Interpreting Biomedical Science

Every discovery brings to our view many things of which we had no intimation before. The greater is the circle of light, the greater is the boundary of darkness by which it is confined. But notwithstanding this, the more light we get, the more thankful we ought to be.

Joseph Priestley (1790)

Interpreting Biomedical Science

Experiment, Evidence, and Belief

Ülo Maiväli
University of Tartu,
Institute of Technology,
Tartu, Estonia

AMSTERDAM • BOSTON • HEIDELBERG • LONDON
NEW YORK • OXFORD • PARIS • SAN DIEGO
SAN FRANCISCO • SINGAPORE • SYDNEY • TOKYO

Academic Press is an imprint of Elsevier

Academic Press is an imprint of Elsevier
125, London Wall, EC2Y 5AS
525 B Street, Suite 1800, San Diego, CA 92101-4495, USA
225 Wyman Street, Waltham, MA 02451, USA
The Boulevard, Langford Lane, Kidlington, Oxford OX5 1GB, UK

Notices
Knowledge and best practice in this field are constantly changing. As new research and
experience broaden our understanding, changes in research methods, professional practices,
or medical treatment may become necessary.

Practitioners and researchers must always rely on their own experience and knowledge
in evaluating and using any information, methods, compounds, or experiments described
herein. In using such information or methods they should be mindful of their own safety and
the safety of others, including parties for whom they have a professional responsibility.

To the fullest extent of the law, neither the Publisher nor the authors, contributors, or editors,
assume any liability for any injury and/or damage to persons or property as a matter of
products liability, negligence or otherwise, or from any use or operation of any methods,
products, instructions, or ideas contained in the material herein.

ISBN: 978-0-12-418689-7

British Library Cataloguing-in-Publication Data
A catalogue record for this book is available from the British Library

Library of Congress Cataloging-in-Publication Data
A catalog record for this book is available from the Library of Congress

For information on all Academic Press publications
visit our website at http://store.elsevier.com/

Typeset by MPS Limited, Chennai, India. www.adi-mps.com

**Working together
to grow libraries in
developing countries**

www.elsevier.com • www.bookaid.org

Publisher: Janice Audet
Acquisition Editor: Jill Leonard
Editorial Project Manager: Pat Gonzalez
Production Project Manager: Lucía Pérez
Designer: Greg Harris

Contents

Part II
The Method

3. Study Design

4. Data and Evidence

Part III
The Big Picture

6. Interpretation

7. Science as a Social Enterprise

Preface

This book is about science as practiced by biomedical scientists. Although written for the working scientist, it is not purely a science book. Instead of providing recipes of how to make an edible soup of tasty results and well-rounded careers, we will take a look into the pot from the outside. There would be no compelling reason to do so but for the increasingly strong complaints about the quality of biomedical science coming from paying customers—that is from the medical and pharmaceutical industries. Things have become so bad that we can credibly suggest that the vast majority of scientific results, published in the best journals, are simply false.

One way of dealing with this problem would be to offer specific methodological advice on experimental design or data analysis, an approach that has indeed been tried many times. Substituting standard deviations with standard errors, standard errors with standard deviations, or both with confidence intervals (CIs), giving up P values for CIs, controlling for multiple comparisons, actually doing biological (as opposed to technical) replications, and lognormal modelling of the of data are only some of the suggestions that have been offered as solutions to the wider problem of reproducibility in molecular biology. Such methodological advice has led to various formal guidelines endorsed by journals and scientific societies, which have two things in common: Firstly, they offer relatively simple, cheap, and easily implementable solutions for the working scientist, who is not an expert in statistics or methodology. And secondly, they seem to have had no great effect on research practices, or on the general quality of science.

Such in-the-soup fixes depend on general assumptions: that the problems facing science are small enough to be rectifiable by quick solutions, that the problems are methodological in nature and thus require methodological solutions, and that we understand the workings of science well enough to be able to offer simple solutions to simple problems. These assumptions can be rephrased as questions: What do scientists do? What is the product of science? What is the connection between the data, results, and truth? What are the roles of truth, logic, rationality, probability, and certainty in science? Is there one science or many?

These questions are not strictly from the domain of science. Yet, dealing with them is not only intellectually fascinating, but we believe necessary to diagnose current ills and, perhaps, offer solutions that stick. Thus, in order to understand the methods of science, we must first open the door to philosophy and let some metaphysics in.

Accordingly, the first part of the book is largely concerned with looking at tacit assumptions that we, who are used to living in the soup, usually take for granted. Perhaps, by shifting perspectives, we will end up with a wider horizon and a clearer nose to sniff out the ingredients that need to be replaced.

The second part deals with methodology of experimental science from the design of experiments and analysis and interpretation of data, all the way to the interpretation of the results. The emphasis here is not on cookbook examples of how to do things (although a curious cookbook approach to cancer studies will be mentioned in Section 2.9), but on the assumptions, interpretations, and limits of the various techniques. The reader won't find many equations, but will hopefully better understand the choices one has (or doesn't have) in designing and making sense of experiments.

The third part of the book concerns itself with the more human aspects of science that have great potential to affect the quality of the product, including, among other things, the sociology of science, career structures, incentives, misconduct, and bibliometrics. All this leads to a discussion on possible ways to address the really important question: How can we improve the scientific enterprise so that its results can once again be trusted?

THIS I BELIEVE IN SCIENCE

When a scientist designs and runs an experiment she cannot, in reasonable time, personally check the validity of every assumption inherent in the experimental scheme. Neither is it possible to run every possible control and to deal with every possible bias. Instead, to occasionally publish, each scientist must take some things on trust. The same goes for thinking about science. Having said that, luckily there is no need or requirement that we all should take the same things on faith. What follows is a list of my personal preconceptions. Although I will later present arguments and data in support of these, I must stress that the following are nevertheless preconceptions. They provide a framework for what I try to accomplish in the book, but will in themselves remain conjectures.

1. Doing good science is extremely difficult. In science we try to bring new knowledge into the world, which is no easy task. There is no successful automated or semi-automated way of doing science. In fact, creating a successful theory of scientific progress could be a game changing achievement for humanity because it would open up the possibility for truly machine-driven growth of knowledge. The potential gains (and harms) of this are literally unimaginable.

2. As scientists, we need a science to study the methods of science, and this need becomes more pressing with every passing day because the global nature of scientific practice is changing. For logical reasons, the science of science cannot itself be truly scientific (but this does not mean that it must

be non-scientific or pseudo-scientific). One can call this meta-science, or the philosophy of science; the name does not matter, only the results do.

3. Methodologically, we are not entirely groping in the dark. There is more than one way of doing science and some ways are better than others.

4. Scientific progress is possible. Our goal is to discover how knowledge can be increased and offer specific advice to the working scientist. We are not aiming to prove that science can, in principle, lead to new knowledge or be useful; this much we know from history. However, just because science worked well in yesteryears, does not prove that it will continue to do so. Neither methodology nor history can provide us with such guarantees and we should accordingly exhibit a measure of humility toward what we do as scientists.

5. Everybody has a choice. In fact, we all make many methodological and ethical choices throughout our careers. These choices are personal: no one can make them for us. The only guidance that can be given is clarification of the possibilities and consequences of our choices. Some choices can lead to conflicts between the interests of individual scientists and the scientific enterprise as a whole. This makes science as much a question of morality as methodology.

Ülo Maiväli
March 2015

Acknowledgments

It is a great pleasure to thank the people who have helped to improve this book by reading parts of the manuscript and/or critically discussing its themes: Aare Abroi, Sille Hausenberg, Vassili Hauryliuk, Peeter Hõrak, Nele Jaanson, Arvi Jõers, Niilo Kaldalu, Kalle Kipper, Veljo Kisand, Eva-Liis Loogväli, Margus Pihlak, Marta Putrins, Taavi Päll, David Schryer, Mait Sepp, Aksel Soosaar, and Helle-Viivi Tolk. Sille Hausenberg gave invaluable help with the figures. I am especially indebted to Tanel Tenson, who not only had the courage to read the manuscript in its entirety, but also gave many pertinent comments and criticisms, and encouraged this project from its inception to completion.

The writing of this book would have neither started nor finished without the enthusiasm and support of Jill Leonard from Elsevier. In addition, from the Elsevier side, I am grateful to Elizabeth Gibson and Pat Gonzalez for their kind and professional editorial support. This work was supported by the Estonian Science Agency grant 9040.

This book was conceived with the help of the graduate students of the University of Tartu who attended the scientific communication and methodology course, provided by its Institute of Technology. In writing it I have tried to keep your interests in mind. While there are not many ready-made solutions to the problems that a modern biomedical scientist faces, there are choices to make, relevant knowledge to grasp, questions to ponder, and educated guesses to make; and granted that one invests into considering them, one will not be choosing blindly. I can only hope that this attempt will help the reader to make some sense of his or her life in science.

Introduction

All science is either physics or stamp collecting.

Ernest Rutherford, Laureate of the
Nobel Prize in Chemistry

There is no question regarding the huge impact that science has had on Western thinking, medicine, and technology. Heliocentric world view, laws of physics, recognition of the age of the universe and solar system, dinosaurs, bacteria, viruses, atoms and molecules, DNA and genetic code, automatic computation, ecology, entropy, evolution, the intelligence coefficient, relativity, Schrödinger's cat, GPS, and The Bomb are all instances of the deep influences that scientific knowledge has on our culture. It is common knowledge that much of the rapid technological progress of the twentieth century was fueled by scientific discoveries. Be it antibiotics, transistors, lasers, or nuclear power, the hand of the scientist was instrumental in their creation. It might therefore come as a surprise that things have not always been so. In the nineteenth century, major technological achievements such as the steam engine were at best catalysts for science and in previous centuries across several civilizations, technological progress has been largely independent of science (Ball, 2004).

Of course, when we talk about science we are talking about much more than technology. While most readers may still be able to imagine their lives without effective medicine, electricity, or computers (which is the life lived by many people today), I find it impossible to fathom what my world view would be without access to scientific concepts such as evolution. The whole notion of data-driven rational argument would be moot without the success of science. Science and its methods have completely taken over some aspects of our thinking.

In 2010 about 22% of Americans expressed interest in international issues in news media. Simultaneously, about 40% were interested in "new scientific discoveries" and 60% in "new medical discoveries" (Anon, 2012). Interestingly, 55% of Americans did not think that "the universe began with a huge explosion" and half did not believe in evolution. At the same time, most educated people (65% of Americans in 2009) do think that they can perceive when somebody stares at their back, apparently making use of a theory of perception, already held by Plato, according to which eyes emit physical "rays" that would bounce off from the backs of their heads. The theory that people are able to feel the stare was actually tested, refuted, and published in *Science*, in 1898 (Chabris and

Simons, 2011). People also overwhelmingly believe that they have an above-average sense of humor, grammatical, ability, and logical capability (in one study 97% of university lecturers stated that they are above-average in their job); people believe that they can predict tomorrow's weather by the severity of their arthritis; and that bad deeds will eventually be punished. So, clearly we are not living in a completely rational world (no great news here!) and in our culture the border between what is scientific and what is not is constantly shifting. While gravitational fields, dark matter, and cancer metabolomes are moving into the light of reason, psychoanalysis, eugenics, and the one gene–one protein hypothesis have gone in the opposite direction.

Great scientific theories can move into the shadows not only because they are proven wrong, but also because they gradually lose their relevance. The lactose operon was once a paradigm for the control of gene expression, thereby earning a Nobel Prize for both Françoise Jacob and Jacques Monod (1965) and encouraging countless new studies on gene expression. Today it is perhaps the best-understood system of gene expression. However, is also not terribly relevant to what goes on in the frontiers of molecular biology. The same can be said about the bacteriophage T4 or the bacterium *Escherichia coli*, which were once the preeminent model organisms in molecular biology.

The currents of science flow differently for different people and are by no means irreversible (think of the shifting fortunes of atomism from metaphysics to physics and chemistry, or of multiregionalism in human evolution). As a consequence, we have to resign ourselves to the fact that the question "Is it scientific?" can, on occasion, receive divergent answers from educated people.

SCIENCE MADE EASY

It is usual to teach the scientific method to science students as a "thing" that we have captured, culled, dissected, fully analyzed, and therefore know the workings of. For instance, the US National Science Foundation (NSF), conducts a longitudinal Survey of Public Attitudes Toward and Understanding of Science and Technology where, among other things, it quantifies "public understanding of the scientific process" (Anon, 2012). Obviously, to be able to conduct this they must first believe that they know what the scientific process is. NSFs views on science should give us a good starting point in distinguishing the most salient points of the scientific method.

First question: *Two scientists want to know if a certain drug is effective against high blood pressure. The first scientist wants to give the drug to 1,000 people with high blood pressure and see how many of them experience lower blood pressure levels. The second scientist wants to give the drug to 500 people with high blood pressure and not give the drug to another 500 people with high blood pressure, and see how many in both groups experience lower blood pressure levels. Which is the better way to test this drug?* and *Why is it better to test the drug this way?* (Correct answer: The second way, because a control group is used for comparison.) About half of respondents correctly answer this question.[1]

The second question: *In your own words, could you tell me what it means to study something scientifically?* (Correct answers: formulation of theories/ test hypothesis, experiments/control group, or rigorous/systematic comparison.) About one in five respondents gave a "correct" answer to this request.

Simple explanations of scientific method seem invariably to suggest that doing science involves testing hypotheses by experiments. Richard Feynman was perhaps the most famous scientist of the 1960s and is widely considered to be one of the best physical scientists ever, in addition to being a great writer, popularizer of science, and an avid surfer. So we might as well take our cue from his famous introductory physics course, taught in Caltech in 1961:

> *The principle of science, the definition, almost, is the following: The test of all knowledge is experiment. Experiment is the sole judge of scientific "truth." But what is the source of knowledge? Where do the laws that are to be tested come from? Experiment, itself, helps to produce these laws, in the sense that it gives us hints. But also needed is imagination to create from these hints the great generalizations—to guess at the wonderful, simple, but very strange patterns beneath them all, and then to experiment to check again whether we have made the right guess. This imagining process is so difficult that there is a division of labour in physics: there are theoretical physicists who imagine, deduce, and guess at new laws, but do not experiment; and then there are experimental physicists who experiment, imagine, deduce, and guess.*

> (Feynman and Leighton 2006)

This process is termed "deductive" to account for the use of formal truth-preserving logic. In this, science is supposed to be rather similar to mathematics (and in Feynman's case, identical to physics), where deductive proof is the standard.[2] The gold standard of true science is widely thought to be replication of experimental results by independent researchers, for which free dissemination and criticism of results and hypotheses are necessary.

Of the status of experiments Feynman has this to say: "Now, how can an experiment be 'wrong?' First, in a trivial way: if something is wrong with the apparatus that you did not notice. But these things are easily fixed, and checked

---

1. A quick control question: explain to yourself, why should a scientist trade half of his study sample for a control group? In the blood pressure example he knows, what the average population values of blood pressure are, and even the values of the members of his study group, before giving them the drug.

2. Besides deduction, Feynman uses a different form of logical inference, namely induction, to explain how experiments lead to the formation of new laws; however, he does not advertise this fact. The concept of "induction" is explained below.

back and forth. So without snatching at such minor things, how *can* the results of an experiment be wrong? Only by being inaccurate."

So, as long as your machines are well oiled and will calculate standard deviations, your experiment really cannot go "wrong." Accordingly, science is often described as objective and impersonal, meaning that as long as you are using the scientific method properly, it does not matter who you are because the results of many individual researchers will eventually converge on the truth.

One mechanical engineer, Valery Fabrikant, forcefully tested this principle while imprisoned for the 1992 killings of four of his colleagues at Concordia University, by regularly publishing his research. Relatives of his victims tried, for understandable reasons, to convince journal editors not to publish his work (Spurgeon, 1996). The editors refused, claiming that for science to work properly, even murderers must be on equal footing when subjecting their results for peer scrutiny. Between 1995 and 2013 Dr. Fabrikant published at least 37 single authored papers from prison, which have been cited 97 times. He will be eligible for parole in 2017.

What is the product of science? One could argue with Feynman that as doctors produce cures, engineers design new gadgets, patent lawyers write patents, and lab technicians mix brightly colored chemical solutions—scientists produce knowledge. More specifically, they produce interesting scientific knowledge, which often concerns causal connections (think of smoking and cancer). Knowledge is often defined as "justified true belief," so the scientific product could be defined as "interesting justified true belief." "Interesting" means firstly "nontrivial," as opposed to trivial knowledge, which everybody takes for granted and secondly, that it matters to society (be it curing cancer or reconstructing the evolution of *Homo sapiens*). "Justified" simply means that if it is true accidentally, it is not knowledge. For example, the railway clock that has been out of order since 1945 is still *exactly* correct twice a day, while no working clock has this privilege. Yet, only a clock that is working can lead to advances in scientific knowledge. So for us "justified" means that only correct uses of the scientific method can lead to real knowledge. Truth and causality have been given their own chapter-length treatments in Chapter 2.

Of course, some scientists and philosophers have long argued that the image of science as a single well-understood thing is misleading. I would argue that explanations of the scientific method as a single "thing" not only lead students to a false confidence in carrying out science, but also to inflated expectations about the nature of the scientific product. Let me illustrate this by the notorious case of the Dutch social psychologist Dietrik Stapel. In 2011, while serving as the dean of the School of Social and Behavioural Sciences of the University of Tilburg (and teaching a graduate seminar on research ethics), he was caught making up experimental data for at least 55 papers over an 8-year period. Interestingly, during that time he successfully supervised 10 PhD students, none of which seemed to worry that they never actually collected the data for their theses; however, the alarm was finally raised by two of his PhD students. Apparently, for these 10 happy students, science consisted solely of methodology and lacked both messy experiments and data.

Stapel later commented on his ethics class, where students were asked as an assignment to describe the errors they committed in their research: "They got back with terrible lapses ... No informed consent, no debriefing of subjects, then of course in data analysis, looking only at some data and not all the data." He didn't see the same problems in his own work, he said, because there were no real data to contend with (*New York Times*, April 28, 2013). Indeed, the committee that studied Stapel's wrongdoings seemed to agree and came to some interesting conclusions about the entire field of psychology, finding that his fraud went undetected for so long because of "a general culture of careless, selective, and uncritical handling of research and data." It identified, in the field of psychology, widespread misuse of statistics, unjustified removal of nonconforming data, and pursuit of hypotheses that are scientifically unjustified. To further cite from the Times:

> In his early years of research ... Stapel wrote papers laying out complicated and messy relationships between multiple variables. He soon realized that journal editors preferred simplicity. "They are actually telling you: 'Leave out this stuff. Make it simpler'... Several psychologists ... admitted that each of these more common practices was as deliberate as any of Stapel's wholesale fabrications. Each was a choice made by the scientist every time he or she came to a fork in the road of experimental research—one way pointing to the truth, however dull and unsatisfying, and the other beckoning the researcher toward a rosier and more notable result that could be patently false or only partly true."

The point here is that people who expect simplicity, elegance, and certainty from their research should perhaps have become mathematicians, not scientists; or, like Stapel, they will have some hard choices to make.

DID THE GREEKS GET THEIR MATH RIGHT BUT THEIR SCIENCE WRONG?

Thinking back on my school days it seems that half of what they taught me in mathematics during 11 years of schooling originates from Greek knowledge. While the mathematics curriculum, after 2,500 years, still begins with Pythagoras, Archimedes, and Euclid, and barely reaches the nineteenth century, the physics curriculum starts with the scientific revolution of the seventeenth century and takes us through relativity, big bang cosmology, and quantum mechanics of the early the twentieth century. The biology curriculum begins at an even later stage in history and progresses further: from the nineteenth (cell theory, Darwinism, Mendel) through to the first half of twentieth century (classical genetics, Neo-Darwinism, biochemistry) up to the 1980s (molecular biology, ecology). So, how is it that the Greeks would have grasped much of my mathematics curriculum, but not physics or biology? Is mathematics actually easier than science?

Rudimentary mathematics has arisen independently many times as a response to astronomical, economic, political, architectural, technological, and the spiritual needs of different societies. Moreover, quite a few branches, as in Babylonia, Egypt, Greece, China around 100 BC, and Mesoamerica managed to reach remarkable sophistication more or less independently. The classical Chinese mathematical treatise, the "Nine chapters of the mathematical art" (compiled from much older material during Han Dynasty, around 210 AD) deals with subjects as different as field measurement, the addition, subtraction, multiplication, and division of fractions, the extraction of square roots, the solutions to linear equations with multiple unknowns, the calculation of the volumes of pyramids and cones (Robson and Stedall, 2009).

Nearly 800 years before and a continent apart, the Eupalinos tunnel was built, which, being a part of a water-supply system for the city of Samos, runs for more than a kilometer under a mountain. Excavations show that the tunnel was made by two teams, which started to dig at the opposite sides of the mountain and met in the middle. To achieve this, the workers must have had to figure out how to start both ends on the same level and, as a separate problem, how to make the two excavations meet in the middle. It is hard to see how this could have been achieved without having a firm command of arithmetic, geometry, and triangulations (Cuomo, 2001).

While Greek mathematicians were adept at geometry, the thing that really set them apart was the axiomatic-deductive demonstration of proof. It is important to note that quite a few sophisticated mathematical cultures, including the Chinese, have flowered independently of the Greeks, without ever discovering or using axiomatic proofs, so this method appears not to be necessary for the development of fairly high levels of mathematical knowledge. However, it seems to have been extremely important for the development of science.

Traditionally, the honor of originating the axiomatic-deductive method rests with another native of Samos, Pythagoras (*ca*. 530 BC). Pythagoras was a mathematical mystic, for whom contemplation resulted in mathematical knowledge, which retained an element of ecstatic revelation. His famous utterance is: "All things are numbers"; and of course, everybody has heard of the "Pythagoras theorem." He founded a politically and intellectually influential mystical order that spread into southern Italy and Sicily. However, in the modern view, Pythagoras himself was not a creative mathematician; there is no good reason to think that he actually proved any theorems.

Pythagoreans were interested in speculative numerology, believing that mathematics reveals the abstract structure of the world, rather than just models it. It is so that Pythagoras "became the midwife of pure mathematics and the founder of the whole mathematical side of scientific theory" (Taylor, 1997). The axiomatic method was perfected by Euclid and married to deductive syllogistic logic by Aristotle. In fact, Euclid's *Elements* systematically axiomatized most of the contemporary mathematics.

Axiomatic-deductive proof starts from true, primary, and necessary axioms. Modern versions of Euclid's five axioms are: (i) A straight line can be drawn between any two points. (ii) Any terminated straight line can be extended indefinitely (iii) A circle with any radius can be drawn around any point. (iv) All right angles are equal. (v) Only one line can be drawn through a point so it is parallel to another line.

In fact, Euclid's fifth axiom did not seem to many mathematicians (and possibly Euclid himself) truly self-evident, and many unsuccessful attempts were made to deduce it from the other four. Still, it wasn't until the early nineteenth century when alternative non-Euclidean geometries were created, often dispensing with Euclid's fifth axiom.

Starting from the five axioms, ancient geometricians deduced theorems, which are still true more than 2,000 years later and seem anything but self-evident. These successes had great impact on Greek science. Axiomatic-deductive demonstrations were attempted in hydrostatics, music theory, astronomy, and medicine. For example, the foremost medical authority of antiquity, Galen (second century AD), tried to introduce mathematical reasoning into medicine by proposing as an "axiom" or indemonstrable principle that "opposites are cures for opposites" (Cuomo, 2001).

The logical route from mathematical proof to scientific proof is to first propose that Euclid's axioms and the resulting theorems not only define a self-consistent mathematical space, but also hold true for actual space (a very reasonable assumption until Einstein's theory of relativity). Then it would be theoretically possible to discover solid truths about the physical world by simple deduction (Russell, 2008). Moreover, if geometry deals with pure forms and if exact reasoning applies to idealized situations, and if results can be interpreted as eternal truths, one might be tempted to argue that the objects of thought must be more real than the messy perceptions we take in through the senses. Mathematical objects are eternal and outside time and space, so, if they are in fact real, the door is opened for all sorts of interesting and mystical things. This is a step that young Bertrand Russell took, but later bitterly regretted when he, while working out what was to become the modern theory of mathematical logic, realized that mathematics can never be anything more than a web of very elaborate trivialities (Russell, 2009). This view is concordant with Aristotle, who, unlike Plato, modestly believed that mathematics studies the mathematical properties of physical objects, not independently existing realities. For him, mathematics was a necessarily imperfect model of reality, not a part of reality itself. People, who believe that it is possible to arrive at absolute truths in science by deductively testing theories, should be made aware of this tidbit of history.

While mathematics seems to be a reasonably common occurrence—in fact it has to be present in every culture that uses a calendar or knows how to count—science is an only child. Arguably, science first arose in Greece as an inseparable part of philosophy ("natural philosophy"). The first great scientist, Aristotle (384–322 BC), is also a founding father of Western philosophy and the inventor of syllogistic logic. Aristotle, the scientist, was at his greatest in biology, especially in zoology. Among other achievements, he introduced the comparative method to biology, earning the praise of being the greatest contributor to biology before Darwin (Mayr, 1982). He also did original research in botany, astronomy, chemistry, meteorology, and in other fields. His physics, although erroneous, was hugely influential until Newton and the Scientific Revolution. In Aristotle's work in physics there are no experiments, no measurements, and no observations other than those of ordinary everyday experience (Hankinson, 1999). It seems that his qualitative and comparative methods, which worked so well in biology, did not take kindly to application into physics.

Aristotle taught us that science must start from empirical observations; but it must also use rigorous deductive arguments that start from axioms and lead to theorems. The axioms are the fundamental facts upon which everything else depends. Aristotle also saw science as a search for causes. For him all biological structures and activities have a biological meaning, which can be elucidated. According to Aristotle, we have scientific understanding, if we know the necessary cause of the structure or process we are studying. From true premises true conclusions can be drawn. The premises of a scientific argument must be "true and primitive and immediate and more familiar than and prior to and explanatory of the conclusion." In Greek philosophical parlance such views are known as "dogmatist," while defenders of the alternative outlook were called "empiricists." According to Diocles of Carystus:

> Those who think that one should state a cause in every case do not appear to understand first that it is not always necessary to do so from a practical point of view, and second that many things which exist are somehow by their nature akin to principles, so that they cannot be given a causal account. Furthermore, they sometimes err in assuming what is unknown, disputed, and implausible, thinking that they have adequately given the cause. You should disregard people who aetiologize in this manner, and who think that one should state causes for everything; you should rather rely upon things which have been excogitated over a long period on the basis of experience [empeiria]; and you should seek a cause for contingent things when that is likely to make what you say about them more understandable and more believable.
>
> (as cited by Hankinson, 1999)

Luckily, Aristotle was not too dogmatic in following his own epistemological advice, as his works in biology are more empirical inquiries than proofs. Aristotle urges students not to be contemptuous of biology, as opposed to "higher" sciences like astronomy, which study eternal things. In biology the researcher is closer to the subject matter and therefore better able to study it. Aristotle saw the world as eternal and essentially unchanging and, accordingly, would probably not have been receptive to the idea of evolution. For the biologist, according to Aristotle, the most important task is thus to understand the final cause of the animal that is studied (Hankinson, 1999). The animal lives in its environment, which determines its survival. Therefore the task of the researcher is to determine how each part of the animal helps it to survive and reproduce. This is the "final cause" why the part exists. In animals, in addition to final causes, efficient causes and formal causes are also worthy objects of study. For example, the semen of the father carries the form of the father and transmits it to the offspring (efficient cause), but it also contains the instructions for the formation of the embryo (formal cause). It looks as if Aristotle would have enjoyed a pint in the company of Watson and Crick.

For the sake of balance, it must be stated that while Aristotle's comparative anatomy deeply impressed even the nineteenth-century greats, like Cuvier, physiology was not among his strengths. For example, he believed that the heart is the seat of the soul and that the blood connects sense organs with the soul and contains food for the tissues of the body (which is not inaccurate). According to Aristotle some animals originate from seawater, some from putrefying matter (spontaneous generation). While a keen observer of animals, Aristotle was apparently not a great admirer of women. According to him the matter of the embryo is from the mother, while the form comes entirely from the father. He also rather revealingly believed that women have fewer teeth than men. One suspects that this last result might indicate something deep and profound about the scientific method and the next example will show us, why. Over 2,000 years after Aristotle, in 1921, a prominent American cytogeneticist, Theophilus S. Painter, published an erroneous human chromosome count, 48 (humans have 46 chromosomes). Notwithstanding the popularity of chromosome studies in the first half of the twentieth century, it took a full 35 years before this error was corrected in print by Tjio and Levan. In the meanwhile, however, "[a] photo of the human karyotype … in a widely read textbook of the day, by the eminent British geneticist Cyril Darlington, clearly showed forty-six chromosomes. The photo caption, however, read forty-eight" (Leaf, 2013, pp. 19–20). After publication of the correct chromosome number by Tjio and Levan, Masuo Kodani reported in *Science* that a third of the Japanese men in fact have 48 chromosomes (Kodani, 1958). This suggests that in the real world it is not only the cold facts that count in disentangling scientific controversies, but also the expectations, previously accepted theories, and the respective scientific authority of the participants.

Ancient biological science did not die with Aristotle but flourished for another 500 years, until Galen in the second century AD. The focus shifted from the description of animals and plants and their parts to physiology and medicine. Herophilos of Alexandria (335–280 BC) was perhaps the first Greek doctor to systematically dissect humans. He apparently went even further and (as did Erasistratus, see below) "cut open criminals provided by kings from prison, and inspecting, while they were still alive, those parts which nature had previously hidden as to their position, color, shape, size, arrangement, hardness, softness, smoothness, connection, and the projections and concavities of each, and whether anything is inserted into something else and whether anything receives into itself a part from some other" (Celsus, as cited in Hankinson, 1999). He gave a detailed anatomical description of the brain and the eye (including isolation of the optic nerve) and may have been the discoverer of the function of nerves and perhaps even the first to distinguish between the motor and sensory nerves (based on the work done on the prisoners).

The second famous Alexandrian, Erasistratus (304–250 BC), described the anatomy of the heart and understood that it functions as a pump. He also held that all fevers are caused by inflammations, which are in turn caused by transfusion of blood from the veins (where it naturally belongs) to the arteries, where its presence is pathological. Erasistratus rejected the view that disease is caused by antecedent causes, such as overheating, overwork, over indulgence in food or sex. He contends that heat and cold cannot be the causes of illness, because they are not invariably followed by it, and do not persist at the time of the illness:

> *Most people, both now and in the past, have sought the causes of fevers, trying to ascertain and learn from the sick whether the illness has its origin in being chilled or exhausted or repletion, or some other cause of this kind; but this kind of inquiry into the causes of diseases yields results neither true nor useful. For if cold were a cause of fever, then those who have been chilled the more should suffer the greater fever. But this is not what happens: rather there are some who have faced extreme danger from freezing, and who when rescued have remained unaffected by fever.... [And] many people who experience far worse exhaustion and repletion than that which coincides with fever in some others yet escape the illness.*
>
> (Hankinson 1999)

We will revisit very similar arguments in Chapter 2, where we discuss the great smoking controversy that raged among the brightest statisticians of the 1950s.

The founding of empiricism is usually attributed to Philinus (fl. 250 BC) and Serapion (fl. 225 BC). Their method requires observation of concurrences of events. Experience of a successful cure is formed from single occurrences of success of some treatment in individual patients. Should a treatment (perhaps administered by accident) be successful once, an empiricist will want to try it again. The idea here is that what worked in the past, is likely to do so in the future. The empiricists were suspicious of grand theoretical schemes because they believed that there are always many possible theories to account for the empirical facts. They were not great believers in the power of deductive logic to explain the workings of the human body. For an empiricist is a more modest creature, for whom all that matters is to find patterns that work; and as soon as they don't, they will be replaced with those that will.

Practitioners of a somewhat extremist form of empiricism in the first century AD called themselves the Methodists. According to them, all that is required of medicine is to know the differences between healthy and pathological states, which should not prove too difficult, because they only need to study what differentiates patients from the healthy—not some theoretical entities. For a Methodist, medicine is something one can learn in 6 months. The obvious weakness of the Methodist position was recognized by Galen, who pointed out that in treating a person who has been bitten by a rabid dog a Methodist would only concern himself with the wound itself, which has nothing to do with the condition of the dog. Thus he would give the same prognosis to people who are bitten by rabid or healthy dogs. Galen (129 to *ca.* 200) strived to unify the dogmatist and empiricist schools and became known as the greatest of the ancient doctors. He agreed with Aristotle that science must start from basic axioms and thereby lead through logic to truths. The axioms can be logical, mathematical, or metaphysical ("nothing occurs causelessly"). The axioms should be self-evident in such a way, that to deny one would simply seem silly in the eyes of reasonable people. However, in addition to the things evident to the reason, which are exemplified in the axiomatic method, there are also many things that are accessible through perception. Here is where the real strength of Galen lies. A strong argument can be made that it was he who introduced controlled experiment to science (Hankinson, 1999). For example, he demonstrated the function of the recurrent laryngeal nerve in voice-production by placing ligatures at different points of the test animal's spinal column and recording how each affected its movements and vocalizations. The major methodological insight here is that when you keep all else equal and change just one variable in the experimental setup (the placement of the ligature), the observed effect (lack of vocalization) was likely caused by that variable.

Not all experiments are fountains of truth. For example, Galen claimed that a boy was able to survive for a day with an ox bladder over his nose and mouth to prevent him from breathing; thus illustrating the importance of due diligence in designing and conducting experiments (not to mention the importance of ethics-committees). In summary, since the time of Galen it is good form to perform an experiment to test a theory.

Before jumping to the origins of modern physics in the seventeenth century, it might be useful to briefly return to the question: If the Greeks were so good at mathematics, why did we have to wait another 1,000 years to witness the birth of physical science? The ancient history of biomedical science may shed some light on this. The Aristotelian emphasis on qualitative causal questions nicely fits into the modern conceptual framework of biology, where adaptations really are there for a purpose and inner programs direct development. Without knowing about evolution and gene expression, the Greeks still asked the right sort of questions in biology (Mayr, 1982). In contrast, qualitative causal theories never took them (or anybody else) far in physics and because experimental approaches entered the methodological toolbox of ancient medical science rather late in the game, they failed to make it into physics before the end of the Mediterranean civilization. This omission left any ancient speculation on the physical world prey to the Pythagorean folly of misunderstanding mathematics as the true description of the world, the shortest repudiation of which that I am aware of, is attributed to Richard Feynman: "Physics is to math what sex is to masturbation."

The early modern success of physical science and relative dormancy of biology during the scientific revolution led to the assumption that the physicists of the scientific revolution had discovered the scientific method and that it would therefore be highly commendable if the quantitative mathematical ideals of physics guided every science to the rigorous form and deductive predictions of physical theories. This belief dominated the views of both scientists and philosophers of science (many of whom were physicists by education) well into the twentieth century. A fair example comes from Lord Kelvin (1891):

> *When you can measure what you are speaking about and express it in numbers, you know something about it; but when you cannot measure it, when you cannot express it in numbers, your knowledge is of a meagre and unsatisfactory kind: it may be the beginning of knowledge, but you have scarcely, in your thoughts, advanced to the stage of science...*

(cited in Westfall, 1973)

Such thoughts are echoed in the media in its adoration of big data approaches within the humanities, where counting the frequencies of words is hoped to revolutionize literary studies. If it works in physics, it must be only a matter of time until it saves literary theory from the evils of subjectivity.

THE SCIENTIFIC REVOLUTION

While Greek proto-science often asked the right questions, its answers were as yet pretty much irrelevant to contemporary technological or medical fixes. The technological marvels of the Ancients (of which there were more than a few) owed more to mathematics and to "human natural knowledge" than to accumulative growth of science (Sivin, 1982). This seems to hold equally to classical Greek and Chinese cultures, which managed to independently develop quite extensive scientific traditions.

In a word, classical proto-science was on the whole speculative rather than empirical, and it took more than half a millennia to rectify this. What we now recognize as "The Scientific Revolution," occurred fairly rapidly and only once in the early seventeenth-century Europe. This poses the Grand Question originally attributed to the British biochemist/Sinologue Joseph Needham: why did modern science in its technologically useful form develop only once and in Europe, but not, for example, in China? Were the reasons economic, political, ideological, cultural, religious, climatological, linguistic, or just due to different paths that human thought can take (i.e., accidental)? Despite more than 60 years of intensive work, nobody really knows the answer (Cohen, 2001). The question of the origin of science is not dissimilar to the one about the origin of life: we do not even know if we are looking at a causally preordained event or are witnesses to (and products of) an extraordinarily lucky chance.

Galilei Galileo is often thought of, and with some reason, as the father of the experimental method in science. He was born in 1564, the same year as Shakespeare, and spent most of his life teaching mathematics in the University of Padua. When 46 years old, Galileo was invited to Florence by his former pupil Cosimo II de' Medici and was suddenly free to do all the science he liked.

He used his time well. In 1616, Galileo was censured by the Pope for abandoning Copernicus's heliocentric theory, which according to a panel of ecclesiastical authorities contradicted Holy Scripture.

That the Church has a long memory was proven 16 years later, in 1633, when the Inquisition ascertained that his *Dialogue Concerning Two Chief World Systems, Ptolemaic and Copernican* had ignored the ruling of 1616, tried the now elderly Galileo, and found him guilty. Luckily for science, the sentence was life imprisonment, which amounted to house arrest that left him once again free to work without hindrance under the wing of powerful patrons. His main work, *Two New Sciences*, was published during this period of imprisonment. Galileo was widely seen as a pioneer of science during his lifetime and this view has not changed. It was he who introduced the experimental method to physics and some of his discoveries have at least in principle survived to this day and can be found in any elementary physics syllabus.

In the science of motion Galileo's insight was to complement the traditional Aristotelian question about the causes of motion with a mathematical characterization of it. Questions about how freely falling objects moved could now be treated as problems of geometry and accurate and predictive models of motion were accordingly developed. This geometrical approach was complemented by carefully designed experiments; and designing experiments in the seventeenth century was by no means easy. For example, while Galileo could measure distances with reasonable accuracy, the best way for him to measure short intervals of time was probably still to whistle a tune and count on the beat (Gower, 1997).

By measuring time and distances of rolling balls on inclined surfaces, Galileo discovered the "times-squared" rule, according to which the total distances traveled by freely falling objects are proportional to the squares of the times taken. This rule can also be derived from the mathematical definition of naturally accelerated motion that Galileo adopted in his *Two New Sciences*. So, according to Galileo, the carefully designed experiment, which taken alone is insufficient to prove some universal truth, can in conjunction with the axiomatic method do just that.

Natural philosophy of the early seventeenth century was concerned with universal truths. It was clear to everybody concerned that an experiment was never to be sufficient to achieve this ideal. Galileo's work, while containing descriptions of experiments, does not list specific results. Indeed, he placed little weight on the results of real experiments. As true experiments were difficult to perform, less messy thought experiments were preferable for him.

For a thought experiment we have no need for complex apparatuses or experimental designs but merely imagine simple arrangements of objects under conditions, which would lead to a single uncontested outcome. Also, thought experiments are democratic: everybody is in equal position to check everybody else's results and devise new and better experiments. Although using very simple designs, some of Galileo's thought experiments had the power to overthrow entrenched common-sense notions of the physical world, and to do so without using any empirical data. For example, it comes naturally to suppose that of two falling objects like a feather and a brick, the heavier brick will travel faster than the feather and will thus reach the ground first. Indeed, this was Aristotle's belief. In an ingenious thought experiment Galileo tied them together. Now the brick will pull the feather more quickly and the feather will try to slow the brick down. And yet the combined object must be heavier than either the brick or the feather alone, which means that according to our original intuition it should have fallen both faster and slower than the brick (which, of course, is a logical impossibility). The only logically possible result is that the feather-brick falls with exactly the same rate as the feather, and the brick. Thus, by the power of logic alone we have reached an unexpected result about the empirical world, and in the process have disproved a long-standing common-sense theory (Gower, 1997). Intriguingly, Galileo's theory was neither deduced from data, nor is it a purely logical truth (Brown, 1994). Nevertheless, the fact that nearly 400 years later Galileo's solution is still taught to every schoolchild seems to indicate that there must exist an abstract realm, which is perfectly real, and from which we can obtain knowledge *a priori*.

While Galileo strove to put physics on a more mathematical footing, his contemporary, Francis Bacon stated in his *Novum organum* (1620) that "natural philosophy…is tainted and corrupted…by mathematics, which ought only to give definiteness to natural philosophy, not to generate it or give it birth." Unlike Galileo, he promoted an inductive experimental method, which was nevertheless supposed to lead to certainty. He clearly saw the perils of simple enumerative induction where, to borrow Aristotle's example, white swan + white swan + … + white swan = all swans are white. A single observation of a non-white swan is in principle enough to overthrow the general conclusion and, because we wouldn't know where to look for a black swan before actually finding one, Aristotle's failure of never visiting Australia is not very instructive for us. Bacon's eliminative form of induction was more nimble than that and tried to reason from single observations or experimental facts to laws of nature, by using only those facts, which enable us to exclude alternative theories. He believed that logical certainty was an attainable goal when using this method.

According to Bacon's *Novum organum:*

> *The true method of experience … first lights the candle, and then by means of the candle shows the way; commencing as it does with experience duly ordered and digested, not bungling or erratic, and from it deducing axioms, and from established axioms again new experiments…. What the sciences stand in need of is a form of induction which shall analyse experience and take it to pieces, and by a due process of exclusion and rejection lead to an inevitable conclusion.*

The importance of Bacon, however, lays not so much in eliminative induction, which had already been put to philosophical use by Plato, but rather in popularizing the experimental method. As the lawyer he was, in Bacon's thinking nature must be put in the dock and questioned vigorously by the scientist, who must take special care to avoid bias, instances of which he grouped into four classes of "idols." Idols of the tribe are common to the human race, idols of the den to the individual, idols of the marketplace come from misuse of language, and idols of the theater from philosophical dogmas. Without destroying these idols, we should have no hope for achieving truth and pure understanding.

The Baconian ideal is personified in the character of Sherlock Holmes, who, after excluding all logical possibilities excepting one, is willing to put his trust in the most preposterous theory, as long as it is the last one standing. The success of eliminative induction in crime detection is perhaps a good indication of why it has been rather precarious in science. In a crime novel we usually have half a dozen suspects with motives, means, and opportunities—making it feasible for the detective to eliminate all but one. In science, as was discovered by the early twentieth-century French physicist Pierre Duhem, the number of possible self-consistent theories, that could explain any set of experimental results, is infinite. As Bacon himself rather eloquently put it: "The universe to the eye of the human understanding is framed like a labyrinth, presenting as it does on every side so many ambiguities of way, such deceitful resemblances of objects and signs, natures too irregular in their lines and so knotted and entangled."

The more important child of Bacon's theories is not Sherlock Holmes but The Royal Society that received its charter in 1662 (Jardine, 2000). The great experimentalist Robert Hooke was made its first Curator of Experiments, thereby setting the tone for the empirical tradition of British science for the next few centuries (Jardine, 2004). Incidentally, Hooke found a more humane way, than the one prescribed by Bacon, to pry into the secrets of nature by the use of the microscope, so that instead of "pry[ing] into [nature's] secrets by breaking open the doors upon her," we "quietly peep in at the windows, without frighting her out of her usual byas" (Ball, 2012). Bacon, who was not a great scientist himself, died in 1626 of bronchitis, acquired during an experiment involving stuffing a chicken with snow (or, according to a different version of events, of experimenting too liberally with opiates).

The fledgling experimental science was kindled by Hooke's collaborator and benefactor, Robert Boyle of the vacuum pump fame, who demanded that experiments be actually conducted (as opposed to imagined) and be witnessed by several competent judges. This constitutes a break with the closed, secretive tradition of alchemy and natural magic, and provides a reason why, starting from the 1660s, experiments were routinely and openly conducted in the halls of the Royal Society, recorded, and signed by witnessing members. The replicability of experiments was also highly desirable, to which end Boyle published several of his experiments as letters to other potential experimentalist: "not barely to relate, ... but to teach a young gentleman to make them" (Shapin and Schaffer, 1985). At the same time, Boyle was not above complaining about the difficulty and slow pace of replication of his vacuum pump experiments.

Boyle was one of the first to draw a distinction between the "confident" philosophers, who build grand systems, and "sober" experimental scientists, who are characterized by their modesty, diligence, and whose proudly voluntary role it is to be "a drudge of greater industry than reason" (a very Baconian sentiment). Boyle also laid the foundation for the modern "functional" style of academic writing, his style being described as "plain, ascetic, unadorned (yet convoluted)" (Shapin and Schaffer, 1985). He was also keenly aware of the gap between experimental facts and scientific conclusions, which to him were always tentative. The object of controversy was not the fact but the interpretation of facts, the way to conduct a controversy was to attack the work, not the person, of your opponent, and the experimenter was supposed to be a disinterested observer of facts, not a builder of great systems of thought. Such are the beginnings of the modern ethos of science, which was codified by Robert Merton nearly 300 years later (Section 7.3).

Without a doubt, the most important figure of the scientific revolution was Isaac Newton, whose principal works are the highly mathematical *Philosophiae Naturalis Principia Mathematica* (1687) and the experimental *Opticks* (1704). (Incidentially, in *Opticks* Newton was probably the first modern scientist to use the arithmetic mean as a summary statistic to present his measurements.) In *Principia* he lists his four "rules of reasoning in philosophy":

> Rule 1: "We are to admit no more causes of natural things than such as are both true and sufficient to explain their appearances."
> Rule 2: "Therefore to the same natural effects we must, as far as possible, assign the same causes."
> Rule 3: "The qualities of bodies, which admit neither intensification nor remission of degrees, and which are found to belong to all bodies within the reach of our experiments, are to be esteemed the universal qualities of bodies whatsoever."
> Rule 4: "In experimental philosophy we are to look upon propositions inferred by general induction from phenomena as accurately or very nearly true, notwithstanding any contrary hypotheses that may be imagined, till such time as other phenomena occur, by which they may either be made more accurate, or liable to exceptions."

While rules I and II show the importance of causal explanations and rule II seems to postulate the uniformity of nature, rules III and IV define Newton's reliance on a particular form of induction. Rule III narrows the scope of induction by stating that we can only generalize from our experimental knowledge when the studied property belongs to all objects. Rule IV has to do with the level of certainty we can assign to our inductions. For Newton the only kinds of reasoning were deductive reasoning, which he successfully used in *Principia* to deduce the inverse square law of gravitational attraction from Kepler's law, using a theorem that he had proven for this reason, and inductive reasoning as described in rules III and IV. In Newton's view, rules III and IV will never lead to absolute certainty, but can establish conclusions beyond reasonable doubt. So, as with Galileo's law of motion discussed above, Newton's law of gravitation seems to rely heavily on abstract logical deduction (although it was later amply confirmed by empirical studies; Brown, 1994).

Newton's theory of gravitation initially drew its fair share of criticism. The great German mathematician and philosopher Gottfried Wilhelm Leibniz, who was sucked into a bitter dispute with Newton over priority for discovering the calculus, noted that the conclusion that everything is attracted to everything else by a gravitational force does not identify a natural agency for this attraction. There is no causal link; instead the theory merely reports what happens. Leibniz concluded that Newton must suppose that gravitation is produced by a continual miracle. Newton's reply was, while conceding the lack of causal explanation, that the important thing here is the great explanatory power of the theory (the power of explaining observations and measurements on heavens and the sea).

As years went by, doubt about the veracity of Newton's law of gravitation dissipated to the extent that for two centuries the theory was (nearly) universally considered literally true. The reason for this change of fortunes was the amazing predictive power of the theory, which was again and again corroborated by new and ever more exact measurements. It had led to predictions of tiny deviations of the orbits of all the planets from Kepler's laws and the orbits of their satellites, which were experimentally corroborated. Moreover, Newton's laws were equally good in predicting movements in both the spheres of heavens and in terrestrial mechanics. The importance that was accorded to accurate measurements, as proof for Newton's laws, is best emphasized by Newton himself, who, so it seems, took great care to fit his data to the theory. For example, when calculating the correlation of two independently measured values: acceleration of gravity (g) on earth and the moon's centripetal acceleration, he claimed a precision of fit to theory of about 1 in 3,000 (Westfall, 1973). This and other instances of uncanny precision were only incorporated into the second edition of *Principia* (1713) and were probably intended to bolster Newton's view of exact quantitative science over science as consisting of speculative causal hypotheses.

Meanwhile, we are still grappling with an explanation of how gravity works, have recently ran a $13.25 billion experiment to get a glimpse of the Higgs boson (according to *Forbes Magazine*, July 5, 2012), which is a manifestation of the Higgs field, which in turn "imparts" mass to particles. We must concede that gravity is still the only type of force, which we cannot hope to learn to manipulate anytime soon.

DEDUCTION AND INDUCTION AS TWO APPROACHES TO SCIENTIFIC INFERENCE

When we look back to the early history of science from Antiquity to Newton, we immediately see that there is no single method that everyone agreed on. Instead we may try, while taking the risk of looking simple-minded, to represent the available choices as a series of dichotomies (Table I.1).

TABLE I.1 Some Choices Available for a Scientist

Goal:	True description	A working model
Means:	Causal explanation	Saving the phenomena
Method of proof:	Deductive logic	Inductive logic
Probability approach:	Probability theory	Statistics
Evidence:	Certain	Uncertain
Philosophical stance:	Rationalist	Empiricist

Different people have used these choices in different combinations. For example Newton, eschewing causal explanations, strove to use both deduction and induction to arrive through quantitative precise measurements at certain true description of the world. His laws, while not considered literally true, are still universally taught and used under many circumstances as excellent approximations of truth. In contrast, the mathematically naïve Charles Darwin used deduction in combination with qualitative observations to arrive at a causal theory of evolution, which, regardless of its minimal predictive power, has become the basis of modern biology (Mayr, 1982). As a third example, Gregor Mendel used statistical arguments derived from the crossings of some 29,000 pea plants to arrive at a corpuscular theory of inheritance (1865) postulating independently inheritable particulate genetic units, of which both parents can have different versions (alleles), which can have either a recessive or dominant effect on the phenotype. It took until the 1940s to fully reconcile Mendelian qualitative inheritance with the continuous (quantitative) variation of most characters that were being studied by evolutionary biologists (Mayr, 1982).

In a remarkable passage in his book "The General Theory of Natural Selection" (1930), another great biologist (and statistician), Roland Fisher, proposed that had any mid-nineteenth-century thinker been so inclined, he could have arrived at the better part of the Mendelian theory purely deductively by conducting the kind of thought experiments that were so loved by Galileo. Knowledge of the "noninheritance of scars and mutilations would have prepared him to conceive of the hereditary nature of an organism" and assuming that genes completely determine (i) heredity and (ii) sexual reproduction: "he would certainly have inferred that each organism must receive a definite portion of its genes from each parent, and that consequently it must transfer only a corresponding portion to each of its offspring." Further, "our imaginary theorist ... would scarcely have failed to imagine a conceptual framework, in which each gene has its proper place or locus, which could be occupied alternatively, had the parentage been different, by a gene of a different kind." From here concepts of homozygosity and heterozygosity and Mendel's laws follow as necessary truths. Fisher concludes that "It thus appears that, apart from dominance and linkage, including sex linkage, all the main characteristics of the Mendelian system flow from assumptions of particulate inheritance of the simplest character, and could have been deduced *a priori* had one conceived it possible that the laws of inheritance could really be simple and definite" (cited in Dawkins, 2008).

These are of course by no means the only methodological choices one could make, and judging by the resilience of their results, not the worst ones either. Here it might be useful to note that molecular biology started out as a highly theoretical enterprise, when Watson and Crick derived the DNA structure from X-ray diffraction patterns, cut out 2D models of its individual constituents (bases, sugars), and also from Chargaff's rule (purine–pyrimidine pairing). The genetic code and the central dogma of molecular biology were likewise researched in a highly theoretical hypothesis-dependent manner, appropriating the concepts of the then budding field of information theory (Judson, 1979). From these august beginnings, however, molecular biology has gradually verged more and more in the direction where hypotheses are often generated post hoc (if at all) and mainly serve as rhetorical means for discussing the experimental results. It is likely that already during the past 5 or 10 years most of the data in molecular biology has been generated by hypothesis-free omics approaches.

Although the public perception of science may lean on the side of truth and certainty, we will in later chapters see that the practice of biomedical science is heavily dependent on inductive statistical inference. It is important to note that while the approach that leads to true descriptions is an offshoot of the axiomatic-deductive proof that was already working well in mathematics in 400 BC, the mathematical basis for the inductive approach, which was sufficiently developed to be of use to the working scientist, did not emerge until some 2,300 years later.

Deduction starts from a general rule or premises and proceeds through well-defined (and well understood) logical steps toward the conclusion. As long as the premises are true and no errors were made in the process of logical inference, the conclusion is also necessarily true. But there is a nasty trade-off. Namely, everything that was revealed in the conclusion was necessarily already present in the premises. So, technically, no new knowledge is created in the process of deduction. There are two complementary methods of deduction: (i) classical logic, which is the method for determining truth-values of propositions, and (ii) probability theory, which allows one to deduce frequencies of events.[3] Both logic and probability theory are well worked out by mathematicians, so the practical problem for the scientist is the truth of the premises.

Using induction we do the opposite of deduction. We start with specific data and try to infer a general rule. Inductive inference can create new knowledge,

---

3. Propositions are sentences, which are so structured that they can only be either true or false (never both or neither).

which is another way of saying that the conclusion is not fully embedded in the premises. But again, there is a nasty trade-off. Unlike deduction, induction does not generate certainty. In other words, induction is not a truth-machine. In contrast to deduction, there never has been a widely accepted and clearly defined propositional inductive logic. Probabilistic inductive reasoning, however, is codified in statistics.

The difference between probabilistic deduction and induction and between probability theory and statistics is best understood when looking at the game of poker. Let us suppose that you have a deck of cards and want to calculate the long-run relative frequency of drawing a full house, assuming the rules of poker and a fully randomized deck. Using probability theory, you will arrive at the answer (1/693) which, as long as the assumptions hold, is *exactly* true over an infinite number of draws. There is nothing mystical about this result, simple combinatorics will provide the only logically possible answer every time.

Now we come to the difficult problem: if shown a number of hands drawn from a deck, what is the composition of the deck? This is logically the reverse of the first problem and the answer must be provided by statistics. Unfortunately, there is no single correct answer. Without being able to observe the actual deck, we will never know for sure. There are in fact an infinite number of possible answers and the most one can do is to show that some are more likely than others.

However, all is not lost. If the assumptions (randomized deck, etc.) hold, the more individual hands you will observe, the more trust you can put in your answer. In Chapter 5 we will learn how to convert this trust into a number and call it a probability.

The trade-offs present in deduction and induction lead to an obvious problem for those who want science to generate novel and true knowledge about the external world. This is exactly how Newtonian physics was seen for 200 years and in Chapter 2 we will discuss a very interesting attempt by Immanuel Kant to solve this problem, the failure of which still resonates today.

REFERENCES

Anon 2012. National science board. 2012. Science and Engineering Indicators 2012, Arlington, VA: nsf.gov.

Ball, P., 2004. Critical Mass: How One Thing Becomes Another. Random House, New York, NY.

Ball, P., 2012. Curiosity. University of Chicago Press, Chicago, London.

Brown, J.R., 1994. Smoke and Mirrors: How Science Reflects Reality (Philosophical Issues in Science). Routledge.

Chabris, C.F., Simons, D.J., 2011. The Invisible Gorilla. Random House Digital, Inc.

Cohen, H.F., 2001. Joseph Needham's grand question, and how to make it productive for our understanding of the scientific revolution. Sci. Technol. East Asia 9, 21–31.

Cuomo, S.S., 2001. Ancient Mathematics. Routledge, London, New York.

Dawkins, R., 2008. The Oxford Book of Modern Science Writing. Oxford University Press, USA.

Feynman, R.P., Leighton, R., 2006. Feynman Lectures on Physics. Pearson.

Gower, B., 1997. Scientific Method: An Historical and Philosophical Introduction. Routledge, London; New York.

Hankinson, R.J., 1999. Routledge History of Philosophy Volume II From Aristotle to Augustine. Routledge, London; New York.

Jardine, L., 2000. Ingenious Pursuits. Random House, New York.

Jardine, L., 2004. The Curious Life of Robert Hooke. Harper Perennial, New York.

Judson, H., 1979. The Eighth Day of Creation. Cold Spring Harbor Laboratory Press, New York, NY.

Kodani, M., 1958. Three chromosome numbers in whites and Japanese. Science 127 (3310), 1339–1340.

Leaf, C., 2013. The Truth in Small Doses. Simon & Schuster, New York.

Mayr, E., 1982. The Growth of Biological Thought. Belknap Press,Cambridge, MA.

Robson, E., Stedall, J.A., 2009. The Oxford Handbook of the History of Mathematics. Oxford University Press, Oxford, New York.

Russell, B., 2008. History of Western Philosophy. Simon & Schuster, New York.

Russell, E.B., 2009. Autobiography. Routledge.

Shapin, S., Schaffer, S., 1985. Leviathan and the Air-Pump. Princeton University Press, Princeton, NJ.

Sivin, N., 1982. Why the scientific revolution did not take place in China—or didn't it? Chin. Sci. (5), 45–66.

Spurgeon, D., 1996. Paper from jailed professor stirs debate over publication. Nature 381, 458.

Taylor, C.C.W., 1997. Routledge History of Philosophy Volume I: From the Beginning to Plato. Routledge.

Westfall, R.S., 1973. Newton and the fudge factor. Science 179, 751–758.

Part I

What Is at Stake:
The Skeptical Argument

Chapter 1

Do We Need a Science of Science?

It is possible to envision two kinds of approaches to describing scientific methodology. In the first, which might be called the textbook approach, you are told how things are or what you need to do to get certain results. This is the most effective way of disseminating knowledge, as myriad textbooks attest. Unfortunately, textbooks are generally not very good at exploring controversies as such activity is often thought to be confusing to young minds. In fields like inferential statistics, where controversy is rife, this can lead to grossly inadequate and misleading expositions.

In writings on scientific method the textbook approach can mutate to a form where an established scientist explains to his younger self, how to reach the sort of pinnacles of fame and power that he himself so effortlessly conquered. Indeed, there are numerous more or less entertaining books and articles falling into the mold of "Advice to a young scientist," often written from the perspective of career advancement. Alas, career advice from a scientific heavyweight presupposes that he actually knows the secret of his success. Furthermore, it presupposes that there is a secret, that the success was not a random event. The same goes for the scientific method. In order for a scientist to give a normative description of science in the sense that a Textbook of Cellular Biology is a normative treatment of cell biology, there must be a set of generally accepted norms in the field of scientific methodology, as there are in cell biology. If this is the case then the main body of experimental biological science must be methodologically fairly homogeneous and, more importantly, it is likely that, given the past successes of biological science, the biology on the whole "works."

"Does experimental biology work?" is the main question for this section. However, it is not immediately clear, what "work" in this context means. As a minimum, one expects that scientific results are repeatable by independent researchers (repeatability of experiments is of course not sufficient for claiming the truth of the interpretations emanating from the experimental results). Also, it would be nice if scientific truths were permanent (as opposed to transitory) and would occasionally lead to technical or medical advances, which would

Interpreting Biomedical Science. DOI: http://dx.doi.org/10.1016/B978-0-12-418689-7.00001-6

directly benefit society. These questions should be answered based on evidence. However, recently the generally assumed high level of long-term usefulness of basic research for clinical practice has been seriously questioned by prominent medical researchers (Chalmers et al., 2014) and the median lag time from discovery to medical application may exceed a quarter of a century (Contopoulos-Ioannidis et al., 2008).

Although the evidence that will be presented below could be construed to claim fundamental irreproducibility and, therefore, irrelevance of modern biomedical science, I would caution against such radical interpretations. This is because, as we shall see, both the quality and availability of evidence are somewhat uneven. However, available evidence certainly seems to point away from a neat and ordered picture of scientific methodology that leads to reproducible results and accumulation of stable scientific theories. Instead it points to the opposite direction of serious maladies in biomedical science.

This opens the door for the second, more informal, approach to scientific methodology; an approach that is comfortable with uncertainty and tries to keep an open mind on different ways of doing science. In this approach we embrace both the known and the unknown and assume that before we can write a textbook on scientific method, we must discover what the scientific method is; and this is why we need a science of science.

This view has the obvious disadvantage over the textbook approach in that it cannot tell you what to do, or even what to think. What it can do is to present alternatives and help to work out some of their consequences. So, in practice, both approaches will be used where appropriate: the second one in presenting and comparing different wider theoretical options with their respective advantages and disadvantages and the first one in less controversial situations, where explaining some practical uses arising within these theoretical frameworks is in order.

1.1 ARE WE LIVING IN THE GOLDEN AGE OF SCIENCE?

The sheer volume of modern science is staggering. According to ISI Web of Knowledge Science Citation Index (SCI), which represents about 70% of all published scientific articles, 2,917,318 papers were published in science and technology in 2013. It has been estimated that about 50 million scholarly articles have been written since the beginning of modern scientific publishing in 1665, with half of these published during the last 15 years or so (Jinha, 2010; Ioannidis et al., 2014). In a 16-year period (1996–2011) more than 15 million people authored a scientific paper, while only about 8% of them published during at least 8 of those years (Ioannidis et al., 2014).

While the growth in global scientific output shows no sign of slowing (Figure 1.1), it hides significant regional disparities. The 2.6% annual growth in publishing during the decade from 1999 to 2009 came overwhelmingly from Asian countries (and Brazil). The fact that increase in submission rates to top journals (*Nature* and *Science*) has been about twofold slower indicates that

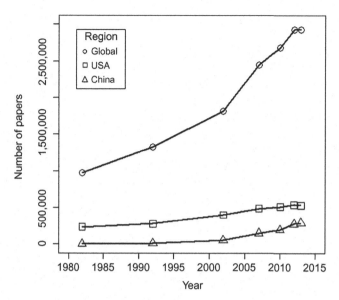

FIGURE 1.1   Growth in the number of papers published annually according to Web of SCI and its US subset.

most of the recent growth has been channeled into journals with lesser impact (Reich, 2013). Indeed, the emerging Asian scientific powers have recently been characterized as "knowledge sinks," because their scientific output gets under-proportionally cited globally (Mazloumian et al., 2013). Interestingly, on this background of solid global growth, the number of papers published by the US and EU scientists remained almost constant over 1999–2009 (1.0% and 1.4% increases, respectively).

This period of relative stasis in the number of US publications was not accompanied by similar stasis in the number of US authors: the number of authors in academic papers originating in the United States increased fourfold from 1988 to 2012 (National Science Board, 2013).[1] This feat was accomplished by increasing the mean number of authors per paper from three in 1988 to nine in 2011 (National Science Board, 2013).

The growth in the number of papers seriously underestimates the growth in scientific data volumes. DNA sequencing has become incredibly fast and cheap since the first genome sequence was determined in the 1970s. Technical advances that allow collecting of huge amounts of RNA and protein expression

---

1. And the number of US scientists, who coauthored an academic paper, increased by 62% from 1999 to 2009, from about 75,000 to nearly 120,000 (Anon, 2012).

data in single experiments have led to big data "omics" approaches, which have greatly increased the availability of biological data during the last 10 years.

For an example let's take microRNAs, which constitute a diverse universe of tiny cellular RNAs, whose function is to fine-tune mRNA's expression, thus constituting one of several superimposed levels of control for the expression of genetic information. During a 10-month period, ending in April 2012, more than 750 research articles reporting effects of different microRNAs on global mRNA levels were published, each containing amounts of data that would have been sufficient to fill hundreds or even thousands of "ordinary" research papers (Witwer, 2013). There should be enough data to go around!

Even normal garden-variety experimental biology papers have become considerably more data-heavy. A quick analysis of biology and chemistry papers in the *Proceedings of National Sciences of USA* (analyzing 29 experimental papers from each June issue of 1965, 1975, 1985, 1995, 2005, and 2012) reveals a steady threefold increase in the number of authors per paper and in the number of different experimental systems used (from 2.3 on average in 1965 to 6.4 in 2012). Even more conspicuous is the fivefold increase in the number of panels in experimental figures *per* paper (Figure 1.2).

One can only surmise that it takes a lot more data to provide convincing evidence for the hypotheses that were tested in 2012, than was the case 30 or 40 years ago. Now, more than ever, different people are responsible for different aspects of a research paper and in many cases nobody is in the position to control or even form an informed opinion of the quality of every experiment, interpretation, and inference.

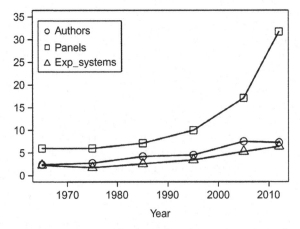

FIGURE 1.2 Growth in the number of figure panels, authors, and experimental system used per paper published in the *Proceedings of the Academy of Science of USA*.

The recent steep rise in the size and complexity of the content of science is in curious dissonance with the dynamics of the institutional growth of science. To better understand this, we must go back to the beginnings.

The flowering of science in the nineteenth and twentieth centuries was the culmination of a long process of germination and growth. Getting from the first non-religious speculations about natural phenomena by the pre-Socratic philosophers to the recognizably modern ways of Galileo and Newton was a slow journey with many false starts and detours. The explosion of the scientific revolution started the exponential growth phase of science, which lasted for nearly three centuries, from about 1700 to (perhaps) the 1970s (Goodstein, 1994). There were about 50,000 professional scientists in 1900, 150,000 in 1950, about 1 million in the early 1960s, but "only" about 3.5 million in 2001 (OECD countries only) (Vinck, 2010). According to the discoverer of this period of exponential growth, Derek de Solla Price, the doubling time of science, and of scientists, was a mere 10–15 years and, therefore, throughout 300 years every scientist that ever lived could have claimed contemporaneity with most of the science ever done. It did not escape the notice of Price that the corollary of the exponential growth of science is that on one fine day the demands of science must exceed the financial possibilities of society. In hindsight, it seems that when he published these thoughts in 1961 (de Solla Price, 1975), the process of slowing down was just beginning. While Price's extrapolation for the number of scientific journals for the year 2000 was around 1 million, the actual number turned out to be closer to 15,000 (Jinha, 2010).

The huge surge of popularity of science among the Deciders of the world during the 1950s and 1960s had a lot to do with the technological nature of modern warfare: in 1940 the United States spent 0.3% of its gross domestic product (GDP) on R&D, by 1965 this had risen to 3% (in real terms the federal spending increased 200-fold) and it has fluctuated between 3% and 2% ever since (Vinck, 2010; www.nsf.gov/statistics/). Today over half of the US federal R&D expenditures come from the Department of Defense.

Also note that the relative stasis over the past 50-odd years in total US R&D spending in proportion of GDP was accompanied by about a twofold relative increase in private sector funding and concomitant reduction in the federal commitment (Figure 1.3). And yet, most of US research (as opposed to development) funds come from the federal government, which in 2009 also directly paid for about 12% of industry R&D activities (not including substantial tax breaks).[2]

To look at it in another way, in 2009 the federal government gave $35 billion to directly support business sector R&D, while the universities got $32.6 billion

---

2. True R&D costs to industry are highly controversial, making it possible that the 12% seriously underestimates the fraction of industry R&D that is directly financed by the taxpayer.

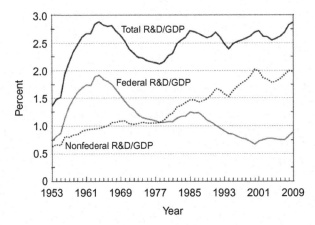

FIGURE 1.3 Ratio of US R&D to GDP, roles of federal and nonfederal funding for R&D: 1953–2009. *Source: National Science Board, Science and Engineering Indicators 2012.*

for their R&D ($54 billion from all sources). The real value of the federal contribution to basic science in the United States has stagnated since 2003, and this trend is set to continue at least in the medium term (Press, 2013). For example, the resources available to the US National Institute of Health diminished in real value by at least 25% between 2003 and 2014 (Alberts et al., 2014). It is not that the industry has picked up the slack in the funding of basic research: in 2009 the business sector provided $247 billion for R&D, only $3.2 billion of which went to universities. Accordingly, in 2009 basic research constituted only 19% of total US R&D (Anon, 2012).

The stagnation of spending on basic research must appear deeply illogical to those who believe that every dollar spent on research will net a handsome return in future economic growth:

> There is no question that when the [US] federal government invests in scientific research there is a tremendous return. Knowledge is gained; discoveries are made with profound implications for our health, safety and quality of life; future scientists, doctors, teachers and leaders are educated; innovations give birth to new technologies, companies and industries; and jobs are created. All of this activity advances our economy and global competitiveness. Yet federal funding for research and development has been on a downward trend for the past decade and has not exceeded 5 percent of the federal budget since 1990. Funding for R&D in FY 2013 accounted for just 3.8 percent of the federal budget, which is a historic low point. Significantly, the budget sequestration assures that research funding will continue to be squeezed for many years to come.
>
> (Anon, 2013)

However, the general quality of results of the branch of economics that studies the economic impact of science is controversial at best (Macilwain, 2010; Lane and Bertuzzi, 2011; Weinberg et al., 2014) and characterization of the wider societal impact of science is mired in conceptual difficulties (Bornmann, 2012). Obviously, there is no good way of conducting controlled experiments on entire societies' long-term investments in science. This leaves us with observational and correlative studies. For instance, we know that there is a strong positive correlation across many countries between investment in physical sciences and future economic growth (Jaffe et al., 2013). Does this mean that we now believe that scientific activity causes economic growth? Here we have (i) a plausible hypothesis, (ii) a strong correlation, and (iii) a timeline, which should exclude as an explanation the backward causation from the riches of society to investment in science. And yet our answer must be "no."

This is because we also know that, unlike for the physical sciences, investments in the sciences of medicine, psychology, biochemistry, genetics, and molecular biology are all strongly negatively correlated with future economic growth. Also, the countries that drive the positive correlation between science funding and GDP growth in the early twenty-first century include Armenia, Azerbaijan, Kazakhstan, Ukraine, Latvia, Belarus, Georgia, Moldova, and China (Jaffe et al., 2013). Simply put, wherever experiments are impossible for practical or ethical reasons, scientific knowledge of causal links is extremely hard to come by (see Chapter 2).

As global economic might shifts to Asia, so does scientific spending. Interestingly, due to fast-growing R&D spending in East Asia, especially in China, global R&D spending is still doubling with every decade or so (www .nsf.gov/statistics/). China's R&D spending is growing at the incredible rate of 18% (inflation-adjusted) and in 2011 it constituted 15% of the world's R&D spending. In 2009 Asia held 34%, the United States 30%, and the EU 22% of the global R&D (National Science Board, 2013). If future manifests itself in the number of first university degrees awarded, then China, which awards 24% of the global total, should celebrate its victory over the EU (17%), and the United States with its 10% share is a spent force.

The R&D expenditures, and by extension knowledge and intellectual property (IP) creation, are extremely lop-sided globally: the top three countries—the United States, China, and Japan—collectively spend over half of the global R&D dollars and, in the other extreme, the continents of Africa and South-America manage to make up 0.8% and 2.5% of global R&D, respectively. As future innovation tends to build on past innovation, whose use is made difficult by IP laws for most of the world's potential innovators, this lop-sidedness of the geography of IP can have important consequences for global trends in history.

The implications and effects of the abovementioned changes in the geography of science are still mostly unexplored and must probably remain so until sufficient distance in time is achieved to enable an unbiased evaluation of the quality of late twentieth- and early twenty-first-century science.

1.2   R&D AND THE COST OF MEDICINE

There are still companies that try to focus on excellent science and that attract first-rate scientists. These are, however, exceptional situations.

(Cuatrecasas, 2006)

The reason why we do research is to discover new knowledge, which, occasionally can be of practical use. Abstract knowledge is rarely directly converted into technological or medical progress. There are two-pronged reasons for this. Firstly, in order to turn a scientific discovery into a practical discovery it must be both technically and economically feasible to produce in a large scale. And secondly, scientific knowledge tends to be less than perfect. This is especially clear in medicine, where transforming a promising biological pathway or target molecule into an effective cure generally proceeds by a series of false starts, trials and errors; first by finding a biologically effective molecule and then by testing its efficacy and safety in animals and humans.

When converting knowledge into products we engage in an activity, called development, which resides in the border of science and technology. While research is all about generating new knowledge, development is (mostly) about engineering new products. For example, we might study the physiological mechanisms and molecular pathways of cancer, which will later be used in the development phase as potential drug targets. In a typical approach, when university researchers identify a protein, whose activation is associated with cancer, this knowledge can be licensed out to a pharmaceutical company, which then tests its huge chemical libraries to identify compounds, which inhibit the target molecule (Angell, 2004; LaMattina, 2008). From there follow clinical trials to determine the safety and efficacy of the potential drug.

Drug development is overwhelmingly done by for-profit entities, the revenues of some of which considerably exceed those of my country. Their critics have pointed out that for taxation and public relation purposes it is in the companies' best interest to inflate their development costs on paper. As the disagreement over the true cost of drug development ranges well over an order of magnitude (Angell, 2004; LaMattina, 2008; Light and Warburton, 2011), pharmaceutical industry R&D statistics should be approached with caution. Indeed, the average cost of new drugs, as cited by the industry, has been increasing exponentially. In the 1950s the industry could get about 70 working drugs for every billion USD invested, by *ca.* 1995 this had fallen to one drug per billion USD (Scannell et al., 2012). Since 1950 the number of drugs approved per billion dollars halved on average every 9 years and by 2010 had fallen about 80-fold in inflation-adjusted terms.

To rub salt into the wound, the history of medicine tells us that the major innovative period occurred between 1930 and 1960, when many of the prototypes of drugs, which we still commonly use, were developed (Horrobin, 1990). For example, each of the five major classes of psychiatric drugs, which we use today, was discovered before 1960. Although many new members have been created and are in current use, we are still confined to these five classes and, ironically, after many decades of intensive study are still far from understanding the mechanisms of their action, not to mention a lack of consensus on whether the current crop of psychiatric drugs actually works (Angell, 2011a,b; Oldham, 2011; Carlat, 2011).

During the last half century improvements in patient care have been decidedly less dramatic than they were before the advent of modern randomized placebo-controlled clinical trials in the 1960s (although nobody wants to suggest that we haven't made many incremental cumulative advances in medicine). It is clear that the huge increases in R&D expenditures have not resulted in commensurate advantages to real patients.

There is no consensus about the causes of this problem, but possibilities seriously discussed in trade press include at least the possibility that the low-hanging fruit have already been picked, the Beatles effect (you have to be better than The Beatles to compete with them in a market where fashion never changes), frequent mergers between pharmaceutical companies, which reduce diversity of the global drug pipeline and sap researcher morale, over-cautious regulators raising the safety bar for new drugs, over-cautious company men who worry more about their pay check than about the long-term prospects of their companies, confounding effects of some subjects in drug trials not taking their medicine, the high degree of removedness of the experimental models of biomedical science from human patients, and putting too much trust in the modern big data approaches to biology (Scannell et al., 2012; LaMattina, 2011; Horrobin, 2003; Mullane and Williams, 2012; Cuatrecasas, 2006; Blaschke et al., 2012).

The explanation for how corporate mergers could lead to decreased rate and efficiency of drug discovery is especially interesting (LaMattina, 2011; Mullane and Williams, 2012). It all starts with a large drug company discovering that its drug discovery pipeline is smaller than its investors might hope for. The reason may well be a general lack of agility of big bureaucratic systems and the quickest solution is, of course, to acquire a smaller, more agile company, whose drug pipeline instantaneously becomes a part of the resulting, even larger, entity. You can probably guess where it goes from there. Every merger reduces the global potential for innovation and thus lays the seed for further mergers, which in turn further reduce the diversity of approaches to drug discovery and thus the long-term prospects of the pharmaceutical industry.

As such self-destruction makes perfect financial sense in the short term of the next shareholder meeting, there is no market-based mechanism to reverse the trend. Accordingly, only a quarter of the members of the pharmaceutical companies trade organization (PhRMA) of 1996 were still around in 2011 (LaMattina, 2011). A 1996 analysis concluded that major pharmaceutical companies needed three new drugs per year per company to maintain reasonable growth, while actually developing 0.25 per year per company, which further underlines that mergers within the industry are inevitable, economically speaking. This continuously unmet productivity goal has been estimated to have reduced the value of the pharmaceutical industry from 2000 to 2011 by $1 trillion (Mullane and Williams, 2012).

Drug trials are typically divided into three successive phases, which precede the approval of the treatment by the regulatory agency (the NIH, in the US case). Phase I trials are small (20–80 participants), often without a control group and involve healthy people on whom the safety and dosage of the experimental drug are tested. Phase II trials are either extensions of the phase I trial with more participants (100–300) or else they look like a smaller version of the controlled phase III trial. Phase III trials are randomized, controlled (either by placebo or by another treatment), and typically double-blinded, involving 1000+ participants. Their goal is twofold: to see whether a treatment works better than the control and to determine the frequency and severity of its side effects. Often at least two phase III trials are needed to win approval for a novel drug.

We will now turn to John LaMattina, an avowed defender of the industry and a former president of Pfizer Global Research and Development. He illustrates the costs of drug discovery by unsuccessful development of the cholesterol-lowering drug torcetrapib (LaMattina, 2008). The story starts in 1990, when academic researchers found that a mutation in a gene for cholesteryl-ester transfer protein (CETP) can increase the levels of high-density lipoprotein (HDL) cholesterol and lower the harmful low-density lipoprotein (LDL) cholesterol levels (Inazu et al., 1990).

Taking their cue, from 1991 to 1997 Pfizer scientists searched for an inhibitor for CETP by screening chemical libraries and synthesizing modified versions of the compound that bound CETP. They selected a strong binder (named torcetrapib), which also worked in animals. Animal safety studies and developing a viable formulation for use in humans were carried out in 1997–1999. The phase I clinical studies on healthy subjects were successfully conducted in 1999. In a subsequent phase II trial in cardiovascular patients it was found that torcetrapib worked best in combination with a statin, so for the phase III trial torcetrapib was combined with atorvastatin (versus atorvastatin alone in the control group). By the end of 2006 the trial was prematurely terminated, because the combination drug was seen to increase mortality over the control.

In terms of time and research effort the preclinical phase (discovering the target, finding the inhibitor, showing its safety and efficacy in animals) was certainly the most costly. In monetary terms the expense for Pfizer for 9 years of preclinical work was estimated between $25 and $30 million. The phase I clinical trials were estimated to cost between $10 and $15 million and the phase II trials $60–100 million. The phase III cost a whopping $800 million (including the investigator fees (patient recruitment, recording clinical results, monitoring adverse events), lab fees (blood tests), study monitors (site visits, etc.), data managers, statisticians, administrative support, and the data and safety monitoring board). This adds up to $32,000 per study participant (of whom there were 25,000).

As a comparison, the 1994–2002 NIH ALLHAT study, which was mostly federally funded and compared the efficacy of four hypertension drugs in 42,000 subjects, "only" cost $130 million or less than $4,000 per subject. Incidentally, the result of ALLHAT was that the older drugs (diuretics) work better and save more lives than the then-new drugs (Cardura and Norvasc). Patients getting Norvasc (Pfizer) had a 38% increase in heart failure rates over those on the diuretic (http://www.nhlbi.nih.gov/health/allhat/qckref.htm).

So, how can government afford to run its studies so much cheaper than industry? According to a knowledgeable critic of the industry, a major component of the investigator costs, which inflate industry's bill, are the fees paid to physicians for patient recruitment (Angell, 2004). They have been claimed to range from $10,000 to more than $30,000 per recruited patient (while the study subjects are generally paid in the hundreds of USDs). This raises two issues.

Firstly, if the true cost of innovative drug development is only a fraction of the total, wasteful (in terms of innovation, not of business) R&D process, then the global R&D costs present us with an erroneous picture of the dynamics of biomedical innovation over the past 50-odd years.

And secondly, if my physician recruits me for a clinical trial, I would naturally expect that our interaction is in my interest, or at least in that of medicine. If I correctly assume that my doctor doesn't need extra financial incentives from drug companies to give me the best advice she can, and that drug firms do not part with their money without good reason, then their motivation in paying my doctor must have something to do with changing her incentives for recruiting me for their trial. Or to put it more bluntly, if for $10,000 my doctor is willing to do to me, what I would do to her, then I have grounds for concern.[3]

1.3 THE EFFICIENCY OF DRUG DISCOVERY

Despite tremendous efforts, pharmaceutical innovations at the patient level are becoming rare events.

(Wehling, 2009)

As estimated global R&D costs grew from $500 billion in 1996 to almost $1,300 billion in 2009 (www.nsf.gov/statistics/), one might expect that in concert with the huge increases in the publication volume of biological science must come commensurate advances in understanding of the workings of cells, organs

---

3. In 2013 the pharmaceutical industry seems to have made payments to more than three-quarters of practicing American medical practitioners, which collectively amounted to about $3 billion (including research payments) (NYT 10.29.2014. What We're Learning from Drug Company Payments to Doctors).

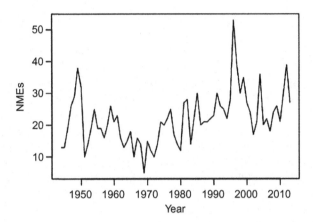

FIGURE 1.4   There has been little annual growth in the number of "NMEs" approved for use by the US FDA.

and organisms, which are in due course followed by new cures in medicine. And yet, since the 1960s there has barely been a positive trend in the number of "new molecular entities" (NMEs) annually approved by the Food and Drug Administration (FDA) for human use, and there is no such trend for the last 30 years (Figure 1.4).[4] The same holds for the so-called priority drugs, which are given the Priority Review status by the FDA because they are expected to work significantly better than available treatments (incidentally, most treatments approved by the FDA are not). It looks as if advances in basic science are rather dissociated from advances in drug discovery!

Also, the efficiency of translating the knowledge of specific drug targets obtained from pre-clinical animal research into successful medicines has been, on the whole, unsatisfactory (Mullane et al., 2014; Mullane and Williams, 2012). The assessment of translatability of a biological target molecule to medical intervention can involve careful weighing of over 20 separate aspects, from *in vitro* evidence to patenting and market issues (Wehling, 2009). In fact, although the vast majority of research effort of drug companies goes into target-based screening, from the 50 first-in-class small molecule drugs approved by the NIH from 1998 to 2008, only a third (17) were discovered through success-ful translation of biological knowledge of drug targets (Swinney and Anthony, 2011). The rest resulted from more old-fashioned trial-and-error methods of drug discovery, which do not require specific knowledge of drug targets.

For example, while in the field of stroke more than 900 treatments have shown benefits in animals, examination of well over a hundred of them indicated

4. Many NMEs contain active moieties that have not been approved by FDA previously, although some drugs are characterized as NMEs for administrative purposes and do not contain novel active moieties.

that success in an animal model adds no value to drugs candidates' future clinical success (O'Collins et al., 2006). Similar results were found with the mouse model of amyotrophic lateral sclerosis where none of the more than 70 promising drug candidates, including about a dozen that had previously been published as very effective in mice, and some that had already led to clinical trial in humans, worked in a well-powered mouse study (Scott et al., 2008).

The proposed reasons for such translational difficulties include low experimental quality of animal studies (lack of randomization, concealment of treatment allocation, and blinded outcome assessment) and sheer complexity of biological systems, but also biological differences between human disease and rodent models of it, using young and healthy animals in lieu of old and sick patients with co-morbidities, studying a homogeneous group of animals versus a heterogeneous group of patients, the use of either male or female animals only, publication bias, and differences in outcome measures and the timing of outcome assessment (van der Worp et al., 2010; Macleod et al., 2008, 2009; Kenakin et al., 2014; Scott et al., 2008; Swinney, 2013).

On top of this, there is growing concern, emanating from industry outsiders, over the quality of evidence collected from clinical trials funded by the pharmaceutical industry and the purity of motivation behind these studies (Goldacre, 2013; Angell, 2004). Indeed, the major players in the industry have in recent years paid billions of dollars in criminal damages and civil liabilities for serious crimes, including knowingly selling ineffective or dangerous drugs, illegally promoting drugs for off-label uses, misrepresenting the results of clinical trials, and using Third-World manufacturing plants resulting in wrongly dosed or ineffective pills. Let me further stress that these examples are not about the generics, but about the major firms, which sell their product dearly. A random look at the *New York Times* drug blog shows:

> *31 July 2013: "Justice Department announces that Pfizer has agreed to pay $491 million to settle criminal and civil charges over illegal marketing of kidney-transplant drug Rapamune; settlement is latest in string of big-money cases involving sales practices of major pharmaceutical companies."*
>
> *27 July 2013: "GlaxoSmithKline agrees to pay Louisiana $45 million to settle lawsuits accusing it of improperly marketing diabetes drug Avandia and other drugs."*
>
> *27 July 2013: "Chinese government releases more details about its accusations involving GlaxoSmithKline, which has been rocked by bribery and corruption scandal at its unit in China."*

The largest settlements are currently in the order of $3 billion: in 2012 GlaxoSmithKline (GSK) pled guilty to criminal charges for promoting its best-selling antidepressants for unapproved uses and failing to report safety data about a top diabetes drug. The agreement also includes civil penalties for improper marketing of a half-dozen other drugs. In 2010 Glaxo had already paid $750 million for knowingly selling contaminated baby ointment and an ineffective antidepressant.

An example of how the results of clinical trials are disseminated to the medical public is described in Turner et al. (2008). They looked at 74 studies of 12 antidepressants, which were mandatorily registered with the US FDA and whose results were re-analyzed by FDA statisticians. From the 38 studies whose results were deemed positive by the FDA, 37 were published in medical journals, all with the same conclusion. From the 12 studies with questionable results according to the FDA, six were published claiming positive results and another six were left unpublished. From the 24 studies, the results of which were deemed negative by the FDA, three were published as negative results, 16 were left unpublished, and five were published as positive results!

This means that in a field where half the studies of specific drugs give negative results, a doctor believing in the evidence-based medicine, who conscientiously reads the medical journals, will see that almost all studies (94%) provide positive evidence for the efficacy of these drugs. Taking into account the unpublished trials reduced the effect sizes for 11 of the 12 antidepressants by 11% to 69% (the one that remained, had no unpublished studies).

A more recent comprehensive study revealed that pharmaceutical companies published only 40% of the results of phase II and III trials that were conducted at least partly in the United States and tested a drug already approved by the FDA (Prayle et al., 2012). The publishing of these trials is mandatory under FDA rules.

There are very few people better versed in the big picture of pharmaceutical research than the executive editor of *The New England Journal of Medicine*. Marcia Angell, who filled this post from 1988 to 2000, summarized her beliefs like this:

> *It is simply no longer possible to believe much of the clinical research that is published, or to rely on the judgment of trusted physicians or authoritative medical guidelines. I take no pleasure in this conclusion, which I reached slowly and reluctantly over my two decades as an editor of* The New England Journal of Medicine.
>
> (Angell, 2009)

This criticism was seconded by Richard Smith, who was editor for 25 years at the *British Medical Journal* and according to whom "Medical journals are an extension of the marketing arm of pharmaceutical companies," whose publication practices of seemingly high-quality double-blinded controlled clinical trials routinely mislead the doctors who try to cure us (Smith, 2005). The current editor of the third top journal of medicine, *JAMA*, was quoted as saying that as drug companies decided 15 years ago to shift dollars from science to marketing, "that has too often led to increased dependence by researchers on industry funding, manipulation of clinical trials, suppression of results, delay of publication, and even deliberate lying. [...] Those outcomes [...] threaten the credibility of biomedical research" (Levin, 2014).

Marcia Angell accuses the industry, among other things, of deliberately designing its trials to yield favorable results. Examples include comparing your drug with a placebo when a better alternative is available, comparing it with a competitor's drug administered in a very low dose, and testing drugs that are likely to be used by the old, on young people, who are less likely to exhibit side effects. And, of course, trying (and often succeeding) to suppress negative results. For patenting reasons it is in the industry's interests to sell newer rather than older drugs. Therefore, its financial interests are best served by not directly comparing new drug candidates with best current treatments and thus avoiding the Better-than-the-Beatles problem. Combined with heavy marketing this means that there is often no reason to think that a new and expensive treatment is better than a cheap generic alternative. If the relevant comparisons have not been made, it doesn't matter if you are the richest man on earth—when you fall sick then the best doctor cannot give you the right advice.

The work of Prasad et al. (2013) may help to shed further light on problems with established medical practices. They reviewed a decade's worth of *The New England Journal of Medicine* and found 363 papers studying established medical practices, which are in wide use as standards of care. Only 38% of these found that the established clinical practices were beneficial to patients and in 40% of studies the examined practice was found to be ineffective or harmful. The list of reversals includes hormone therapy in postmenopausal women, high-dose chemotherapy and stem cell transplant for breast cancer, and intensive glucose lowering in type 2 diabetes patients in intensive care.

Could the low rate of drug discovery be explained by methodological problems in clinical trials? Indeed, a systematic study of clinical trials deposited with the ClinicalTrials.gov mandatory registry revealed that most trials have less than 100 participants and exhibit a "significant heterogeneity in methodological approaches, including reported use of randomization, blinding, and data monitoring committees" (Califf et al., 2012). As a case study on the importance of methodological rigor on the results of clinical trials, Nowbar et al. (2014) studied the effect of elementary logical and mathematical discrepancies in reporting of bone marrow stem cell therapy in heart disease patients on the effect size of the therapy.

They identified 49 randomized clinical trials of stem cell therapy for heart disease, the meta-analysis of which showed a clear benefit of the therapy. The number of identified discrepancies in papers describing individual trials ranged from 0 to 89, with the largest number of discrepancies associated with the largest study. All discrepancies—ranging from dead patients still taking drugs and reporting symptoms to suspicious or impossible statistical results and simple conflicts of numbers in different tables and figures—were found simply by reading published papers; something that the reviewers and editors were supposed to have already done before publishing. What is really shocking is that the authors found a strong linear correlation between the number of discrepancies and the effect size. The five trials with zero discrepancies showed on average zero effect size, and on the other side the five trials with more than 30 discrepancies exhibited a robust 7.7% mean effect size of stem cell treatment.

This is how Nowbar et al. describe their efforts to resolve the discrepancies with the original journals and authors:

> *We have asked for resolution of over 150 discrepancies through journals. None were resolved, although we found it triggered correspondence from lawyers.*
> (Nowbar et al., 2014)

Considering that many companies around the world are already offering bone marrow stem cell therapy to cardiology patients and several large phase III trials are currently (as of spring 2014) on the way or starting, this news couldn't have come at a more interesting time (Abbott, 2014).

To conclude in a more general vein, from 108 reported phase II clinical trial failures in 2008–2010 only 19% were due to safety concerns and 51% were caused by lack of efficacy (Arrowsmith, 2011). A comprehensive study of the success rate of more than 2,500 drug candidates found that more than one-third of the drugs that successfully passed phase II trials never made it through phase III (DiMasi et al., 2010). Things are even worse in the field of oncology where 62% of phase III trials of new treatments failed, mostly due to lack of efficacy (Gan et al., 2012).

1.4 FACTORS THAT ENDANGER THE QUALITY OF MEDICAL EVIDENCE

In his equally informative and scary book, *Bad Pharma*, Ben Goldacre lists a number of ways in which evidence-based medicine is being distorted at the level of evidence collection and dissemination to doctors and patients. To my knowledge, this book is currently the best single review on the subject; it relies heavily on systematic reviews and is up-to-date as of August 2013 (2nd edition).

So, how to best skew the results of clinical trials? It is surprisingly easy if you know what you are doing.

Firstly, the results depend more than anything on study design (Chapter 3). The first thing to do is therefore to ensure that your test subjects would not be representative of the patient population that will later be using the treatment. If you study your drug in younger and healthier people than your average patient, then the chances are that you will see fewer side effects and consequently have a better benefit–harm ratio and a better chance of winning approval for your drug. But it also means that nobody knows if the benefit–harm ratio is still favorable for the patient population in which the drug is actually used.

As most medical treatments have fairly small beneficial effects (but remember that a drug that shaves even a 1% off mortality of a very common disease can save many lives), this leaves doctors guessing if, for example, a bone-fracture prevention drug, which was tested in only women who were mostly in their 70s, would work in the patient population consisting of both sexes and whose mean age is 80 years (Järvinen et al., 2011). The problem of external validity, as it is called in medical parlance, seems to be nearly ubiquitous and current evidence shows that for many treatments more than 90% of real patients would not have been eligible for participation in any of the clinical trials, whose results determined the course of their treatment (Rothwell, 2005). The percentage of patients excluded from clinical trials is rarely expected to be less than 50%.[5]

Another method of distorting the evidence is to design the study so that your new drug is compared against an existing drug, which is given either in an unusually low dose (to reduce efficacy) or high dose (to increase side effects); or to use a placebo instead of an existing treatment (for this you might want to go to some parts of Africa, where you can claim that the usual treatment for most diseases is nothing) (Safer, 2002). There is little doubt about the effectiveness of this strategy, but I am unaware of any reviews quantifying the prevalence of the problem.

Because of cost considerations it can happen that a trial is long enough to discover a short-term positive effect of the tested drug, but not long enough to see the real medical relevance of the drug. This is connected with using surrogate outcomes, like cholesterol levels or blood pressure, which can be easily measured in short-term trials, instead of outcomes that really matter to the patients, like death and disability. As the regulators are increasingly demanding mortality data before approving a new drug for a serious disease, there is hope on this front (LaMattina, 2012).

The above presents a bag of tricks that one can choose from in the study planning stage. Next comes a brief list for the data analyst (see also Safer, 2002).

1. You could peek at your data as it is collected and stop the trial as soon as a formal criterion of statistical significance is achieved. In so doing you harness the normal variation present in your data and allow blind chance to declare victory over the control group (see Chapter 3). Such behavior is nowadays fairly universally recognized as fraudulent, plus there are statistical tools that enable you to peek at data at pre-determined intervals and yet make scientifically meaningful decisions (see Chapter 4).

   A systematic review of these questions identified 217 randomized controlled clinical trials, which were stopped early because of strong results (Bassler et al., 2010). After looking at 91 of them, for which there exist comparable full-length trials of the same treatment, the authors concluded that stopping early led to reporting of effects sizes, which were inflated by about 30%. Also, 76% of the truncated studies had a pre-planned stopping rule, suggesting that such rules can be ineffective in real-life research situations. Interestingly, fully 68% of the truncated trials had been published in six journals with very high-impact factors, while for the control group this number was 30%.

---

5. Damaging in the opposite direction is careless recruitment of patients with similar symptoms but differing underlying pathologies. This dramatically reduces the potential of the study to identify real benefits of tested drugs, and it has been proposed to play a major role in persistent failure of trials involving hard-to-diagnose conditions, such as sepsis (Fink and Warren, 2014).

2. In addition to stopping early, you might want to study a smaller sample and thus leave chance more room to flex its muscles.
3. You could analyze your data in several alternative ways and pick for presentation the one that works the best.
4. You could run many small studies and pick for presentation the ones that work.
5. You could manipulate results by choosing and picking measuring scales or by using open-ended questionnaires wherein patients can choose which side effects to mention, for reporting side effects.
6. You could ignore the patients that drop out from the trial, who are likely to be the people with more side effects and lesser treatment efficiencies.[6]
7. After the trial finishes, you could change the main outcome of your study to present a positive story.
8. You could publish the positive results more than once.

The above is more than a list of possibilities. The use of every trick in it by the industry has been documented (Safer, 2002; Goldacre, 2013). Of course, one should not assume that academic research is clean of misdeeds, and there are other possibilities of obfuscation, some quite complex and hard to detect.

Also, it might be interesting to note that a recent study of statistical re-analyses of already published clinical studies—most of which were done by the original authors—revealed that 35% of the re-analyses (13/37) changed the conclusions of the original studies (Ebrahim et al., 2014).[7]

The methods listed above have a distinct scent of fraudulence about them and, maybe for just this reason, it is unclear how frequently they are used. Be it as it may, there is a method, which is not (yet) widely considered fraudulent and for which we know that it is both widely used and very effective in distorting the knowledge base of medicine. Namely, if nothing else works and your trial refuses to give positive results, simply ignore it and do another one. In your defense you could always claim that you failed to see any treatment effect because the placebo effect was too strong.

The problem of missing trials is currently under extensive academic attention and it can be divided in two. Firstly, there is the problem of withheld raw data behind the published trials. There are now good indications that this may be changing very soon (we have Ben Goldacre, among other people, to thank for it). If and when the raw data behind approved medicines becomes available

---

6. In the situation where most drugs have relatively small effects, ignorance of non-compliance can not only reduce side effects but it can also inflate effect sizes. By some estimates 5–10% of subjects of trials of central nervous system drugs have simultaneously enrolled into more than one trial and are thus in the position to choose which treatment seems to work the best for them and to secretly drop the others, while collecting stipends from all of them.
7. But note that re-analyses of published data are rare and that we do not know what caused these particular ones to be conducted.

for independent analysis, the quality and integrity of the initial analyses can and will be re-checked by independent academic types.

The second problem is non-publishing of negative or harmful clinical results. There is reason to believe that about half of clinical studies will never be published and these are greatly enriched for negative results (Goldacre, 2013). Although compulsory regulation for publication of all phase II and III data is in place, it is widely ignored. (A fine of $10,000 per day is in effect, but has never been levied.)

Although there appears no quick fix in sight, a recent promising trend is for pharmaceutical companies, like Johnson & Johnson and GSK, to increasingly open up their clinical data for independent researchers.

How pervasive is the problem of distorted medical evidence? To be able to make an informed guess, we will look at the most recent systematic review that looks at the available literature on differences between company-sponsored and otherwise sponsored drug trials (Lundh et al., 2012). It identified 48 papers comprising several thousand individual studies, which had compared industry-funded and otherwise funded clinical studies. The results are summarized in Table 1.1.

It seems that drug companies are quite effective in demonstrating the relative effectiveness of their product over the competitor. They are more likely to get and report positive effects by about a third and almost twofold less likely to find harm to patients resulting from their treatments. That is certainly indicative of trouble, but the real magnitude of the problem is unlikely to be revealed by comparing company-sponsored research with academic research. To see why, look at the last row of Table 1.1. If nothing else, lack of concordance between results and conclusions is surely a sign of ethically challenged authors, and 24% non-concordance in non-industry-sponsored trials seems a little high for comfort.

Djulbegovic et al. (2013) published a comparative analysis of all published and unpublished phase III cancer randomized controlled trials (RCTs) conducted

TABLE 1.1 The Success Rates of Industry-Funded and Non-Industry Funded Clinical Studies (Lundh et al., 2012)

	Industry Sponsored (%)	Non-Industry Sponsored (%)
Favorable results	69	51
Favorable harms results	65	33
Favorable conclusions	76	65
Concordance between results and conclusions	70	76

by GSK and the not-for-profit Canadian NCIC Clinical Trials Group (CTG) from 1980 to 2010. This study was made possible by the 2004 legal settlement of GSK with the State of New York, by which GSK agreed to set up a comprehensive online clinical trial registry, thus making results from all trials publicly available. To cite directly from the abstract of Djulbegovic et al.:

> GSK conducted 40 cancer RCTs accruing 19,889 patients and CTG conducted 77 trials enrolling 33,260 patients. 42% (99% CI 24 to 60) of the results were statistically significant favouring experimental treatments in GSK compared to 25% (99% CI 13 to 37) in the CTG cohort (RR = 1.68; p = 0.04). Investigators concluded that new treatments were superior to standard treatments in 80% of GSK compared to 44% of CTG trials (RR = 1.81; p < 0.001). Meta-analysis of the primary outcome indicated larger effects in GSK trials (odds ratio = 0.61 [99% CI 0.47–0.78] compared to 0.86 [0.74–1.00]; p = 0.003). However, testing for the effect of treatment over time indicated that treatment success has become comparable in the last decade.

In fact, the downward trend in treatment success in GSK cancer trials has been dramatic over time. Cancer treatments studied by GSK from 1980 to 1990 seemed to work on average 4.5-fold better than controls. In the period 1991–2000 this fell to 2.0-fold and the trials conducted after 2001 showed on average a 1.2-fold advantage over the control (CTG trials from the same period showed a 1.1-fold advantage).

The saddest thing is that what you just got is the good news. Things may be slowly improving in cancer research.

1.5   THE STABILITY OF EVIDENCE-BASED MEDICAL PRACTICES

Your physician is not supposed to treat you according to the wishes of the pharmaceutical industry. Instead, if she has faith in science, she usually follows clinical practice guidelines, which are developed to assist in patient care. The purpose of these is to convey scientific evidence behind efficacy of specific treatments to medical professionals and thus increasingly base the medical practice on scientific evidence. Evidence-based medicine de-emphasizes intuition and unsystematic clinical experience as grounds for clinical decision-making and stresses the examination of evidence obtained from clinical research.

This kind of systematic scientific approach to medicine is a surprisingly recent invention. The first modern randomized clinical trial, which studied streptomycin for patients with tuberculosis, was conducted in 1948 in the United Kingdom (the therapy was beneficial). The first systematic review (of perinatal interventions) was done only in 1986 at Oxford by Iain Chalmers and colleagues. Systematic reviews and meta-reviews have since become the standard in interpreting medical evidence. Systematic reviews combine multiple studies that look at the same clearly defined research question. Usually the objective is to critically evaluate all the available evidence that a single treatment works.

This is very different from the conventional review format in biology, where the object is to give a review of the results (but usually not the methodology or biases) of recent studies in a wider field of science. Biological reviews usually assume that the individual studies under review are correct, while medical systematic reviews make no such assumption and try to critically compare all relevant independent studies. Biological reviews rarely attempt to be exhaustive of the relevant literature and are not "critical" in the sense that systematic reviews in medicine are. Because, unlike in medicine, very few independent research groups in biology address exactly the same research questions or outcomes, systematic reviews in biology would be very different, if not impossible.

A meta-analysis differs from a systematic review in that the results of multiple studies on the same outcome are re-examined statistically, which "transcends" simple analysis and can be used to reconcile studies with different results. The gold standard of evidence-based medicine is The Cochrane Collaboration (founded in 1993), which currently has more than 28,000 contributors and has so far published over 5,000 reviews taking together evidence in support, or otherwise, of specific medical interventions (http://www.cochrane.org).

Clinical evidence is by common consent divided into quality grades by type of experimental design. It is important to note that the grading reflects the likelihood of obtaining biased results in each design, not the ease with which each design can be perverted by interested parties to give desired results. Therefore the reader of medical literature should consider not only study design as a factor of bias, but also the interests and biases of study authors. This is why ghost authoring is such a big problem in medicine (Goldacre, 2013).

The Hierarchy of Clinical Studies

1. The highest-grade studies in medicine are RCTs, preferably double-blinded. RCT design maximizes the chance that the study group and control group are equal with respect to all the other (confounding) characteristics, which could affect the outcome under consideration. Double blinding minimizes the chance that unconscious biases of the researchers or expectations of patients could change the study outcome. RCTs are the least likely to introduce bias (unless bias is designed into them), but are the most expensive and hardest to conduct. They are often unsuitable for studying extremely rare outcomes.

2. The next best are cohort studies, where subjects are identified based on their exposure or lack of exposure to the risk factor being studied. The fates of the two groups (smokers and non-smokers, say) are followed over time and compared with each other. Because the subjects are not randomly assigned to the two groups, they may well differ on other important characteristics besides the studied exposure. On the positive side, cohort studies tend to cost less than RCTs and allow to study issues for which randomization would be unethical or difficult.

3. The next step down the ladder is the case-control study, which begins with cases who have and controls who don't have the outcome of interest. For example, one might compare patients with lung cancer with an equal number of healthy people to look at the proportion of smokers in each group. Matching of patients and controls for age, education, or whatever confounding factors one could think of is usually performed to make the two groups as similar as possible.
4. On the lowest rung we find case series, which are simply descriptions of the characteristics or clinical course of sets of individuals with a given exposure or outcome. Here a success or harm of a medical practice is described on real patients without using a control group.

In the simplest model of scientific progress, science grows cumulatively as true facts or evidence accumulate. In this scenario clinical practice guidelines are relatively stable and accumulate cumulatively: once a fact is discovered, it is here to stay. In an extreme alternative, evidence is essentially random noise, unchecked by the real world, and science-based clinical practices should accordingly fluctuate randomly, as chance dictates ever-new changes to existing practices.

In this context it would be interesting to know the lifetimes of science-based medical recommendations, which is exactly what is offered by Shojania et al. (2007). They did a survival analysis for a random sample of 100 systematic reviews published between 1995 and 2005, restricting their interest to randomized controlled clinical trials concerning efficacies of specific drugs, devices or procedures and excluding reviews of alternative or complementary medicine. A result was considered overthrown when adding the data from future work changed the statistical significance status of the original (except for trivial changes in P values in the gray zone between 0.04 and 0.06) or when the effect magnitude changed by at least 50% for a primary outcome. In addition, qualitative signals that were used to reverse original findings included new information about harm, important caveats to the original results, emergence of a superior alternate therapy, and important changes in certainty or direction of effect.

The result was that median survival time for systematic reviews seems to be just 5.5 years. Seven percent of the reviews were actually stillborn (main results already obsolete at the time of publication) and only 43 reviews out of 100 published between 1995 and 2005 were still in good health on 1st of September 2006. Shojania et al. provide detailed descriptions of 22 medical treatments that did not survive their analysis, from which the original results were invalidated in 14, and in eight the original results were changed to demonstrate increased effect sizes or benefits to new patient groups. The fluidity of science-based clinical practices was deemed very similar by Shekelle et al. (2001), who studied in depth the validity of 17 clinical practice guidelines set by the US Agency of Healthcare Research and Quality. Their estimated half-life for the guidelines was 5.8 years.

The Shojania et al. study is limited to systematic reviews of randomized controlled clinical trials, which are generally believed to provide the most reliable (and thus stable) medical evidence, but these are far from the only source of medical evidence. Much of the medical research is conducted by observational studies, where the assignment of subjects into a treatment group versus a control group is outside the control of the investigator. A study by Young and Karr provides evidence of how well the observational study designs hold up if compared with presumably better randomized designs (Young and Karr, 2011). They picked 12 randomized clinical trials, which re-tested a total of 52 claimed effects emanating from previous observational studies. The result was that none of the claimed effects were reproduced in randomized studies, and five gave opposite effects to the observational study.

If observational studies are untrustworthy, the obvious question is: How much trust is actually placed in them? To search for an answer, Tricoci et al. looked at clinical practice guidelines released by the American College of Cardiology and the American Heart Association, to find that only 11% of specific recommendations emanate from multiple randomized trials or meta-analyses, while 48% are based on "expert opinion, case studies, or standards of care" (Tricoci et al., 2009). The proportion of recommendations, for which there is no conclusive evidence, had actually grown over time. Very similar results were obtained in an analysis of Infectious Diseases Society of America guidelines (Lee and Vielemeyer, 2011).

In conclusion, although the quantity of scientific evidence pertaining to medical practices has been growing exponentially for many decades, the stability of many specific evidence-based recommendations is consistently low. A lot of the scientific evidence behind medical treatments thus seems to be contradictory and potentially misleading.

1.6 REPRODUCIBILITY OF BASIC BIOMEDICAL SCIENCE

With millions of research papers published every year, instead of the "mere" tens of thousands in the 1950s and 1960s, does it mean that scientific knowledge accumulates accordingly faster? Growth of knowledge is not easy to quantify, but it does in fact seem that more real scientific breakthroughs were achieved in the olden days of physics, chemistry, biology, and medicine. Of course, it is entirely possible that the grass was greener then because of the benefit of hindsight: we simply don't know yet, which of today's discoveries will help to build the future. There is also the possibility that precisely because science works so well, the low-hanging fruit have already been picked and thus, as time goes on, progressively more effort will be required for ever more meager scientific progress (Horgan, 1996). This scenario was predicted to end with a "golden age" where technological bliss is accompanied with the stagnation and slow death of science (Stent, 1969). To prove or disprove this theory, we should presumably need extra-scientific knowledge about the hidden reservoirs of possibilities, waiting to be tapped by scientists. Barring that, the believers can only wait for the hour when the costs of doing science exceed the tolerance of society and the universities will be shut down.

Luckily, there is an empirically tractable approach to the problem of the progress of science. This concerns the quality of research. It is possible to ask and answer, what is the frequency of errors in scientific data? How does existing quality of research affect the conversion of data into scientific knowledge? Are scientific results reproducible? Furthermore, in addition to the above micro-level questions, we could build macro-level models describing the workings of science, and compare them with the facts on the ground.

For example, if science is working perfectly well and we have no reason for concern, we may expect (at least in the confines of a single experimental paradigm) cumulative growth of knowledge. A prediction from this is that when we start to describe a novel protein, after the initial flurry of discoveries, as we have exhaustively described the limited number of interactions a protein can have, the scientific interest should gradually taper off. (This is what molecular biologists do: they describe the interaction partners, measure the strength of individual interactions, determine the time and place for each interaction, and then build models, which use this information.) Plotting the published literature on a protein over time, one would then expect an inverted U-shaped graph. A subfield of biology is born, flourishes briefly and then, as the front of science moves on, gradually withers.

From time to time we indeed see births of new fields (like the RNAi field in the late 1990s), but do we ever see the exhaustion and death of a field of biology?

As a case study, we will take the p53 protein, which is known to be important in many (or most) cancers. Starting from tens of papers published annually in the early 1980s there was a period of exponential growth in the 1990s, followed by a relative stasis from 1999 to 2007 at a level of about 4,200–4,300 publications each year, after which started a new rise in publication rates, which shows no signs of abating (Figure 1.5). In fact the last year a scientist interested in p53 could have reasonably been expected to read everything that was published on this protein was 1990 (398 papers published). From there we have reached an annual output of more than 5,200 papers that list p53 as a "topic" in ISI WoS (for 2012, 395 of them are listed as reviews). The picture gets even stranger when we look at the growth of cumulative knowledge on this protein (Figure 1.5). To date more than 80,000 scientific publications on p53 have been listed in WoS (excluding conference proceedings). This implies hundreds of thousands of scientific facts, from which one is expected to construct a full picture of the role of p53 in normal and cancerous cells.

Now, perhaps it is worth pondering, what are the implications of this for the cumulative growth of science? Or in a more practical vein, on our ability to comprehensively model even a single cellular pathway, let alone a cell?

What is really surprising about all this is not the large numbers *per se*. It is, to cite David Lane who is a co-discoverer of p53, that "our certain knowledge

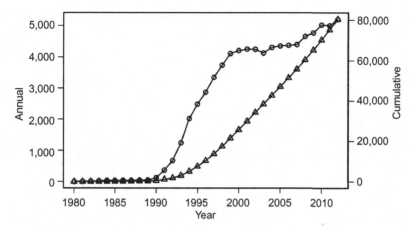

FIGURE 1.5. Annual and cumulative growth in the number of papers listing p53 as a topic in ISI Web of Science (WoS).

of the p53 system is surprisingly incomplete, and internationally, we still have neither effective p53-based therapies nor diagnostics in approved clinical use except in the People's Republic of China" (2005, cited in Leaf, 2013). So the several billions of dollars spent on research into this protein over a 30-year period have led to a less than satisfactory understanding of its role in health and cancer.

In fact, there seems to be a more general problem with ubiquitous signal transduction and transcriptional activator proteins (NF-κB, Akt, transcription factor AP-1, STAT proteins, Bcl2), which tend to pop up in a multitude of studies as potential drug targets to many different diseases (Kenakin et al., 2014). Unfortunately, because of the ubiquitousness of their cellular roles, the specificity of any potential drug that targets them must be questionable. To cite from a recent review:

In the interim, reviewers supporting the peer review process have to deal with a deluge of articles that document the effects of drugs … on the expression and presumed activity of various signaling pathway members that inevitably conclude that NFκB, AP-1, Akt, etc. are targets for the treatment of Alzheimer's, cancer, diabetes, obesity, hypertension, atherosclerosis, RA, depression, etc., etc. Given the finding that the effects of so many different natural products appear to converge on the very same battery of signaling pathway targets, it becomes difficult to reconcile a natural and very necessary level of skepticism with the naive hyperbole surrounding the reporting of these findings in the context of viable targets, the majority of which lack any pretense of replicability, specificity, concentration/ dose dependence…

(Kenakin et al., 2014)

As mentioned in the above citation, the quality of experimental science is often assessed by the reproducibility of its results. This is why journals require (at least as a stated principle) that all published experiments be sufficiently described to allow their independent replication. The idea is that when other scientists successfully repeat the original work, it is unlikely that the original results were caused by chance, and they are therefore expected to give insight into the real world. Conversely, the failure to replicate results by competent colleagues is taken as a strong indicator that something must be amiss in the original work. In the next chapter we will examine some important limitations to these assumptions, but for now we will resign ourselves to looking at the evidence for reproducibility of biomedical research.

1.6.1   Genome-Wide Association Studies

Reproducibility of big science "omics" experiments has been an active field of study for over a decade. The reasons for this interest are manifold: in omics we use novel and fast-developing techniques; the machines that generate data tend to be complicated, expensive, and quickly driven to obsolescence; experimental designs and data analysis are often very different from what "small science" experimenters are used to; and the quality of results cannot be intuitively assessed. On the other hand, the sheer volume of data creates possibilities for statistical quality control, which would be impossible in small science (Leek et al., 2010).

For example, when a genome-wide association study (GWAS) shifts through 3 million single nucleotide polymorphisms (SNPs) in 20,000 people, to end up with half a dozen SNPs that weakly correlate with a disease, intuition will be useless in assessing the veracity of results. This has been aptly described as looking for a needle in a stack of needles (Cooper and Shendure, 2011). While today most GWASs focus on polygenic diseases and traits where the genetic contributions of most associations are small, the method has a chequered record of reproducibility even when used for single severe disease mutations (MacArthur et al., 2014).

By the end of 2012 more than 8,000 genetic loci had been associated with more than 250 diseases or traits using GWAS (Panagiotou et al., 2013). The number of GWASs annually published has risen from around a dozen in 2006 to more than 5,000 in 2012 (http://www.genome.gov/gwastudies/). However, the biological and medical importance of much of this work has been questioned and there are reasons to think that most observed associations are either not real or are so weak as to make the question of reproducibility moot (Ioannidis, 2005; Cooper and Shendure, 2011; Manolio et al., 2009; Mullane et al., 2014). (Although there is a growing number of reproducible associations, which tend to be enriched in gene-rich regions and depleted in intergenic areas (Schork et al., 2013).) Meanwhile, a meta-meta-analysis of GWAS meta-analyses by Panagiotou et al. (2013) revealed that in only 7 out of 62 meta-analyses (ankylosing spondylitis, Crohn's disease, rheumatoid arthritis, multiple sclerosis, ulcerative colitis, type I diabetes, and Parkinson's disease) were the discovered associations able to explain more than 20% of the genetic variance, and that there was only a weak correlation between the number of SNPs and the proportion of variance explained.

To make things worse, replication studies of many disease and gene associations often demonstrate very poor reproducibility, even within the originally studied populations. In several attempted replications of GWASs for type 2 diabetes mellitus, only about 15% of the claimed associations could be reproduced by any of the groups that tried (Hong et al., 2012). In schizophrenia GWAS we have more than 900 reported associations, 25 of which were successfully replicated in a large study ($N = 37,000$) (Ripke et al., 2014; Flint and Munafo, 2014).

Such a poor record of reproducibility can reflect either a high rate of misidentification of associations or a lack of statistical power to find existing associations. In accordance with the latter possibility the analysis of (Dudbridge, 2013) indicates that for GWAS to characterize the majority of relevant genetic associations with complex phenotypes, the sample sizes must be greatly increased (but see Section 5.6 for a discussion of why this might not work). This means that to meaningfully measure the associations of, for instance, breast cancer, with a million SNPs, it would be necessary to collect and analyze at a minimum hundreds of thousands of cancer patients and as many healthy controls. Current sample sizes tend to be at least an order of magnitude lower (Dudbridge, 2013).

There is an advanced variant of GWAS, called epigenome-wide association studies or EWASs, which does not look at genetic polymorphisms, but at epigenetic polymorphisms (e.g., DNA methylation patterns). A comprehensive critical review of this field identified 257 EWASs published from 2008 to mid-2013 (Michels et al., 2013). About half of them searched for methylation patterns associated with cancer. Because, unlike for DNA mutations, changes in genome methylation give continuous signal (most DNA positions are not 100% methylated and the signal is an average from many cells), and the methylation patterns change over time, one would expect that required sample sizes for EWASs must greatly exceed those of GWAS (Michels et al., 2013). And yet the actual median sample size used in EWAS was 46 (range: 6–2,442). If GWAS results are anything to go by, this number seems to be low by several orders of magnitude! In addition, about two-thirds of the EWASs did not try to validate their results with a second independent sample, thus greatly increasing their chances of publishing false-positive results.

The potential good that is expected to come from GWAS can be seen in three ways. In one view the goal is to predict disease risks from individual genotypes. This approach has been largely thwarted by the missing heritability problem (Eichler et al., 2010; Manolio et al., 2009). By design the GWAS approach is restricted to common DNA variants and it cannot find genomic insertions, deletions, inversions and copy number variants, which as we know can be associated with diseases like schizophrenia and autism (Mullane et al., 2014). The missing heritability problem is that for many common illnesses, like heart disease, where we know from classical genetic studies that a substantial genetic component exists, GWAS has not been able to add much to risk prediction by conventional environmental and behavioral risk factors and family history (Do et al., 2012). Even for rare diseases where family history has practically no predictive power, like Crohn's disease, SNPs were predicted to be merely "potentially useful as supporting evidence alongside other types of clinical information" (Do et al., 2012)—signifying very unsatisfactory current predictive power of GWAS results. In addition, 90% of GWAS-based studies of genetic risk prediction, published in 2010, failed even to describe their results and methodology precisely enough, to make external validation of their risk prediction models feasible (Iglesias et al., 2014).

For reasons that are unclear, GWAS have been more successful in pharmacogenomics, where clinically relevant predictions for individual patients about efficacy and toxicity of at least 14 drugs can already be made (Meyer et al., 2013).

In the second view, the reason behind GWAS is to generate new hypotheses concerning new molecular pathways, which will be subsequently tested by other, presumably more traditional, means. While the first approach is personalized and requires the associations to have a reasonably high predictive power, the proponents of the second approach are willing to accept tiny effect sizes, as long as they keep finding new and interesting candidate genes for their favorite traits. However, given the very small predictive power of most GWAS results, it looks like that by the time we are able to explain a sizeable proportion of the genetic component of any complex trait, we may well have associated most human genes with every complex trait studied. Or, alternatively, we may discover that most of the genetic contribution for complex traits is hidden into very rare SNPs or into complex interactions between genes, neither of which is currently amenable to GWAS (Eichler et al., 2010).

It is becoming clear that the vast majority of identified associations cluster into gene regulatory regions, making it very likely that their biological significance involves plenty of complicated transcription-factor-mediated networking (Maurano et al., 2012). This means that the biologically causal genes may not be genetically linked at all with the SNPs that reveal themselves in GWAS. Thus, replicability of specific GWAS results may not say much about their (short term) scientific worth.

Another problem with the view of "GWAS as biological hypothesis generator" is that it seems to be factually wrong: A comprehensive analysis of 90 papers from 2011, which cited the ArrayExpress database revealed that none of them used its data troves to test specific biological hypotheses (the common use of data was to help in generating new computational methods, analysis of bulk data for biological insight, or for independent GWA replication studies) (Rung and Brazma, 2012).

The use of GWAS in the generation of biological hypothesis is ultimately a question of cost-effectiveness; the question "was it worth it?" will no doubt be eventually settled by the users of these hypotheses, that is, by traditional small-science biologists.

The third view is that the ultimate goal of omics is not so much in helping traditional experimental biology as in supplanting it with an integrated model, which entails the totality of omic data collected at every level from DNA polymorphisms to small metabolites. When the model is released, a biologist will have two choices: either to keep on studying the wet and fickle cells, or to switch to studying the virtual cell by virtual (and instantaneous) manipulations. There can be no doubt that the second approach will provide faster results, but the answer to the question, if studying a model will be relevant to the study of biological organisms, will surely depend on the coverage and quality of the omic data that will be incorporated into the model. The tolerable level of errors that can be incorporated into the model is currently unknown—as are the actual error levels in various omic approaches.

A much-publicized example of attempts to integrate large-scale omic data into a single picture of the working of the cell is the very large ENCODE (Encyclopedia of DNA Elements) project, whose main paper listed 594 authors (The ENCODE Project Consortium, 2012). The goal of the project was to map regions of transcription, transcription factor association, chromatin structure and histone modification in human cells. An early result was a matrix of assays versus cell types, encompassing 1,640 datasets. The main conclusion from this matrix, published with great fanfare, was assignment of biochemical "functions" to 80% of the genome.[1] ENCODE claimed that 62% of the genome is transcribed, 56% is associated with modified histones, 15% is found in open-chromatin areas, 8% bind transcription factors, and 4.6% consists of methylated CpG nucleotides.

Strangely enough, the authors chose to label any nucleotide that was positive in any of the above-mentioned assays as "functional," thus promoting a completely new paradigm of human gene expression. Their critics were quick to point out that such a framework is in striking disagreement with the evolutionary framework of biology (Doolittle, 2013; Graur et al., 2013). Because only about 3–9% of the nucleotides in the human genome appear to be under purifying selection, it seems that the vast majority of ENCODE results represent remnant actions of pseudogenes, organismically non-significant actions of genomic parasites (transposons), random biological noise and/or false-positives (Eddy, 2013; Graur et al., 2013).

If this interpretation holds, then the ENCODE data can be no more than a first step on the road to create a useful map of gene expression, which could subsequently be used either in traditional hypothesis-driven small science or in constructing big science models of the functioning of the cell (a subsequent greatly enlarged dataset did not change this conclusion, Muerdter and Stark, 2014). The adoption of a concept of "function" that ignores evolutionary considerations may be chiefly responsible for turning a potentially useful resource for hypothesis generation into a media circus (Graur et al., 2013).[1]

---

1. And even this number should be an underestimate, because only a few cell lines were heavily studied by the ENCODE consortium.

A common explanation for why it is prudent to collect large hypothesis-free datasets is that they may come in handy in the future when, presumably, we will have the hypotheses to test them by. However, this idea presupposes both good enough accuracy of the present data and the stagnation of technological innovation between now and "the future." Should it turn out—and by our present experience it seems likely—that in 5 years, say, we can produce more accurate datasets more cheaply, then the ones that are collected today at great cost and poor interpretability will be studiously forgotten.

There is suspicion whether accurate prediction of genetic susceptibility is ever going to be achievable for most common diseases. This is because of statistical power considerations the genetic disease associations with the strongest effect sizes were likely to be revealed in the first and smaller studies, leaving the smaller effects for later and larger studies. Therefore, since the first crop of GWAS was able to collectively explain a very small proportion of genetic contributions to most diseases, there is little reason to think that any common DNA variant, which could predict the onset of a common disease, still waits for discovery (Kraft and Hunter, 2009). This could mean that as the GWASs grow ever larger, more expensive, and more sophisticated, their results will become ever more significant statistically and ever less significant biologically. As concluded in an authoritative review by the Evaluation of Genomic Applications in Practice and Prevention Working Group: "Genetic risk profiling based on GWAS study results is unlikely, as currently developed and implemented, to translate into management decisions that improve individual health or provide significant additional information as compared with traditional risk assessment" (EGAPP Working Group, 2014).

1.6.2 Microarray Studies

Another popular rapidly developing omics approach is transcriptional profiling. Here microarray studies, where cellular levels of thousands of RNAs are measured by massively parallel hybridization to artificial DNA probes, were until recently all the rage. To ensure their replicability, mandatory policy of data deposition to public databases, based on the Minimum Information About a Microarray Experiment (MIAME), was adopted by most journals around 2001. Following its requirements not only allows people to independently repeat microarray experiments, but also facilitates re-using of the original data by independent researchers, thus giving the society a better return for its buck.

Indeed, from a comparison of two large-scale microarray experiments we know that if they are done correctly, then the correlation between mRNA expression profiles obtained in different labs using the same cell lines, but different assay platforms, can be very high (Haibe-Kains et al., 2013).

You might think that when a journal demands deposition of raw data as a prerequisite for publication, the authors have no choice but to comply (especially as data deposition provides clear benefits in terms of added peer trust to the results, increased number of citations to your paper should other researchers re-use your data, and better chances of spotting errors). You might well think so, but then you are wrong! Six years after the official adoption of the policy, a review of microarray studies across 20 journals found that the rate of deposition of datasets was below 50% (Ochsner et al., 2008).

In another effort to assess quality of microarray studies, Ioannidis et al. (2009) tried to assess data for 18 primary studies. They successfully retrieved data for ten, but could reproduce quantitative results for just two studies. In a heroic forensic bioinformatics effort Baggerly and Coombes (2009) tried with the help of original authors to reconstitute several microarray study designs, whose results had already led to changes in cancer clinical trial protocols. Their conclusion was that poor description of original data in all five studied papers masked many simple errors and the ultimate falseness of the findings. Common errors included running all controls before treatment samples, mixing up the gene labels (off-by-one errors), and mixing up the group labels (sensitive/resistant). These mistakes are easy to make and have the potential to completely invalidate the results. A simple one-cell deletion in your Excel spreadsheet, coupled with shifting up of all values in the affected column is a common source of error that is hard to detect. Unlike in traditional hypothesis-driven research, in omics studies such errors can easily escape detection by intuitive grasp of the absurdity of the ensuing biological conclusions.

A dozen years after adoption of the MIAME policy by most journals, Kenneth Witwer checked whether things had changed by looking at MIAME compliance of 127 articles, all containing microRNA profiling by DNA array techniques (Witwer, 2013). He found that almost three-quarters of the papers did not follow the MIAME rules. Only 34 papers out of 127 deposited full datasets and described their study details adequately. In addition, Witwer found that the papers, which were MIAME-compliant, had on average much higher experimental quality, suggesting that perhaps people, who do not like to share their data, are acting rationally after all.

Another interesting question arising from the low technical quality of published papers is: What is the role of journals in selecting what gets published and what does not? The Witwer sample was collected from the four journals that published the most miRNA array papers (*PLoS One*, *Blood*, *Oncogene*, and *Journal of Biological Chemistry*). While none of these are considered top journals, all are respectable outlets for experimental biology.

It very much looks like journals are not enforcing their own publicly stated rules for the description and sharing of experimental data, and the reason cannot not be lack of resources. Moreover, if the authors really believed that providing accurate descriptions of experiments and depositing raw data are required for consideration of the manuscript, they would surely comply. Describing the experiment accurately is not an impossible requirement, and is probably often easier to do than fairly discussing the importance and novelty of your results.

Every submitted paper goes through peer review and is checked by an editor. Raw data submission and full description of experiments are indispensable for a competent assessment of the technical quality of submitted research by reviewers and editors; and every editor surely knows this. Thus the question becomes: If it is possible to sail through peer review without giving the reviewer even a chance of understanding the technical aspects of the research, what is the purpose of the peer review process?[8]

1.6.3 Proteomics

Starting from DNA sequence variations and going through RNA expression studies, we are finally arriving at the biologically most important level of gene expression: the protein level. It is here that fates of cells, organs, and organisms are decided. For example, no cancer is deadly because of the genes it has, but its cells proliferate because their protein complement is different from those in normal tissues. Because the mRNA levels are generally poor predictors of protein levels (Schwanh a usser et al., 2011), proteins must be measured directly. This can be done with high sensitivity and massive parallelity by mass spectroscopic methods.

Overexpression of an oncogene and underexpression of a tumor suppressor (TS) gene can lead to inappropriate synthesis of proteins needed for cell division and removal of cell cycle blocks that can both lead to uncontrolled cell division. There are many ways by which cell proliferation can go haywire, some curable with specific treatments, and all exhibiting different patterns of protein expression. Therefore, classifying cancers by their proteomes (a proteome consists of all proteins expressed in a given cell or tissue) promises a prognosis for individual patients based not on the location of their cancer, but on the underlying biology of each disease. To achieve this, researchers must first compare the proteomes of healthy and cancerous tissues by mass spectroscopy and define the patterns of protein levels that would be of predictive value in diagnosing cancer.

This is exactly what was done in the (so far) most highly cited paper of the proteomics field (Petricoin et al., 2002). Petricoin et al. compared the serum proteomes from 50 healthy women and 50 ovarian cancer patients and used a searching algorithm to identify a proteomic pattern that completely discriminated cancer patients from healthy controls. This pattern was then used to classify an independent set of serum samples with amazingly good results: they claimed a sensitivity of 100% and specificity of 95% for ovarian cancer diagnosis, which surpassed existing diagnostic methods by a wide margin. Although

---

8. Another serious issue is that reviewers are not generally given the software that authors have created to calculate their models. This alone makes thorough reviewing of many big data biology papers a moot question—even if academic reviewers had the time to engage in such an activity for free.

the plan was to turn these results into a marketable test by 2004, in 2013, and more than 2,000 citations later, this hope had still not been realized (Baggerly, 2013; Ransohoff, 2005).

The same holds for many other cancer proteomics studies that report incredible predictive power, but the vast majority of which has turned out to be clinically useless (Ransohoff and Gourlay, 2010). In fact, omic techniques, including proteomics and microarray studies, have during a relatively short time generated an extensive literature on more than 150,000 biomarkers, less than 100 of which had made it into clinical practice by 2011 (Poste, 2011). In 2010 alone more than 20,000 papers on biomarkers were published, including 8,000 potential cancer biomarkers (Prensner et al., 2012).

Things are equally bleak when correlating mRNA expression levels with effects of various anti-cancer drugs on cell lines. A metastudy of results of two large consortiums found—after comparing data from 15 drugs, 471 cell lines, and 12,153 genes—little correlation between results, which were obtained by the two consortiums (Haibe-Kains et al., 2013). In 2010 there were no validated molecular biomarker tests for the early detection of any cancer (Diamandis, 2010; Leaf, 2013).[9]

Behind these dashed hopes lurks the inherently poor reproducibility of proteomics work. This is not due to limited accuracy of measurement. In fact, the opposite may be true. Mass spectrometers are extremely accurate and sensitive, which means that if they are sensitive enough to detect minute changes in proteomes, they are also sensitive enough to do the same for different reagent lots used in preparing biological samples. Small differences in treatment of samples from the control and study groups, confounding influences of differences in age, gender, etc. between the groups, even the order in which samples are loaded to a mass spectrometer, can each make the difference between a true effect and an artifact that looks *exactly* like a true effect (Ransohoff and Gourlay, 2010).

While true effects have only one source (reproducible differences in the proteomes of cancer and normal tissue), the confounding errors come from an almost infinite supply of seemingly insignificant experimental trivia (e.g., did you use the same batch of plastic for storing of controls and cancer samples?). These diversely sourced confounding effects tend to group together and result in differences between experiments run on different dates or by different people. In an ingenious study Leek et al. (2010) made use of the very difficulty inherent in the big data omic approach to assess the prevalence of this type of error. These kinds of systematic errors ("batch effects") are usually undetectable in small data experiments and must therefore be dealt with in the planning

---

9. There are several approved biomarker tests for detecting recurrence or prognosis of cancer—here the accuracy of the test can be lower than when testing the general population (Anderson and Kodukula, 2014).

stage of experiments. However, high-throughput technologies provide enough data to detect batch effects. The trick is to use a statistical method called the principal component analysis to find biological variables, which are strongly correlated with technical variables. Principal components are estimates of the most common patterns that exist across features. For example, if most proteins in serum samples are differentially expressed with respect to cancer status, the first principal component will be highly correlated with cancer status. On the other hand, if most proteins are differentially expressed with respect to run date, as, incidentally, it turned out to be in the Petricoin et al. study (Baggerly et al., 2004), any biological conclusions drawn will become problematic.

In gene expression studies the greatest source of differential expression seems nearly always to be across batches rather than across biological groups, which can lead to incorrect biological conclusions (Leek et al., 2010). Leek et al. analyzed in depth nine experimentally very different high-throughput omics studies, including Petricoin et al. (2002), only to find strong evidence of batch effects in every single one of them. Across the nine case studies 32–99% of measured features showed statistically significant associations with processing date, irrespective of biological phenotypes.

Should most of the data generated by omics result from batch effects, rather than genuine biological differences, it follows that most hypotheses generated by omics are effectively random noise, here only to mislead and confound. This means that many confirmations of existing hypotheses by omic methods will be spurious, lulling researchers into a false confidence and from there to a complex maze of theories, where both true and misleading paths are treaded with equal vigor, and will both get experimental confirmation.

Examples of such spurious confirmation of theories can be found in Baggerly and Coombes (2009), who discuss microarray studies in which gene lists were supplied and "explained" in light of biology, despite the fact that the lists referred to the wrong genes owing to a simple indexing error.

Ioannidis and Panagiotou (2011) evaluated the effect sizes of 35 disease biomarker association studies, selected due to their popularity (each had received more than 400 citations). Only 15 of the associations turned out to be statistically significant based on the largest study available, and only seven had a relative risk estimate greater than 1.37. This means that 80% of the most-cited biomarker studies were red herrings, likely to lead the reader down the wrong path.

1.6.4   Small Science

Even if the fast-moving omics field has trouble getting a grip on the quality of its product, most biologists still work in traditional hypothesis-driven small data settings, which have been around for some time, are less complex, less expensive and should therefore be easier to understand and implement. Small science is not expected to progress in big leaps, but rather in small but secure steps. The process of peer review can be viewed as the guarantee for the right direction of this walk of science. Indeed, successfully passing through peer review is often considered to put a piece of research into the domain of scientific truth.

While the critics of omics have usually come from within the field, the wake-up call for classical experimental molecular biology came from outside of academic science. In 2011 and 2012 two large companies, Bayer and Amgen, took the unusual step of publishing summaries of a decade's worth of their in-house efforts to reproduce preclinical research published in academic journals by academic scientists (Prinz et al., 2011; Begley and Ellis, 2012). Their interest was largely directed at cancer and on papers whose results promised new and potentially medically important knowledge in cancer biology. A reason for going public was stated in the Bayer paper: "success rates for new development projects in Phase II [clinical] trials have fallen from 28% to 18% in recent years, with insufficient efficacy being the most frequent reason for failure." Because the pharmaceutical industry depends heavily on academia for basic science, the quality of research published in academic journals is paramount for progress in drug development. For knowledge transfer from universities to companies to work, the results that are published in academic journals must (at a minimum) be reproducible in independent settings.

In the Bayer effort 23 in-house labs tried to reproduce a total of 67 papers, working on average 6–12 months on each, only to find that the main results of only 20–25% of these could be reproduced (Prinz et al., 2011). Reproducibility did not correlate with journal impact factors of the original studies, the number of publications on the respective drug target, or the number of independent groups, which authored the publications. The authors further state that:

> Our findings are mirrored by "gut feelings" expressed in personal communications with scientists from academia or other companies, as well as published observations. An unspoken rule among early-stage venture capital firms is that "at least 50% of published studies, even those in top-tier academic journals, can't be repeated with the same conclusions by an industrial lab."

In a separate effort Amgen scientists tried to confirm 53 landmark cancer studies, which were defined as papers describing something completely new, such as fresh approaches to targeting cancers or alternative clinical uses for existing therapeutics (Begley and Ellis, 2012). They fared even worse by confirming the scientific findings in only 11% of the cases, leading them to contact authors of the irreproducible papers. This is what they concluded from these interactions:

> Investigators frequently presented the results of one experiment, such as a single Western-blot analysis. They sometimes said they presented specific experiments that supported their underlying hypothesis, but that were not reflective of the entire dataset. [...] Some non-reproducible preclinical papers had spawned an entire field, with hundreds of secondary publications that expanded on elements

of the original observation, but did not actually seek to confirm or falsify its fundamental basis. More troubling, some of the research has triggered a series of clinical studies ....

Comparable replicability issues have been described in a survey of academic cancer researchers where most respondents had at some point in their careers unsuccessfully tried to replicate somebody else's findings (Mobley et al., 2013). In two-thirds of the cases the issues remained unresolved even after contacting the original authors. Furthermore, when researchers tried to publish their unsuccessful replications, only a third of the attempts were successful (Mobley et al., 2013).

A failure to replicate specific clinically relevant quantitative PCR results led to a study of more than 1,600 papers on the quality of experimental design and reporting of qPCR experiments, including reporting on RNA quality, reverse transcription conditions, PCR assay details, and data analysis methodology (Bustin et al., 2013). It documented the dismal state of affairs in most studies regarding good practices and formalized guidelines, which should by now come as no big surprise. More interestingly, Bustin et al. found a strong negative correlation between the quality of reporting qPCR data and the impact factors of the journals, which published the papers. Whereas 72% of journals with a low-impact factor had at least a single paper that used a validated reference gene as an internal control, this portion fell to 27% in high-impact journals. Although papers in high-impact journals have on average longer supplementary texts and should thus be better equipped to describe their experimental methods, they tend to contain more insufficiently described and badly conducted qPCR experiments than lower-impact journals.

It seems that the editors, who prioritize high-impact research, are less demanding on its technical quality. This apparent link between novelty and low experimental quality of published results may reflect an inherent difficulty in fast-moving areas of science. When individual career advancement is dependent on extraordinary results, then the winner will be the contestant with the least scruples. Producing false-positive results should not be a difficult task for anyone with a passing knowledge of statistics and experimental design, while producing extraordinary true discoveries is entirely another matter.

Considering that none of the above leads to much happiness, it is perfectly natural for the reader to next want to know the specifics: which papers were not reproduced by the Amgen and Bayer scientists, and what were the problems in each case? Whole fields would presumably need to reassess their knowledge bases, many papers would be modified or retracted, top journals would need to change their policies, pointless clinical trials would be terminated and their participants saved from unnecessary inconvenience and risk. (Accordingly, there are indications of increased reluctance of pharmaceutical companies to invest in academic pre-clinical research (Hayden, 2014).) Conversely, publication of the relevant data would be a boost for the credibility of the minority of successfully replicated papers. However, neither the Bayer nor Amgen papers provide any evidence in support of their claims; neither do they give us information on which papers they tried to reproduce. Here the interesting point is not the companies' reluctance to provide assistance to their competitors and to open themselves up to lawsuits in the process; but rather the enthusiasm of learned journals to publish claims, according to which most of their content could be actively harmful to science—and do this without asking for any evidence. Luckily, a transparent and well thought-out project to reproduce high-impact preclinical cancer studies (and by doing so, the abovementioned papers) is in the works (Errington et al., 2014).

If you are as curious about these things as I am, be aware that *Nature* editors do not share our enthusiasm. An editorial from the issue of *Nature* that published the Amgen paper (Begley and Ellis, 2012) describes it thus:

> *the overall impression the article leaves is of insufficient thoroughness in the way that too many researchers present their data. [...] It is usually the case that original data can be produced, mistakes corrected, and the findings of the corrected research paper still stand. At the very least, however, there is too little attention paid and too many corrections, which reflect unacceptable shoddiness in laboratories that risks damaging trust in the science that they, and others, produce.*

The reading from the editorial staff of *Nature* seems strangely unworried, and is in part diametrically opposite to the claims by Begley and Ellis, but in one thing everybody seems to agree on: irreproducibility is caused by simple and relatively easily avoidable mistakes. If this were true (and this book will argue the opposite), then the solution to maladies of biomedicine would be relatively easily taught methodological fixes. For example, in 2013 Begley published six easy rules, following of which should improve matters: (i) blind your experiments; (ii) repeat experiments; (iii) present all results; (iv) include both positive and negative controls; (v) validate your reagents (are the antibodies specific?, etc.); and (vi) use statistics appropriately (Begley, 2013). On the face of it, these are not complicated commandments. Unlike for the high-throughput omics, where experimental systems are so complicated that confounding batch effects are thought to be practically guaranteed to happen (Baggerly, 2013; Leek et al., 2010), in here anybody, who learns a few rules and abides by them, can apparently be a scientist, and honest too.

1.7 IS REPRODUCIBILITY A GOOD CRITERION OF QUALITY OF RESEARCH?

Reproducibility as a criterion of success has more than a few problems. Firstly, there is the technical problem of how to measure reproducibility. An excellent recent initiative to transparently measure the reproducibility of cancer research proposes four measures of reproducibility: (i) P values of both sets of measurements have to be statistically significant; (ii) to use the original effect size as the

TABLE 1.2 Possible Causes for Replication and Non-Replication of Research Findings

Successful Replication of Results	Failure to Replicate
Effect is not true—replication is due to sampling effect (solution: larger sample)	Original effect not true—was due to sampling effect
Effect is true	Original effect is true—non-replication due to low statistical power of studies (solution: larger sample)
Effect is real but not true—the studies replicated bias	Original effect was due to complexity of experimental system—replication will occur given enough effort, but the experimental system is prone to bias

null hypothesis (which means testing for no change of effect via non-significant *P* values); (iii) to compare the effect sizes of both series of experiments and to see whether their respective confidence intervals overlap; and (iv) to combine both datasets in a meta-analysis (Errington et al., 2014).[10] As we haven't even considered Bayesian and likelihood methods, these options obviously do not exhaust the possibilities and, clearly, there is no consensus on how to measure replicability.

To put the concerns over poor reproducibility of experimental results into wider context, it is helpful to think of the various possible reasons for each outcome of experimental replication. As it turns out, it is possible to see true effects, biased effects, and ghostly sampling effects both under replication and non-replication conditions (Table 1.2).

When the effect is in fact true and the researchers conduct several independent studies to verify it, intending to publish them in a bundle, then, unless the statistical power is very high, it becomes exceedingly likely that at least one replication will fail (Murayama et al., 2014). Thus the requirement to replicate findings in independent experimental systems as a prerequisite for publishing has the potential to both exclude true effects from the literature and to include false ones (due to biases, see below).

We will now take a closer look at the possible interpretations for a failure to replicate.

1. The effect is not real and the original result was an instance of sampling error. If the effect is not real and the scientist works at the significance level 0.05, then he should be able to spuriously reproduce his irreproducible results in

---

10. *P* values and null hypothesis testing will be explained in Sections 4.5–4.7; see also the glossary.

about every 20th try (and this low rate assumes perfectly met assumptions of the statistical test, plus no systematic error). Every result that came out once will do so again, if only the researchers try hard enough. This could be a factor in explaining why some professors find that, as the years pass by, it takes their students ever longer to reproduce the work of last-years' crop of postdocs.

2. The effect is real, but the sensitivity of the experiment is low (Stanley and Spence, 2014). This means that the sample sizes and effect sizes are too small to provide good evidence for a real effect on most tries (but not so in a few lucky tries). The results of replication experiments can still be useful as a basis for meta-analysis.

3. The effect is real, but the experimental system doesn't travel well. If a system only works well in the lab that developed it, then the likeliest reason is that even the people who created it understand its workings poorly. Its performance is at the mercy of unknown factors, which cannot be described in papers or lab protocols. For example, when a cell line changes in random unpredictable ways, the results of experimental treatments can also change randomly. As technological progress in science seems to be accelerating and the new techniques are increasingly expensive and hard to master, poor control over experimental systems has been an increasingly serious problem in experimental biology for at least the past two decades (Bissell, 2013). This means that seeing or not seeing the effect depends ever more on uncontrollable aspects of the experimental system. Careful researchers may be able to control their system for most of the time, but they will still have poor control over the interpretation of its output. In other words, if the experimental system is not robust, it is likely that researchers don't understand its workings well enough, and therefore it is easy to misinterpret its results. This results in increased frequency of bias and decreased external validity of the results (see Chapter 3). Therefore, it often makes little difference that the observed effect is real, as its scientific interpretation is likely to be wrong anyway (but the experimenter will probably never know, how likely!).

To increase the chances that replication of a study will work as intended in discriminating good science from the bad, the samples must be large enough to convey a low rate of false-negative results, experimental systems should be simple enough for the researchers to understand their workings and the replication of the original study should be "straight," meaning that the results of "conceptual replications," which employ similar but not identical experimental systems to the original work, usually fail to convey strong evidence for or against the original result. Replication of a scientific result is by no means an easy task; it should not be combined with novel research into the effects, whose very existence is investigated.

Currently publication of a straight replication of other peoples' results is a rare occurrence (at least outside medicine); for example, in psychology replications of barely 1% of publications are published—and these include replications by original authors (Makel et al., 2012). In psychology, published replications by independent authors are successful about 65% of the time and replications by original authors in more than 90% of the cases (Makel et al., 2012).

1.8 IS BIOMEDICAL SCIENCE SELF-CORRECTING?

Maybe things aren't as bad they look? Sure, the published results are often unreliable and this can be annoying, but many scientists, not to mention philosophers of science, would tell you that this is exactly how science is supposed to work. The secret ingredient is not to be right every time, but to expose one's results to critical scrutiny. By repeating the experiments we can weed out errors and end up with a quality product, which has withstood rounds of criticism and attempted experimental falsification. Thus it is your willingness to subject your work to criticism and learning from mistakes that makes science self-correcting and holds out a promise of eventually converging on truth.

This admirable theory of scientific progress was expounded by arguably the greatest twentieth-century philosopher of science, Karl Popper (1902–1994) (Popper, 1963). We will take a closer look on the philosophy of Popper in Section 2.5, but for now the question is not "does it make sense?" but rather "how often do scientists actually correct each others' mistakes?"

Here you can do a little experiment. First ask yourself, what is a realistic error rate for your own research? If you do a lot of statistical significance testing then you are committed to incorrectly rejecting at least 5% of the tested null hypotheses (see Section 4.7) and—due to unmet assumptions of the tests, biases, experimental errors, and the actual fraction of false null hypotheses—the fraction of false conclusions can be much greater. If you are a mid-career scientist, you should now open your publication list and count the number of experimental conclusions that you have published. If mutant A grows slower than the wild-type, while mutant B grows at a similar rate, this already makes two results, so your list is likely to consist of hundreds of entries. Now comes the interesting part. Please count the number of published results that have been later invalidated in print by you or by others, calculate the ratio of invalidated to total number of results, and compare this number to the error-rate that you just postulated for yourself. The result will give you a somewhat inflated (because you are likely to underestimate your error rate) estimate of the power of self-correction of science.

Sadly, none of my published results have yet been invalidated in print, which a critical rationalist would probably take as a measure of the importance of my work, not the quality of it. I can only hope that you fared better.

A critic of this little experiment might point out that maybe I'm overestimating the rates by which errors pass through the peer review sieve and therefore the very low numbers of publicly contradicted results reflect the efficiency, not the inefficiency, of the scientific enterprise. To address this, we shall look with Stroebe et al. (2012) at a sample consisting solely of fraudulent papers. Stroebe et al. analyzed the output of 36 notorious scientific cheats (caught between 1974 and 2012), who had published a total of 804 certifiably fraudulent papers. They selected for analysis only high-profile cases involving prominent scientists and popular media attention, which makes it likely that published works, which were later deemed fraudulent, must have initially commanded a certain level of interest in their respective fields.

Of those frauds only two were discovered because of a failure to replicate findings. One of the two involved a 1982 failed self-replication of his results by an ex-postdoc, asked for by his suspicious coauthors. The other involved a fraudster, who must have been really desperate as he chose to "discover" a new element for the periodic table (element 118). In the new element discovery business the rule is that another group must reproduce any discovery, before it can be accepted. In the event three groups tried and failed.

> The results of Stroebe et al. (2/804) imply that there is about a 0.25% chance that a high-profile false result will be retracted by virtue of independent replication. Fang et al. showed that out of 25 million papers registered in Medline, about 2,000 had been retracted by 2012, and that almost half of those retractions were caused by fraud or suspected fraud (Fang et al., 2012). Taken together with the systematic review of Fanelli (2009), according to which 2% of scientists would admit to having fabricated, falsified or modified data: out of those 25 million papers we may expect about 50,000 to be fraudulent, and therefore for the fraudster the chances of feeling the negative consequences of their acts would be about $10^3/5 \times 10^4 = 0.02$, or 1 in 50 per fraudulent paper published. If it were assumed that not all fraudsters would admit to fraud when asked about it, and that on average each would publish several fraudulent papers, the estimation of the 1 in 50 risk of discovery becomes inflated, and the 50,000 fraudulent papers a deflated, number.

Incidentally, the Stroebe et al. analysis revealed that fraudsters are most often caught due to whistle-blowers (56%) and due to inconsistencies spotted in their papers by outsiders (15%). Journal editors and peer reviewers were responsible for the uncovering of only three cases of fraud in their sample (Stroebe et al., 2012).

This does not bode well for the efficiency of fraud detection. If the main possibilities of being unmasked are being caught red-handed at the bench or being careless enough to write inconsistent papers, and the gain for an individual fraudster is the ability to control the direction and speed of his or her career, it doesn't take much imagination to arrive at an interesting mental image of any good university's faculty.

Of course, there are instances of independent invalidation of published studies, and of the ensuing controversy. For a case study we will turn to the field of TS genes.

The first gene whose germline mutations predispose to human cancer, RB or retinoblastoma, was discovered in 1987. By the turn of the millennium there were just 24 TS genes, but by 2003 a database (TSGDB) lists 174 TS genes, and by June 2013 the list had been expanded to 716 human TS genes (http://bioinfo .mc.vanderbilt.edu/TSGene/). Of the 716 "TS genes" the notation of most (504) originates from a single data source, indicating the extent of fluidity in the field. As there are only about 20,000 genes in humans, more than 3% of them are currently candidates to protect us from cancer, and this number will no doubt increase in the near future.

One of the better-cited (739 citations as of January 2015) discoveries of a new TS emerging in the explosive growth of TS research of the early 2000s was RUNX3 transcription factor, which was described by a collaboration of a Japanese and a South Korean group as being expressed in normal gastric epithelium (Li et al., 2002), while the absence of this protein was strongly associated with gastric cancer. Deletion of RUNX3 also increased the proliferation of mouse gastric mucosa, suppressed apoptosis (programmed cell death) in the gastric mucosa, and suppressed the proliferation of a cell line. Moreover, Li et al. were able to show that mutating the RUNX3 gene led to increased tumor growth in a mouse model, and, vice versa, overexpression of the RUNX3 led to reduced tumors. So, presented in this paper are at least five independent lines of evidence, all supporting the hypothesis that RUNX3 protein can protect us against gastric cancer. Accordingly, their paper is titled "Causal Relationship Between the Loss of RUNX3 Expression and Gastric Cancer."

Over the next decade hundreds of studies from other labs confirmed the role of this gene as a TS and linked its inaction to numerous other types of cancers including bladder, brain, breast, colorectal, liver, lung, pancreas, and prostate cancers. However, in 2011 serious doubt was cast over all this work by a paper that aimed to overthrow the crucial result of the original 2002 work that RUNX3 is expressed in the gut epithelium (Levanon et al., 2011). The title of this work is "Absence of Runx3 Expression in Normal Gastrointestinal Epithelium Calls into Question Its Tumour Suppressor Function."

Correcting somebody else's mistakes can be arduous work. Levanon et al. did the following experiments to counter just two experiments in Li et al. (2002) showing RUNX3 expression in the gut: (i) they tested RUNX3 expression in mice by inserting a marker gene into the chromosomal RUNX3 locus (result: no expression in the gut), (ii) they tested expression of wild-type RUNX3 with eight different antibodies (result: in each case, no expression in the gut), (iii) they showed that RUNX1, a close relative, is indeed expressed in the gut, while RUNX3 is absent, (iv) they fused the RUNX3 protein with green fluorescent protein (GFP) to show in three independent assay systems the lack of GFP in the gut, (v) they did the same with a red fluorescent protein, and (vi) they repeated the original Japanese experiment exactly, but with opposite results. In addition, two independent reports, which analyzed more than 100 patients, failed to detect RUNX3 mRNA or protein in human gastric epithelium (Carvalho et al., 2005; Friedrich et al., 2006).

Nevertheless, Levanon et al. counted 285 papers on the potential involvement of Runx3 in a variety of cancers, none of which had carefully examined whether Runx3 is actually expressed in the tissue in which it was reported to be expressed. Instead, on the basis of the original 2002 work by Li et al., the majority of this literature assumed that Runx3 must indeed be expressed in the normal gut epithelium and that it is therefore likely to act as a TS in the particular epithelial cancer investigated.

The papers that did look for RUNX3 expression, but did not result in the correction of the original result, were divided into three groups by Levanon et al. Group 1, which consisted of 145 papers, used RUNX3 promoter methylation status in various cancers as a proxy for its expression. Unfortunately, hundreds of genes are known to get methylated in tumor cell genomes, most of which are not expressed in the normal tissue of origin of these cancers.

Members of group 2 used poorly characterized (or fully invalidated) antibodies to detect RUNX3 protein, which in several cases stained the cell cytoplasm instead of the nucleus, where the RUNX3 transcription factor resides.

Group 3 used either RT-PCR or validated antibodies, failed to detect Runx3 expression in the gut epithelium, and never made a big thing out of it. Considering the trouble that Levanon et al. must have gone through to publish, this is perhaps understandable.

As a response, the authors of the original 2002 *Cell* paper presented new evidence of RUNX3 expression in human and mouse gut and concluded that there was no reason to doubt their original results (Ito, 2012). Also, the publication of Levanon et al. (2011) apparently did not reduce public enthusiasm for the work it set out to falsify; from August 2011 to December 2014 the original Lee et al. paper has been cited 146 times, and the paper by Levanon et al. 11 times.

What might be the reasons behind a poor track record of independent replication of experiments in ensuing the self-correction of science? For one, outside the relevant departments of pharmaceutical companies whose members are sworn to silence by non-disclosure agreements (Begley, 2013), nobody really gets paid for repeating other people's experiments. In science even people who have tenure need to constantly compete for grants, where premium is put on originality and expansion of knowledge, not on one-to-one replication of the work of others.

And secondly, as discussed above, lack of reproduction hardly means lack of validity for the original results. There are a million reasons for differences in experimental protocols, which can cause failure to replicate true results. Also, there is the possibility of getting non-replications of true results by pure chance, the frequency of which is measured by the concept of statistical power. The power to detect true effects is rarely formally estimated in experimental biology, and in many cases it is likely to be very low (Button et al., 2013).

1.9   DO WE NEED A SCIENCE OF SCIENCE?

Philosophers and scientists as different as Richard Feynman, Richard Rorty, and Karl Popper have claimed that no theory is required to explain the success of science; precisely because science was so obviously and tremendously successful (Brown, 1994). A friend of mine, who is a first-rate biologist, recently put it like this: if we want to know how many legs a dog has, it is best to count them. Similarly, nobody doubts that DNA is made of four nucleotides and is a double helix; so you should do more science and worry less about its methodology.

There are a couple of objections to this, however. Firstly, we cannot actually be sure that DNA even exists. This may sound strange, coming from a molecular geneticist, but it is entirely within the realm of possibility that we have created our scientific concepts because of their usefulness, not their truth-likeness. This means that there may be other unexplored and very different conceptual systems, which can be as useful in generating scientific and technological progress, as are the ones that we have. Hasok Chang has explored this radical idea in the context of the history of chemistry (see Chapter 8) (Chang, 2014).[11]

Secondly, even if DNA really does exists, the fact that we were able to correctly deduce its structure 60 years ago doesn't imply that we know how to get truthful results today. This is because science gradually moves away from the simpler questions, which were naturally asked first, to more complicated ones. This means that the divide between what is accessible to common sense and what is actually studied by scientists can only widen over time (Wolpert, 1993). Thus the methods that were so successful yesteryear may yet fail us today.

And thirdly, the empirical evidence presented above suggests that, notwithstanding its past successes, there is presently something seriously wrong with biomedical science. At a minimum we must say that general reproducibility of results in experimental biology seems to be very low, that its ability to self-correct has been put into doubt, and that the efficiency of translating biological knowledge into medicine is becoming an embarrassment.

This is indicative of widespread problems in biology, which in turn raises the questions: "how can we know for sure?" and "what can we do about it?" If we want to return to the glory days of science, we must first explain where good theories come from. To make science work again, we must first begin to understand it as a process, which leads to growth of knowledge. And for that we need a science of science.

If you suddenly discovered that your house was in fire, dealing with the crisis would surely take precedent over whatever else you might have been doing there. Similarly, if a scientist suspected that there was something amiss with science, finding some answers would be expected to be a pretty high priority.

---

11. By the way: would you accept the literal truth of an untestable theory—like the one that there exists a multitude of parallel universes, which cannot interact with each other—if doing so would make working with your existing testable theories much easier (Lewis, 1986)?

As scientists, our instinct is to use the scientific method to try to answer questions about the world. Even if we are unsure of what exactly the scientific method is, most biologists would probably associate it with learning about the world through controlled experiments; and if these are impractical, through systematical observations. However, there are logical limits to this approach when the object of study is science itself.

When we do a controlled experiment on the scientific method, we are comparing the abilities of different versions of the method to produce true knowledge. But, since it is the scientific method itself that is under test, we must have an independent way of recognizing truth (a criterion of truth) to be able to decide on the winner. Clearly, our criterion of truth must come from outside of science; otherwise we would have decided on the winner before the race. On the other hand, if indeed we discovered an extra-scientific way of recognizing truth, what would we still need science for? This dilemma naturally leads to questions like: What is truth anyway? What are the attributes of scientific knowledge? How to recognize good science from bad? Science from pseudo-science? Physics from metaphysics?

Such questions are from the domain of philosophy, not science. A science of science should therefore combine an empirical scientific approach with the methods of philosophy. The ingredients needed for this particular sauce were already put together by Ernst Mayr in his 1982 classic, *The Growth of Biological Thought*:

> *Increasingly often one reads references to a "science of science". … It relates to an evolving discipline that would combine the sociology of science, the history of science, the philosophy of science, and the psychology of scientists with whatever generalizations one can make about the activities of scientists and about the development and methodology of science.*

To this list I would add the economics and politics of science, as well as empirical study of the dynamics of its product (publications, patents, etc.).

Whatever this new science of science will eventually be, a look at the above list makes abundantly clear that it won't be easy to get there. This brings us again to Richard Feynman, who famously quipped that philosophy of science is about as useful to scientists as ornithology is to birds. Now, the 1960s and 1970s were a time when many scientists actually read contemporary philosophers of science like Karl Popper, Thomas Kuhn, Imre Lakatos, and Paul Feyerabend. (Care to name a living philosopher of science?) While these philosophers didn't agree on most things, they presented the reader with alternative theories of the scientific method in physical science (or in Feyerabend's case, about the lack thereof).

What Feynman probably had in mind, was that it really doesn't matter who is right and who is wrong, as long as science keeps on producing new knowledge. And this was a state of affairs he had little reason to doubt. In addition, being a physicist, he might have not been fully aware of the sorry state of many species of birds. A robust knowledge of ornithology might actually be a good thing for a bird these days.

REFERENCES

Abbott, A., 2014. Doubts over heart stem-cell therapy. Nature 509 (7498), 15–16.

Alberts, B., et al., 2014. Rescuing US biomedical research from its systemic flaws. Proc. Natl. Acad. Sci. USA. 111 (16), 5773–5777.

Anderson, D.C., Kodukula, K., 2014. Biomarkers in pharmacology and drug discovery. Biochem. Pharmacol. 87 (1), 172–188.

Angell, M., 2004. The Truth About the Drug Companies. Random House Inc., New York.

Angell, M., 2009. Drug Companies & Doctors: A Story of Corruption. The New York Review of Books, New York.

Angell, M., 2011a. The Epidemic of Mental Illness: Why. The New York Review of Books, New York.

Angell, M., 2011b. The Illusions of Psychiatry. The New York Review of Books, New York.

Anon, 2012. National Science Board. 2012. Science and Engineering Indicators 2012. NSF, Arlington, VA.

Anon, 2013. Sparking Economic Growth 2.0. Available at: http://www.sciencecoalition.org/reports/Sparking%20Economic%20Growth%20FINAL%2010-21-13.pdf.

Arrowsmith, J., 2011. Trial watch: phase II failures: 2008–2010. Nat. Drug Discov. 10 (5), 328–329.

Baggerly, K., 2013. More data, please! Clin. Chem. 59 (3), 459–461.

Baggerly, K.A., et al., 2004. High-resolution serum proteomic patterns for ovarian cancer detection. Endocr. Relat. Cancer 11 (4), 585–587.

Baggerly, K.A., Coombes, K.R., 2009. Deriving chemosensitivity from cell lines: forensic bioinformatics and reproducible research in high-throughput biology. Ann. Appl. Stat. 3 (4), 1309–1334.

Bassler, D., et al., 2010. Stopping randomized trials early for benefit and estimation of treatment effects: systematic review and meta-regression analysis. JAMA 303 (12), 1180–1187.

Begley, C.G., 2013. Six red flags for suspect work. Nature 497 (7450), 433–434.

Begley, C.G., Ellis, L.M., 2012. Drug development: raise standards for preclinical cancer research. Nature 483 (7391), 531–533.

Bissell, M., 2013. Reproducibility: the risks of the replication drive. Nature 503 (7476), 333–334.

Blaschke, T.F., et al., 2012. Adherence to medications: insights arising from studies on the unreliable link between prescribed and actual drug dosing histories. Annu. Rev. Pharmacol. Toxicol. 52 (1), 275–301.

Bornmann, L., 2012. What is societal impact of research and how can it be assessed? A literature survey. J. Am. Soc. Inf. Sci. Technol. 64 (2), 217–233.

Brown, J.R., 1994. Smoke and Mirrors: How Science Reflects Reality (Philosophical Issues in Science). Routledge, London, New York.

Bustin, S.A., et al., 2013. The need for transparency and good practices in the qPCR literature. Nat. Drug Discov. 10 (11), 1063–1067.

Button, K.S., et al., 2013. Power failure: why small sample size undermines the reliability of neuroscience. Nat. Rev. Neurosci. 14 (5), 365–376.

Califf, R.M., et al., 2012. Characteristics of clinical trials registered in ClinicalTrials.gov, 2007–2010. JAMA 307 (17), 1838–1847.

Carlat, D., 2011. The Illusions of Psychiatry: An Exchange. New York Review of Books, New York.

Carvalho, R., et al., 2005. Exclusion of RUNX3 as a tumour-suppressor gene in early-onset gastric carcinomas. Oncogene 24 (56), 8252–8258.

Chalmers, I., et al., 2014. Research: increasing value, reducing waste 1. How to increase value and reduce waste when research priorities are set. In: Chang, H. (Ed.), Is Water H$_2$O? Springer, Heidelberg.

Chang, H. (Ed.), 2014. Is Water H$_2$O? Springer, Heidelberg.

Contopoulos-Ioannidis, D.G., et al., 2008. Medicine: life cycle of translational research for medical interventions. Science 321 (5894), 1298–1299.

Cooper, G.M., Shendure, J., 2011. Needles in stacks of needles: finding disease-causal variants in a wealth of genomic data. Nat. Drug Discov. 12 (9), 628–640.

Cuatrecasas, P., 2006. Drug discovery in jeopardy. J. Clin. Invest. 116 (11), 2837–2842.

de Solla Price, D.J., 1975. Science Since Babylon. Yale University Press, New Haven, CT, London.

Diamandis, E.P., 2010. Cancer biomarkers: can we turn recent failures into success? J. Natl. Cancer. Inst. 102 (19), 1462–1467.

DiMasi, J.A., Feldman, L., Seckler, A., 2010. Trends in risks associated with new drug development: success rates for investigational drugs. Clin. Pharmacol. Ther. 87, 272–277.

Djulbegovic, B., et al., 2013. Treatment success in cancer: industry compared to publicly sponsored randomized controlled trials. PLoS One 8 (3), e58711.

Do, C.B., et al., 2012. Comparison of family history and SNPs for predicting risk of complex disease. PLoS Genet. 8 (10), e1002973.

Doolittle, W.F., 2013. Is junk DNA bunk? A critique of ENCODE. Proc. Natl. Acad. Sci. USA. 110 (14), 5294–5300.

Dudbridge, F., 2013. Power and predictive accuracy of polygenic risk scores. PLoS Genet. 9 (3), e1003348.

Ebrahim, S., et al., 2014. Reanalyses of randomized clinical trial data. JAMA 312 (10), 1024.

Eddy, S.R., 2013. The ENCODE project: missteps overshadowing a success. Curr. Biol. 23 (7), R259–R261.

EGAPP Working Group, 2014. The EGAPP initiative: lessons learned. Genet. Med. 16 (3), 217–224.

Eichler, E.E., et al., 2010. Missing heritability and strategies for finding the underlying causes of complex disease. Nat. Drug Discov. 11 (6), 446–450.

Errington, T.M., et al., 2014. An open investigation of the reproducibility of cancer biology research. eLife 3.

Fanelli, D., 2009. How many scientists fabricate and falsify research? A systematic review and meta-analysis of survey data. PLoS One 4 (5), e5738.

Fang, F.C., Steen, R.G., Casadevall, A., 2012. Misconduct accounts for the majority of retracted scientific publications. Proc. Natl. Acad. Sci. USA. 109 (42), 17028–17033.

Fink, M.P., Warren, H.S., 2014. Strategies to improve drug development for sepsis. Nat. Drug Discov. 13 (10), 741–758.

Flint, J., Munafo, M., 2014. Schizophrenia: genesis of a complex disease. Nature 511, 412–413.

Friedrich, M.J., et al., 2006. Lack of RUNX3 regulation in human gastric cancer. J. Pathol. 210 (2), 141–146.

Gan, H.K., et al., 2012. Assumptions of expected benefits in randomized phase III trials evaluating systemic treatments for cancer. J. Natl. Cancer Inst. 104 (8), 590–598.

Goldacre, B., 2013. Bad Pharma: How Medicine Is Broken, and How We Can Fix It. Fourth Estate, London.

Goodstein, D.L., 1994. The Big Crunch. In: NCAR Symposium. Portland, OR, p. 23. Available at: http://www.its.caltech.edu/~dg/crunch_art.html.

Graur, D., et al., 2013. On the immortality of television sets: "function" in the human genome according to the evolution-free gospel of ENCODE. Genome. Biol. Evol. 5 (3), 578–590.

Haibe-Kains, B., et al., 2013. Inconsistency in large pharmacogenomic studies. Nature 504 (7480), 389–393.

Hayden, E.C., 2014. Universities seek to boost industry partnerships. Nature 509 (7499), 146.

Hong, H., et al., 2012. Technical reproducibility of genotyping SNP arrays used in genome-wide association studies. PLoS One 7 (9), e44483.

Horgan, J., 1996. The End of Science. Basic Books, New York.

Horrobin, D.F., 1990. The philosophical basis of peer review and the suppression of innovation. JAMA 263 (10), 1438–1441.

Horrobin, D.F., 2003. Modern biomedical research: an internally self-consistent universe with little contact with medical reality? Nat. Rev. Drug Discov. 2 (2), 151–154.

Iglesias, A.I., et al., 2014. Scientific reporting is suboptimal for aspects that characterize genetic risk prediction studies: a review of published articles based on the Genetic RIsk Prediction Studies statement. J. Clin. Epidemiol. 67 (5), 487–499.

Inazu, A., et al., 1990. Increased high-density lipoprotein levels caused by a common cholesteryl-ester transfer protein gene mutation. N. Engl. J. Med. 323 (18), 1234–1238.

Ioannidis, J.P.A., 2005. Why most published research findings are false. PLoS Med. 2 (8), e124.

Ioannidis, J.P.A., et al., 2009. Repeatability of published microarray gene expression analyses. Nat. Genet. 41 (2), 149–155.

Ioannidis, J.P.A., Boyack, K.W., Klavans, R., 2014. Estimates of the continuously publishing core in the scientific workforce. PLoS One 9 (7), e101698.

Ioannidis, J.P.A., Panagiotou, O.A., 2011. Comparison of effect sizes associated with biomarkers reported in highly cited individual articles and in subsequent meta-analyses. JAMA 305 (21), 2200–2210.

Ito, Y., 2012. RUNX3 is expressed in the epithelium of the gastrointestinal tract. EMBO Mol. Med. 4 (7), 541–542.

Jaffe, K., et al., 2013. Productivity in physical and chemical science predicts the future economic growth of developing countries better than other popular indices. PLoS One 8 (6), e66239.

Järvinen, T., et al., 2011. The true cost of pharmacological disease prevention. BMJ 342, d2175.

Jinha, A.E., 2010. Article 50 million: an estimate of the number of scholarly articles in existence. Learned Publishing 23 (3), 258–263.

Kenakin, T., et al., 2014. Replicated, replicable and relevant–target engagement and pharmacological experimentation in the 21st century. Biochem. Pharmacol. 87 (1), 64–77.

Kraft, P., Hunter, D.J., 2009. Genetic risk prediction—are we there yet? N. Engl. J. Med. 360 (17), 1701–1703.

LaMattina, J.L., 2008. Drug Truths: Dispelling the Myths About Pharma R&D, 1st ed. Wiley, Hoboken, NJ.

LaMattina, J.L., 2011. The impact of mergers on pharmaceutical R&D. Nat. Rev. Drug Discov. 10 (8), 559–560.

LaMattina, J.L., 2012. Devalued and Distrusted. John Wiley & Sons, Hoboken, NJ.

Lane, J., Bertuzzi, S., 2011. Measuring the results of science investments. Science 331, 678–690.

Leaf, C., 2013. The Truth in Small Doses. Simon & Schuster, New York.

Lee, D.H., Vielemeyer, O., 2011. Analysis of overall level of evidence behind Infectious Diseases Society of America practice guidelines. Arch. Intern. Med. 171 (1), 18–22.

Leek, J.T., et al., 2010. Tackling the widespread and critical impact of batch effects in high-throughput data. Nat. Drug Discov. 11 (10), 733–739.

Levanon, D., et al., 2011. Absence of Runx3 expression in normal gastrointestinal epithelium calls into question its tumour suppressor function. EMBO Mol. Med. 3 (10), 593–604.

Levin, A. Conflicts of Interest Said to Threaten Credibility of Medical Profession. Psychiatric news. <http://dx.doi.org/10.1176/pn.43.24.0001a> ; 2014 [accessed 10.10.14].

Li, Q.L., et al., 2002. Causal relationship between the loss of RUNX3 expression and gastric cancer. Cell. 109, 113–124.

Light, D.W., Warburton, R., 2011. Demythologizing the high costs of pharmaceutical research. BioSocieties 6 (1), 34–50.

Lundh, A., et al., 2012. Industry sponsorship and research outcome. Cochrane Database Syst. Rev. 12, MR000033.

MacArthur, D.G., et al., 2014. Guidelines for investigating causality of sequence variants in human disease. Nature 508 (7497), 469–476.

Macilwain, C., 2010. Science economics: what science is really worth. Nature 465, 682–684.

Macleod, M.R., et al., 2008. Evidence for the efficacy of NXY-059 in experimental focal cerebral ischaemia is confounded by study quality. Stroke 39 (10), 2824–2829.

Macleod, M.R., et al., 2009. Good laboratory practice: preventing introduction of bias at the bench. Stroke 40 (3), e50–e52.

Makel, M.C., Plucker, J.A., Hegarty, B., 2012. Replications in psychology research: how often do they really occur? Perspect. Psychol. Sci. 7 (6), 537–542.

Manolio, T.A., et al., 2009. Finding the missing heritability of complex diseases. Nature 461 (7265), 747–753.

Maurano, M.T., et al., 2012. Systematic localization of common disease-associated variation in regulatory DNA. Science 337 (6099), 1190–1195.

Mazloumian, A., et al., 2013. Global multi-level analysis of the "scientific food web." Sci. Rep. 3.

Meyer, U.A., Zanger, U.M., Schwab, M., 2013. Omics and drug response. Annu. Rev. Pharmacol. Toxicol. 53, 475–502.

Michels, K.B., et al., 2013. Recommendations for the design and analysis of epigenome-wide association studies. Nat. Methods 10 (10), 949–955.

Mobley, A., Linder, S.K., Braeuer, R., Ellis, L.M., Zwelling, L., 2013. A survey on data reproducibility in cancer research provides insights into our limited ability to translate findings from the laboratory to the clinic. PLoS One 8 (5), e63221.

Muerdter, F, Stark, A, 2014. Hiding in plain sight. Nature 512, 374–375.

Mullane, K., Williams, M., 2012. Translational semantics and infrastructure: another search for the emperor's new clothes? Drug Discov. Today 17 (9–10), 459–468.

Mullane, K., Winquist, R.J., Williams, M., 2014. Translational paradigms in pharmacology and drug discovery. Biochem. Pharmacol. 87 (1), 189–210.

Murayama, K., Pekrun, R., Fiedler, K., 2014. Research practices that can prevent an inflation of false-positive rates. Pers. Soc. Psychol. Rev. 18 (2), 107–118.

National Science Board, 2013. Science and Engineering Indicators 2014. NSF, Arlington, VA.

Nowbar, A.N., et al., 2014. Discrepancies in autologous bone marrow stem cell trials and enhancement of ejection fraction (DAMASCENE): weighted regression and meta-analysis. BMJ 348 (April 28), g2688.

O'Collins, V.E., et al., 2006. 1,026 experimental treatments in acute stroke. Ann. Neurol. 59 (3), 467–477.

Ochsner, S.A., et al., 2008. Much room for improvement in deposition rates of expression microarray datasets. Nat. Methods 5 (12), 991.

Oldham, J., 2011. The Illusions of Psychiatry: An Exchange. New York Review of Books, New York.

Panagiotou, O.A., et al., 2013. The power of meta-analysis in genome-wide association studies. Annu. Rev. Genomics Hum. Genet. 14, 441–465.

Petricoin, E.F., et al., 2002. Use of proteomic patterns in serum to identify ovarian cancer. Lancet 359 (9306), 572–577.

Popper, K., 1963. Conjectures and Refutations. Routledge, London.

Poste, G., 2011. Bring on the biomarkers. Nature 469 (7329), 156–157.

Prasad, V., et al., 2013. A decade of reversal: an analysis of 146 contradicted medical practices. Mayo Clin. Proc. 88 (8), 790–798.

Prayle, A., Hurley, M., Smyth, A., 2012. Compliance with mandatory reporting of clinical trial results on ClinicalTrials.gov: cross sectional study. BMJ 344, d7373.

Prensner, J.R., Chinnaiyan, A.M., Srivastava, S., 2012. Systematic, evidence-based discovery of biomarkers at the NCI. Clin. Exp. Metastasis 29 (7), 645–652.

Press, W.H., 2013. What's so special about science (and how much should we spend on it?). Science 342 (6160), 817–822.

Prinz, F., Schlange, T., Asadullah, K., 2011. Believe it or not: how much can we rely on published data on potential drug targets? Nat. Rev. Drug Discov. 10 (9), 712.

Ransohoff, D.F., 2005. Lessons from controversy: ovarian cancer screening and serum proteomics. J. Natl. Cancer Inst. 97 (4), 315–319.

Ransohoff, D.F., Gourlay, M.L., 2010. Sources of bias in specimens for research about molecular markers for cancer. J. Clin. Oncol. 28 (4), 698–704.

Reich, E.S., 2013. Science publishing: the golden club. Nature 502 (7471), 291–293.

Ripke, S., et al., 2014. Biological insights from 108 schizophrenia-associated genetic loci. Nature 511 (7510), 421–427.

Rothwell, P., 2005. External validity of randomised controlled trials: "to whom do the results of this trial apply? Lancet 365 (9453), 82–93.

Rung, J., Brazma, A., 2012. Reuse of public genome-wide gene expression data. Nat. Rev. Genet. 14, 1–11.

Safer, D.J., 2002. Design and reporting modifications in industry-sponsored comparative psycho-pharmacology trials. J. Nerv. Ment. Dis. 190 (9), 583–592.

Scannell, J.W., et al., 2012. Diagnosing the decline in pharmaceutical R&D efficiency. Nat. Rev. Drug Discov. 11 (3), 191–200.

Schork, A.J., et al., 2013. All SNPs are not created equal: genome-wide association studies reveal a consistent pattern of enrichment among functionally annotated SNPs. PLoS Genet. 9 (4), e1003449.

Schwanh a usser, B., et al., 2011. Global quantification of mammalian gene expression control. Nature 473 (7347), 337–342.

Scott, S., et al., 2008. Design, power, and interpretation of studies in the standard murine model of ALS. Amyotroph. Lateral Scler. 9 (1), 4–15.

Shekelle, P.G., et al., 2001. Validity of the agency for healthcare research and quality clinical practice guidelines: how quickly do guidelines become outdated? JAMA 286 (12), 1461–1467.

Shojania, K.G., et al., 2007. How quickly do systematic reviews go out of date? A survival analysis. Ann. Intern. Med. 147 (4), 224–233.

Smith, R., 2005. Medical journals are an extension of the marketing arm of pharmaceutical companies. PLoS Med. 2 (5), e138.

Stanley, D.J., Spence, J.R., 2014. Expectations for replications are yours realistic? Perspect. Psychol. Sci. 9 (3), 305–318.

Stent, G.S., 1969. The coming of the golden age. Am. Mus. Nat. Hist.

Stroebe, W., Postmes, T., Spears, R., 2012. Scientific misconduct and the myth of self-correction in science. Psychol. Sci. 7 (6), 670–688.

Swinney, D.C., 2013. The contribution of mechanistic understanding to phenotypic screening for first-in-class medicines. J. Biomol. Screen. 18 (10), 1186–1192.

Swinney, D.C., Anthony, J., 2011. How were new medicines discovered? Nat. Rev. Drug Discov. 10 (7), 507–519.

The ENCODE Project Consortium, 2012. An integrated encyclopedia of DNA elements in the human genome. Nature 488 (7414), 57–74.

Tricoci, P., et al., 2009. Scientific evidence underlying the ACC/AHA clinical practice guidelines. JAMA 301 (8), 831–841.

Turner, E.H., et al., 2008. Selective publication of antidepressant trials and its influence on apparent efficacy. N. Engl. J. Med. 358 (3), 252–260.

van der Worp, H.B., et al., 2010. Can animal models of disease reliably inform human studies? PLoS Med. 7 (3), e1000245.

Vinck, D., 2010. The Sociology of Scientific Work. Edward Elgar Publishing, Northampton, MA.

Wehling, M., 2009. Assessing the translatability of drug projects: what needs to be scored to predict success? Nat. Drug Discov. 8 (7), 541–546.

Weinberg, B.A., et al., 2014. Science funding and short-term economic activity. Science 344 (6179), 41–43.

Witwer, K.W., 2013. Data submission and quality in microarray-based microRNA profiling. Clin. Chem. 59 (2), 392–400.

Wolpert, L., 1993. The Unnatural Nature of Science. Harvard University Press, Cambridge, MA.

Young, S.S., Karr, A., 2011. Deming, data and observational studies. Significance 2011, 116–120.

Chapter 2

The Basis of Knowledge: Causality and Truth

Now that we have raised doubts on the general veracity of many (or most?) scientific results in biology and medicine, it is time to take a step back and ask: How secure are the basic ingredients of scientific knowledge? We will be looking at the philosophical basis of some of the commonly held concepts that life scientists tend to take for granted as the foundation of scientific progress. Namely that science proceeds by getting ever closer to truth (scientific realism), that a major goal of scientific theories is to describe causal interactions as they exist in the world (preferably, but not exclusively, by experiment), and that in formulating those theories we generalize from observable facts or evidence (induction).

In what follows we hope to reveal some of the assumptions behind those concepts, some of their alternatives, strengths, and weaknesses. In so doing we will develop an understanding of the strength of foundations of scientific knowledge as seen from a philosophical vantage point. We will also get insight into how philosophers think, why do they get such strange results, and perhaps even why all this might be relevant for a working scientist.

In this chapter we will deal mostly with qualitative aspects of scientific reasoning, thereby creating a context for the more quantitative views, which will be discussed in Part 2. As a matter of course we will assume that there exists an external world and that we can, at least in principle, obtain knowledge of it. Remember, we are talking of knowledge in the sense of justified true belief. Knowledge is certain: once obtained it is forever true.

From here, however, trouble begins.

2.1 SCIENTIFIC REALISM AND TRUTH

Scientists tend to believe that they are in the truth business. When I discover that proteins A and B bind to each other and I have ran the relevant controls, measured the binding constant, and this has been repeated in other labs, my natural inclination is to call the interaction of A and B "true." Or I might say that it is very likely or probably true. Or I might say that it is a fact that A and

Interpreting Biomedical Science. DOI: http://dx.doi.org/10.1016/B978-0-12-418689-7.00002-8

B interact. The philosophical stance, according to which science accumulates true knowledge of the external world, as it stands independently of us, is known as "realism." If you, as most scientists, happen to be a realist, you are likely to believe that it is in the purview of science to provide true descriptions not only of observable facts visible to the naked eye, but also of things for whose viewing we need specialized instruments (like the microscope) and, most importantly, of the unobservables that are never "seen" but are inferred or postulated (electron, biological species, quantum tunneling). A scientific realist believes (i) in the existence of independent reality, she believes (ii) that our theories about it can be either true or false (but not both at the same time), and (iii) that when true they give us knowledge about the external world (Chakravartty, 2007). For example, if upon hearing of philosophical doubts about the potency of science to reveal deep truths, you exhort: "But we know truly and well that DNA is a double helix! Of that at least there is no doubt." Then you are definitely in the realist camp.

An obvious question raised by scientific realism is: "What does it mean that something is true?" Without having a theory of truth we cannot have a meaningful theory of knowledge (although we could have plenty of knowledge without being able to talk about it). Moreover, should we in our inquiries on the meaning of truth discover that scientists and their theories are constitutionally incapable of approaching truth, realism would be refuted.

The simplest view on truth is that a truth is identical to a fact. This was the thinking of G. E. Moore and Bertrand Russell around the turn of the twentieth century. To be more precise, identity theories of truth state that a true proposition is identical to a fact. Propositions are sentences, which are objects of belief and give content to beliefs. A true proposition is identical to a fact and facts are simply those propositions that are true. This means that there is no difference between truth and the reality to which it corresponds. If this simple identity relation holds, there really would be no point in discussing the concept of truth any further. Unfortunately, around 1910 Moore and Russell realized that if true propositions could explain true facts, then false propositions should be identical to "false facts," which would bring into the world a shadowy host of fact-like things. This in turn would make the false propositions true, which means that there really could not be any false propositions.

Such considerations led Russell and Moore to adopt a correspondence theory of truth, which postulated that a true belief must *correspond* to a fact, instead of being identical to it. In other words, a true belief requires a corresponding entity in the world (a fact). If there is no such fact, the belief is necessarily false. A correspondence theory of truth can be thought of as a form of realism. The world exists independently of us and our theories, and our theories are about that particular world. This implies that they are objectively true or false. Again, if true, for a scientist this would be a highly satisfying theory of truth in that it would allow understanding what truth is, then to forget all about it and to get on with his real job of discovering new truths.

However, since our beliefs are essentially states of mind, and facts are states of the world (often independent of our presence), we must now explain how a state of mind can correspond to a state of dumb matter. How can an idea and a thing have similar structures or mirror each other? An explanation would conceivably entail our beliefs somehow mirroring reality by entering into correspondence relations with parts of it. As Gottlieb Frege put it:

A correspondence, moreover, can only be perfect if the corresponding things coincide and so are just not different things. … It would only be possible to compare an idea with a thing if the thing were an idea too. And then, if the first did correspond perfectly with the second, they would coincide. But this is not at all what people intend when they define truth as the correspondence of an idea with something real. For in this case it is essential precisely that the reality shall be distinct from the idea. But then there can be no complete correspondence, no complete truth. So nothing at all would be true; for what is only half true is untrue. Truth does not admit of more and less.

(Frege and Geach, 1977)

Frege concluded that truth was indefinable.

An important advance in understanding the concept of truth (although perhaps not in explaining the nature of the correspondence of true proposition to the physical world) was achieved in the 1930s by the Polish logician Alfred Tarski. His work concerns truth in formal languages, whose grammar and rules are precisely determined, where all sentences have unique meaning and a truth value, and which contains no context-specificity or other ambiguities. Tarski's indefinability theorem (1936) states that truth of a sentence in a language cannot consistently be defined within that same language. The solution is to define a higher-order language, called "metalanguage," in which we can consistently speak of truth of the sentences of the lower-order "object language." A sentence in metalanguage can legitimately refer to object language, but not the other way around. A good illustration of the problem is the Liars Paradox, in which I simply state that "I am lying." Now, if it is true that I am lying, then the statement must be false and I am in fact not lying. If, on the other hand, the statement is false, then obviously I lied in making it and the statement "I am lying" is actually true. Either way the statement seems to be true and false at the same time, which is a paradox. Tarski's solution involves separating the object language ("I am lying") and metalanguage (it is true). Now we have a statement: it is true that "I am lying," which can be successfully analyzed without logical contradiction. As Russell succinctly put it, "The man who says, 'I am telling a lie of order n,' is telling a lie, but a lie of order n + 1."

Correspondence theories of truth are the most popular truth theories among philosophers, but that doesn't mean that they are overwhelmingly accepted, or that there aren't many alternatives, or even that a correspondence theory should necessarily be desirable to a scientist. For example, you may believe that a claim should be taken as scientifically "true" as long as it is in principle verifiable (and of course, not yet falsified). Such an "innocent until proven guilty" approach entails that you must deny the principle of logical bivalence (a proposition can only be either true or false) since statements that are both unverifiable and unfalsifiable will constitute counterexamples to bivalence. Indeed, under a verificationist theory of truth we cannot say that any statement about the vast majority of the universe, from where light never reaches us, is either true or false.

Another way of looking at truth is as a measure of coherence of a theory with the body of human knowledge (Young, 2013). For example, when we identify a new hormone, and the truth of our claim depends on a complex bioassay, the interpretation of the bioassay in turn depends on a huge web of scientific facts, theories, and suppositions extraneous to our original scientific question. If any of the myriad major assumptions behind our assay turn out to be untrue, the existence of the hormone in our test tube will be rendered untrue as well. Thus we can argue that it is the coherence of our belief system that is the best guarantee for the success (truth) of our hormone purification effort.

According to coherence theory of truth, a belief is true if and only if it is part of a coherent system of beliefs. This holistic theory makes truth a property of a whole set of beliefs and opens the possibility that many world systems can be equally coherent and therefore equally true. Truth is thus no longer a belief-to-world relation, but a belief-to-belief relation. Moreover, as the coherent web of beliefs grows over time, a scientific fact can gradually become more "true" (Brown, 1994). Idealism is the natural choice for metaphysics, which can make the world amenable to this sort of analysis. It was the rationalist philosophers like Spinoza, Leibniz, and Hegel, along with many idealists, who believed in some version of coherence theory of truth.

A very different, very American, and more science-friendly theory of truth, pragmatism, was developed between 1870 and around 1900 by three illustrious men of letters, Charles Peirce (a hugely original thinker and developer of probability theory and inductive methodology of science), William James (a founding father of modern psychology), and John Dewey (the famous educator). Pragmatism remained a dominant direction in American philosophy until the transformation of American academic life by the influx of German-speaking philosophers and scientists, courtesy of Adolf Hitler (Menand, 2001). More recently it has made a remarkable comeback, thanks to Richard Rorty (Pettegrew, 2000).

There is no single definition of pragmatism, but usually it is taken to mean that a true theory is the one that has proved its effectiveness in actual practice. In effect, pragmatism substitutes truth with the method. Truth is nothing but an expedient way of thinking. For Peirce truth is the end of inquiry for a community of thinkers, meaning that true beliefs will remain settled at the end of prolonged inquiry. True beliefs should therefore not conflict with subsequent experience. Peirce's concept of truth does not entail correspondence with the facts but rather stability of conclusions over time and continuous enquiry

(Hacking, 1983). For Peirce, truth is a long-run thing. In contrast, James' and Dewey's version of pragmatism postulates that what is useful here and now can be viewed as simply true. In the long run we will all be dead—together with our once so well-justified theories. Rationality is what we agree on today. It is about us, not the external world. Stability of our theories cannot therefore be a marker for external (and eternal) truth.

A pragmatist is someone who is fallibilist (he must assume that his theories are not literally true; while in the idealized long run they may approach truth), who nurtures critical discussion and plurality of practical approaches and opinions, and who believes that scientific knowledge is rooted in scientific practice (experimentation, observation, measurement, etc.) (Chang, 2014).

If we further assume that scientific method deals with the objective "real" world, then the pragmatist version of truth may also involve a correspondence with reality, making this theory seductive to scientists. Even if we do not know, which of our theories are really true, a pragmatist can avoid making a fuss about scientific progress by noting that the success of our theories is caused by at least some of them being at least partly true.

The pragmatic argument can be stretched further, by arguing that, strictly speaking, there cannot be a final, universal, truth, unless we re-introduce God into our scientific world view:

> *The suggestion that truth, as well as the world, is out there is a legacy of an age in which the world was seen as the creation of a being who had a language of his own.*

> (Rorty, 1989)

In other words, in the absence of a supreme being, who both made the world and described it in His own words, there can be no strictly true description of the world.

Should you believe Rorty's claim and still have a need for a concept of truth, then it might fit the bill to tie in the concept of truth with that of evidence, so that truth becomes an *idealization of rational acceptability*:

> *… a "true statement" is a statement that a rational being would accept on sufficient experience … "Truth" in any other sense is inaccessible to us and inconceivable by us.*

> (Putnam, 1981)

Thus, roughly speaking, if the evidence points to theory A, then theory A must be true. In Section 4.11 we will see how the concept of evidence has meaning only in the context of more than one competing theories. As long as there is a danger that the true theory is not among the alternatives that are compared with the evidence, Hilary Putnam's theory of truth seems to be in some trouble.

If you still haven't found a suitable theory of truth (and it is easy to criticize pragmatism on the grounds of arbitrariness of its "truth"), there is yet another alternative. The minimalist or deflationary theories of truth claim that we actually have no need for the concept of truth. Saying that "$2 + 2 = 4$ is true" is equivalent to saying that "$2 + 2 = 4$." "True" is taken to be a convenience of language, not something that requires further analysis. In a 2009 survey of professional philosophers one in five supported a deflationary theory of truth (the first place was won by correspondence theories with 45% of respondents). Yet, one suspects that for a scientist to reject the concept of truth wholesale, he should be able to first offer a truth-less explanation of what he is doing, and why.

The above shopping list of philosophies, where each item has some merit and none is perfect, reflects some wider choices. Firstly, there is the choice of realism versus empiricism. Realism claims that well-tested theories should be considered true, while empiricists say that they are merely empirically adequate (van Fraassen, 1980).

The most famous argument for scientific realism is Hilary Putnam's "no miracle" argument, which states that realism is the only philosophy that doesn't make the success of science a miracle. Thus, if we wish to avoid invoking a miracle each time we purify a protein with desired properties after an act of DNA cloning, we better accept the (approximate) truth of concepts, like DNA, restriction enzyme, etc. A counterargument could go like this:

Q. But how do you know that electrons or DNA or natural selection truly exist?

A. Because assuming that they do exist makes us so successful in practice.

Q. Gee, but haven't there been many other successful theories in the past that are now thought to be false anyway? In fact, doesn't the current crop of successful theories claim that the previous crop, which they supplanted, consisted entirely of false theories? If two theories contradict each other they cannot be both true, although they can be both successful. This means that success of a theory cannot be used for inferring its truth (Chang, 2014). Only if we could talk of achieving complete and lasting success (as opposed to relative success in relation to some other theory or practice), could we use such an argument. And anyway, even if your theory is true, this fact alone cannot guarantee the practical usefulness of it: for this all the auxiliary theories needed for its implementation need to be true as well! Success doesn't belong to a theory, it belongs to the use of this theory alongside other theories needed for a particular application.

Empiricists deny that we need to believe that theories provide true descriptions of underlying reality (at least when concerning directly unobservable facts). Theories instead provide us with research programs and explanations, which can be satisfying in many ways, but for which we have no reason to think that they converge on truth. For an empiricist the fact that a theory perfectly explains all the facts in no way implies its truth (Chapters 4 and 5 further touch on this point).

The empiricists do not deny that reality exists; they just believe that realist versions of truth go further than evidence allows. For an empiricist observations and experiments are the only means of testing whether the theory is empirically adequate. In contrast, for a realist they determine whether the theory is true, but there are also other considerations. For example, it is relevant whether a theory is simpler and more explanatory than another. Empiricists tend to dismiss these considerations as pragmatic or aesthetic.

Clearly the empiricist carries less metaphysical baggage than the realist, as he has no wish to infer from empirical phenomena to their underlying causes (the real structure of the world) (Dilworth, 2006). On the downside, the relatively metaphysics-free empiricism has a harder time in explaining the growth of knowledge—which, for a scientist, can be something of a requirement for a useful theory of science. The empiricist has no qualms admitting the increase in the number of scientific observations (and phenomena), as in technological and methodological skills, but he just cannot get into his head how these must lead to convergence of the increasingly complex web of scientific theories on truth (Hacking, 1983).

Some more philosophical theories of truth and reality (adapted from Chakravartty, 2007).

	Independent Reality?	Do Theories Have Truth Values?	Do Theories Lead to Knowledge?
Realism	Yes	Yes	Yes
Empiricism	Yes	Yes	No (for unobservables)
Logical positivism (logical empiricism)	Yes	No (for unobservables)	Yes
Instrumentalism	Yes	No (for unobservables)	No (for unobservables)
Fictionalism	?	No	No
Operationalism	Yes	No	Not to the classical true knowledge

Fictionalism asserts that scientific theories are always false. If you believe that theory is always under-determined by data and that there is an infinite number of potential interesting theories, then factionalism should be a rational choice for you.

Not far from a fictionalist stands the instrumentalist, who believes that theories do not have truth-values but are simply instruments for making predictions. A better theory is nothing but a better tool (in Section 5.5. we argue that Bayesianism is a instrumentalist theory of scientific progress).

Close by is the operationalist who believes that in the final analysis it matters the most, what kind of manipulations we can do to our objects of curiosity (while for the empiricist the main point is, how can we "see" them). If all we do to atoms is to measure their weight (as was the case in the mid-nineteenth century), then for us the atoms are their weights, and very little else (Chang, 2014). Where there exist different parallel traditions of manipulating atoms (by weight, by volume, etc.), then we may concede that the word "atom" means different things to different people, and it may not even be sensible to strive for a single unified concept. By operationalization of a scientific concept, like the atom, we learn to do potentially useful things with it. Speaking of real existence of the concept is apparently not one of them.

Then there are logical positivists or logical empiricists, who (unlike instrumentalists and operationalists) believe that theories concerning unobservables are meaningful, but only if their unobservable terms are linked to observable terms. For them an unobservable term is simply another name for an observable event (electron = a streak in the cloud chamber). Thanks to such dictionaries, which translate unobservables into the language of observables, logical positivists believe that even abstract theories can add to our knowledge about the world (Chakravartty, 2007).

This is clearly not a place to give the reader instructions, what position to adopt. Philosophical theories are not dissimilar to pets: when you take one into your family, certain responsibilities ensue, some of which become apparent to you only after you have left the shop. Also, like it is with pets, you cannot adopt them all. Their differing requirements would simply make the owner's life too difficult.

Metaphysical beliefs matter—even if you are unaware that you have them. For example, in Section 4.3 we will see how the choice of truth (realism) over expediency (empiricism) led some scientists and philosophers to develop an "objective" concept of probability (frequentism) to supplant the "subjective" Bayesian probability. This originally rather metaphysical choice—which did have a practical component, computability—had a huge practical influence on the biomedical science of the twentieth century. There is little doubt that the near-universal denial of Bayesianism as a practical scientific tool, largely due to its perceived subjectivity, had a negative impact on the methodology of science from at least the mid-twentieth century until the 1990s (McGrayne, 2012).

It is time for us to move from the abstractions of truth to the more earthly concept of causality. Causal theories are something that most biologists since the days of Aristotle really cannot live without, while, as we shall shortly see, causality has presented exactly the opposite problem for many a philosopher.

2.2 HUME'S GAMBIT

There are many more people in the world than possibilities to be the first in something noteworthy. David Hume (1711–1776) not only had the good fortune to be the first author in history to get reasonably wealthy on royalties on his books (a multi-volume history of England), but he is arguably the first philosopher since Plato, who has achieved an important philosophical result, which has withstood 250 years of continuous attempts at refutation. He was also a Scotsman renowned for his even temper and generosity toward friend and foe.

What Hume did in one of his lesser-selling books, *Enquiries Concerning Human Understanding* (1748), was to clearly define a problem, pose a simple question, and show that a simple answer is not forthcoming. His problem is how to get from specific data to more general knowledge. This is, essentially, what modern statistical inference is all about, so Hume's shadow still looms large. Hume believed that "causation is the basis of all our reasoning concerning matters of fact, and in our reasonings ... it is constantly supposed that there is a connexion between the present fact and that which is inferred from it." So when the cue hits a billiard ball, there will be a "connection" and this we see as the cause for the sudden movement of the ball.

Hume next showed that this connection is not logical, that is, it does not arise from pure reason, because it is possible to imagine the cue hitting the ball without effect (Figure 2.1). This means causality must be an empirical matter. Hume then offers claims that we think of cause and effect only when two objects are constantly conjoined with each other. Although causal thinking is a deeply ingrained habit, as he points out: "an unexperienced reasoner could be no reasoner at all."

This leaves us with probable reasoning, that is, induction from specific examples, as the way to connect the cue with the movement of the ball. But induction doesn't cut it either, because induction itself is based on the connection between cause and effect. Namely, the connection is based on experience and any inference from experience is in turn based on the supposition that the future will be like the past.

Therefore, causal connection could be established by adding a premise stating that nature is uniform: "For all inferences from experience suppose, as their foundation, that the future will resemble the past ... If there be any suspicion, that the course of nature may change, and that the past may be no rule for the future, all experience becomes useless, and can give rise to no inference or conclusion."

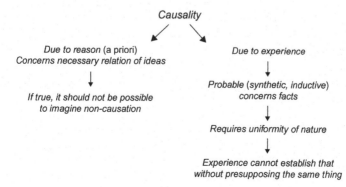

FIGURE 2.1   David Hume's skeptical argument.

Hume's "sceptical doubts about the operations of the understanding" were not meant as "discouragement, but rather an incitement ... to attempt something more full and satisfactory." So you are invited to do better and try to prove uniformity of nature without presupposing the same. Philosophers have been busy doing just that for the past 250 years and, as of yet, there is no winner.

To conclude the argument, Hume finds an unbridgeable gap between induction from specific examples and the general conclusion. He concludes that there is no valid inference from one to another. His somewhat reluctant explanation for why we use causal thinking and induction is habit:

> *But there is nothing in a number of instances, different from every single instance, which is supposed to be exactly similar; except only, that after a repetition of similar instances, the mind is carried by habit, upon the appearance of one event, to expect its usual attendant, and to believe that it will exist. This connexion, therefore, which we feel in the mind, this customary transition of the imagination from one object to its usual attendant, is the sentiment or impression, from which we form the idea of power or necessary connexion.*

This, in a nutshell, is Hume's problem: The way to infer something useful about the world presupposes causality, which in turn presupposes induction, which in turn presupposes uniformity of nature, which again presupposes induction. Therefore, the basis of human understanding is a habit, or perhaps a psychological need, of expecting things not to change.

And this, as Hume clearly saw, opens the door for corrosive skepticism about the ability of sciences to discover truths about the external world. For "truth," seen as "a product of habit," doesn't look beckoning. It is not that Hume dispenses with the existence of causes as such: he simply puts true causes of things into the metaphysical domain, which is hidden to science. In this he takes his cue from Newton's laws where exactness of description is matched by the lack of any possibility of causal explanation of gravitational action from distance. In his *History of England* Hume describes Newton as "the greatest and rarest genius that ever arose for the ornament and instruction of the species" and his contribution: "While Newton seemed to draw off the veil from some of the mysteries of nature, he showed at the same time the imperfections of the mechanical philosophy, and thereby restored her ultimate secrets to that obscurity in which they ever did and ever will remain" (cited in Hacking, 1983). Thus it was the very success of Newton's laws that undermined scientific realism for Hume.

Next we will see how another great philosopher, Immanuel Kant, drew very different inspiration from Newton to offer a striking solution to Hume's problem.

2.3 KANT'S SOLUTION

There are many more people in the world than there are possibilities to fail in a spectacularly useful manner in something noteworthy. Immanuel Kant

(1724–1804) is arguably the most important philosopher since Plato, who has done so. He was also a Prussian renowned for his even temper, kindness of spirit, and modesty. In addition, Kant was a terrible writer of several very important books, which is a state of affairs not conductive to leaving a clear understanding of one's work to posterity. Luckily he was a clear-minded lecturer who tailored his lectures to the less able of his students (as the good ones need no help in learning) and did manage to write a decent semi-popular overview of his philosophy, *Prolegomena to Any Future Metaphysics* (1783). In there we find the famous passage where Kant explains how learning of Hume's work "interrupted his dogmatic slumber." He further claims to have achieved "a complete solution of the Humean problem."

Before we look at Kant's solution, we need to understand the (mostly) unstated assumption behind his work. This assumption is that Newton's laws of motion are literally true. This conjecture was extremely well supported by nearly a century's worth of accumulating scientific evidence, which led to ever more exact correspondence between data and theory. So it was entirely reasonable to believe that for Newton, for the first time in human history, the true writing in the book of nature was revealed by means of experiment and reasoning. Furthermore, not only had Newton achieved absolute and everlasting truth solely by the use of scientific method, but he also discovered that the writing in the book is in the language of mathematics, and of great beauty and simplicity. That would presumably make the other book, which was given to Moses on Mount Sinai, a *Reader's Digest* edition and the one Newton deciphered, the true words of god (Newton accordingly spent the latter half of his life studying the occult sciences and looking for connections between the Bible and the Book of Nature). It would take another century after Kant's death and Einstein's theory of relativity to change things.

DEF: In Kant's terminology judgments can be either analytic or synthetic. Analytic judgments are tautological (all bodies are extended), which means that their truth can be ascertained independent of experience. In contrast, to ascertain a synthetic judgment (all bodies have mass), you need to take a look at the outside world. Secondly, knowledge can be either *a priori*, or *a posteriori*. *A priori* knowledge is independent of experience, necessary and eternally true. *A posteriori* knowledge is dependent of experience and therefore contingent on everything that can go wrong in experimental science.

When Hume showed that causality is not analytic, that is, a logical truth, by showing that he could self-consistently imagine non-causation, his natural conclusion was that it must be either *a posteriori* synthetic, or then simply a habit of thought. The process starts from empirical experience of co-occurrence of things, which leads to induction, which in turn leads to a causal explanation. And, of course, since the use of induction cannot be justified, at the end of the tunnel is darkness (Figure 2.2).

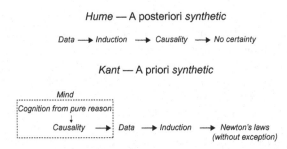

FIGURE 2.2   Immanuel Kant's solution to Hume's problem.

Kant's solution basically turns Hume on his head as it "rescues the *a priori* origin of the pure concepts of the understanding and the validity of the general laws of nature [...] not in such a way that they are derived from experience, but that experience is derived from them, a completely reversed kind of connection which never occurred to Hume." To do that, Kant had to show that there indeed exist *a priori* synthetic concepts. These would be concepts that are not logical truths (we can imagine a world without them), but which are nevertheless strictly universal. For Kant such concepts were space, time, causality, and mathematical truths $(2 + 3 = 5)$. Admittedly, as Hume so thoroughly established, learning from experience is not associated with strict universality. So Kant had to postulate that, "where strict universality essentially belongs to a judgement, this [universality] indicates a special source of cognition for [the judgement], namely a faculty of *a priori* cognition." Kant's scheme therefore starts not in the experience, but in the mind, which forms the *a priori* synthetic concepts. Here causality as a necessary connection is established, and it can then be combined with experience to be eventually able to reach true laws of nature. In this way, starting from the assumption of truth of Newton's laws, Kant could work backward to save causality and deduce the existence of the faculty of *a priori* cognition! In effect Kant suggested that without the ability for *a priori* cognition we would be in grave danger of never being completely able to make sense of the world. Strangely, this remarkable faculty could grant us knowledge of truths about the world, without the need for us to actually take a look at the world.

Our story of Kant is a nice example of how philosophers work. They take a thought (or a system of thoughts) and follow it through rigorously. This means that a philosopher does pretty much what we all do when we talk in prose; only he does it more carefully and the best ones are not afraid to follow their arguments to the bitter end. Philosophical tools include logic, analysis of meaning, and thought experiments.

Consequently, if the product of philosophical analysis seems silly, then perhaps there is something wrong with the premises used. Remember, for Kant a major and at the time practically unquestionable premise was that Newton's laws of motion are true, and consequently that scientific method is able to reach for the literate truth.

Of course, the problem is that Newton's laws are not true (although under non-relativistic conditions they are still widely used as a convenient instrument or model of reality). Thanks to early twentieth-century developments in mathematical logic, we also know that $2 + 3 = 5$ is really a tautology (not synthetic *a priori*). This means that the kindest modern interpretation of Kant's theory would be that synthetic *a priori* knowledge is a psychological fact about the makeup of human brain. In other words, we find causality necessary because we are built to do so (the usual evolutionary parable to explain why this might be so involves tigers and tall grass).

The behavior of reacting to any movement in the grass as there was a tiger lurking in there may have been brought along by natural selection as a good survival strategy, but there is little to recommend it as a means of establishing objective truth. It stands to reason that evolution has endowed us with various coping strategies for situations that our ancestors were likely to encounter in their Paleolithic world. However, there is no reason to suppose that we have developed a sense for recognizing abstract truths in nature or that the actual laws of physics would confirm our inbred expectations of constancy and simplicity. This is a great pity, because otherwise science could be viewed as an adaptation, a sort of destiny for the human race. And so our tiger just ate the remains of Kant's solution to Hume's problem. RIP.

2.4 WHY INDUCTION IS POOR DEDUCTION

So here is the problem. Science is supposed to progress by finding new truths about the world. To reach a general and true scientific theory, we must either (i) deduce it from a set of premises as an unequivocal truth (assuming the truth of said premises), or (ii) we must be able to verify every particular instant generated by the theory, or (iii) we must draw an inference from a limited number of observations to the general theory—that is, we must use induction. (i) Doesn't fly because the structure of the world is not self-evident and therefore we would initially need a separate science to erect the axiomatic structure, from which deduction could start. (ii) Is no good for the obvious reason of the size of the universe, and for (iii), according to Hume, we need to assume the uniformity of nature.

Elliot Sober (1988) by a neat logical trick cast doubt on Hume's contention that induction would be saved, if only we could find a rational way of accepting the principle of uniformity of nature.

Let's suppose that the probability of it raining tomorrow is 20%. Therefore, by the laws of logic, the probability of it not raining tomorrow must be 80%, or to put it more generally: $P(A) + P(notA) = 1$. In other words, the probability of the sum of mutually exclusive and exhaustive events is one. Now, armed with this knowledge we will ask: "What is the opposite of induction?" This would be a principle of counter-induction, according to which the way things are within our experience should be taken as a guide of how they will not be outside our experience. In induction, nature is uniform and the fact that the sun keeps on rising every morning vindicates this option. According to counter-induction, nature is not uniform. The fact that the sun rises every morning shows that counter-induction was unreliable in the past since nature was uniform. But this means, according to the principle of counter-induction, that we can expect nature to be not uniform in the future! Therefore, using the uniformity of nature as a test of induction, we have increased the probability of induction and counter-induction at the same time, which is a logical fallacy P(induction) + P(counter-induction) ≠ 1. Therefore, we cannot use uniformity of nature to justify induction.

In 1943 Carl Hempel formulated his famous raven paradox. In the propositional logic implication "if A then B" is equivalent to "if notB, then notA."[1] This is the same as to say that the statement "if raven then black" is logically equivalent to "if not black then not raven." Classical induction says that by observing a black raven we have added evidence to the hypothesis that all ravens are black ("if raven then black"). Then, by equivalence, when observing a non-black non-raven—a white shoe, say—we have also added evidence to the hypothesis that all ravens are black.[2] But how can a white shoe serve as evidence for a theory about black ravens? Again, it looks like induction is not compatible with traditional deductive logic.

By the way, Hempel accepted his paradoxical conclusion in spite of its counter-intuitiveness. In Section 5.5 we will dissipate his paradox into a whiff of logic by Bayesian analysis.

In 1955 Nelson Goodman proposed what was to become known as his "New Riddle of Induction" (Goodman, 1983). Let "green" be a property of being green. Now suppose that "grue" is a property of being green until the year 2037 and blue afterward. "Brue," on the other hand, is a property of being blue until 2037 and green afterward. From there two conclusions follow. Firstly, any induction would equally support both something being green and grue. And secondly, "green" is a composite of "grue" and "brue," meaning that we should probably look at it as the less basic and thus less interesting hypothesis.

One could argue that the biological concept of "species" is similar to "grue," as essentialist or typological view of the species was supplanted by the population thinking by the Darwinians (Mayr, 1982). For those to whom introducing strange predicates seems too gruesome, the problem can also be illustrated by drawing curves through a finite set of data points. There are an infinite number of curves, which fit the data equally well. Inductively, all are equally confirmed and yet we tend in practice to prefer one curve over all the others.

---

1. Two statements are logically equivalent if they have the same truth value in every model.
2. Our white shoe would equally confirm the alternative hypothesis "if raven then blue"!

The above examples are meant to lead to the intuition that there is something seriously wrong with induction from the viewpoint of logic. To develop this further, we will look at a simple diagnostic example. We know that people with a cold nearly always have a fever. From there comes the hypothesis:

"If a cold, then fever"
the fact is: "I have fever"
and the conclusion is: "I have a cold"

or more generally:

if p then q
q
therefore, p

The fallacy of such reasoning was eloquently put by Bertrand Russell (1945):

"if p then q; now q is true; therefore p is true." E.g. if pigs have wings then some winged animals are good to eat; now some winged animals are good to eat; therefore pigs have wings. This form of inference is called "scientific method."

(Russell, 2009, p. 173)

In deductive logic this is called *post hoc ergo propter hoc* (after that, therefore because of that) type of error, and if Russell is right in describing it as "scientific method," then we surely are in trouble.

The deductively valid form of the argument goes like this:

if p then q
p
therefore, q
(Modus Ponens)

or,

if p then q
not-q
therefore, not-p
(Modus Tollens)

Note that the structure of the argument does not imply a causal connection, only co-occurrence of p and q.

2.5 POPPER'S SOLUTION

Karl Raimund Popper (1902–1994) was a philosopher of science who was for many years better loved in scientific circles than in philosophical ones. His idea of demarcating science from non-science by the criterion of falsifiability is often taken as an established truth by scientists. However, as we shall see, falsifiability carries some philosophical baggage, which, if you choose to use it for making sense of science, you cannot really avoid tugging around with you. In fact Popper's philosophy of science, developed over 70-odd years, is highly original, complex, and controversial; in a word, a thing of beauty.

Like Kant, Popper claimed to have solved Hume's problem of induction. When Kant's solution was to base induction on a novel and somewhat spooky human faculty of synthetic *a priori* judgments, Popper, seemingly anticipating Stalin's famous utterance: "No man, no problem," did away with induction altogether. Nevertheless, his view on the nature of science was remarkably similar to Kant's: both men saw science as a generator of objective knowledge and their aim was accordingly to provide a rational explanation for this remarkable process.

Popper's main work, *Logik der Forschung* (1934) was published in English as *The Logic of Scientific Discovery* (1959) (Popper, 1992). As the title implies, Popper tried to provide a logical base, which would explain the growth of all science. He believed that induction couldn't be saved as a form of logic and therefore that use of induction couldn't possibly lead to growth of knowledge. Thus, being left with deduction as the only form of acceptable logic, he turned to Modus Tollens to save the rationality of science.

MT: if p then q
not q
therefore, not p

Now, let's substitute p with a "general law" and q with a "testable implication" deduced from that law. If we succeed in showing by testing that the implication doesn't hold, then by logic the general law must be false as well. Unfortunately, the opposite doesn't hold (think of Russell's winged pigs).

It is important to note, what Popper's philosophy cannot address. This is the provenance of the law or hypothesis that is used to generate testable predictions. For him, generating the general hypotheses is a matter of creativity, not of logic. However, once the hypothesis is in place, logic takes over; and this can only happen if the hypothesis is self-consistent and we are able to distinguish between its logical and empirical elements (as only the latter can generate testable predictions). Then we deduce from it a narrower test hypothesis and test it by observation or experiment. If it fails the test, we will discard the original wider hypothesis (Figure 2.3). If, on the other hand, the tests succeed, we are stuck in a kind of limbo with the original hypothesis. We couldn't falsify it, but we didn't verify it either. A theory, which survives repeated testing, cannot even be assigned a probability of being true. In Popper's terminology, such a hypothesis is merely corroborated, and we should tentatively accept corroborated hypotheses until their eventual falsification.

There are two reasons why some corroborated hypotheses should be preferred over others. Firstly, since obviously false hypotheses are unlikely to survive stringent tests, the hypothesis that does survive more tests is less likely to be patently false. And secondly, we should prefer hypotheses with larger truth

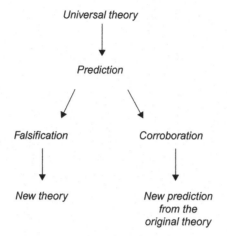

FIGURE 2.3   Karl Popper's theory of scientific progress.

content and smaller falsity content. The truth content of a hypothesis is the set of true deductive consequences of the hypothesis, and the falsity content is the set of its false deductive consequences. Together, they make up the content of the hypothesis. Verisimilitude, or truth likeness, is defined as truth content minus falsity content; greater verisimilitude means that a theory is closer to truth than an alternative. Experimental corroboration is therefore an indicator of verisimilitude and can thus lead to growth of knowledge.

Having said that, in response to his critics Popper conceded that the degree of verisimilitude could not be in most cases numerically determined, leaving the concept of verisimilitude hanging on a limb. Corroboration is now purely a pragmatic act. Although we cannot predict the future performance of our current best theories, what could be more rational than acting on them? One can even work with a falsified theory if it has passed more tests than its competitors and is therefore the best theory available.

This presents an obvious problem: if we cannot speak of truth of theories, or even probabilities thereof, what becomes of the progress of science? Popper's view was that even though theory change in science is revolutionary—the falsified theories are substituted by entirely new theories—the growth of science as a whole is evolutionary; this because of the increasing verisimilitude of each successive crop of accepted theories. There is still continuity in theory-change, as the new theory should explain why the old theory was successful to some extent.

As a methodological rule, a scientist should revise his theories so that the degree of falsifiability is increased. A revision that decreases falsifiability is called *ad hoc* and can remove the hypothesis from the domain of science. Science, like evolution, proceeds through trial and error and leads to ever-better fit between theories and reality. Although this philosophy needs a concept of truth to be workable, truth itself is unattainable. Even if you accidentally stumbled onto a true theory, you would have no means of recognizing this fact.

According to Popper, it is impossible to prove a general theory; but it is possible to falsify one. Therefore the only efficient way of doing science is to deduce predictions, which are likely to fail in experimental testing. The more unlikely your hypothesis (the further from common sense or accepted knowledge), the easier it is to test it. In other words: the more rigorous the test, the more sense in doing it. It follows, rather surprisingly, that the aim of science is not to achieve high probability: if it were, science would have to base itself upon a large number of trivialities with an equally high degree of probability. Rather, science is interested in theories with a high content, whose probability decreases in proportion to the rise in content.

An example of the effect of Popper's theory on scientists can be found with the Nobel Prize–winning biologist John Eccles, who for 10 years championed a theory of electrical synaptic transmission. He was, of course stymied by the eventual acceptance of the chemical nature of synaptic transmission. His subsequent meeting with Popper led him both out of depression and to the following statement: "scientists have often been loathe to admit the falsification of [a] hypothesis [they have published] ... whereas according to Popper falsification is the anticipated fate of all hypotheses, and we should even rejoice in the falsification of a hypothesis we have cherished as our brain child. One is thereby relieved from fears and remorse, and science becomes an exhilarating adventure..." (Dienes, 2008). Eccles went on to coauthor with Popper a very adventurous book on the body–mind problem (Popper and Eccles, 1977).

Falsifiability can do far more than to set free a scientist's inner sense of adventure; according to Popper it is the criterion of demarcation between science and the rest. The easier a theory is to falsify, the easier it is to discard it. And discarding theories is the only thing Modus Tollens is good for. In contrast, if a theory is not falsifiable in principle or in practice, the scientific method is incapable of getting rid of it and we must therefore exclude it from science by fiat. Therefore, an unfalsifiable theory cannot be considered scientific.

Popper's favorite examples of unfalsifiable theories were Marxism, psychoanalysis, and astrology. His specific criticism was that these theories couldn't generate falsifiable predictions. For example, Freud's claim that "If an otherwise highly intelligent patient rejects a suggestion on not too intelligent grounds, then his imperfect logic is evidence for the existence of a ... strong motive for his rejection" suggests that he would take both patients' acceptance and rejection of therapists' suggestion as evidence that his theory is right (Dienes, 2008).

However, if psychoanalysis is not science today, that doesn't mean it cannot become one in the future. The border between what is scientific and what is not is constantly shifting and a theory can move in either direction. A sure way of straying from the domain of science is to modify a falsified theory with the intention of banishing from it the falsified prediction. For example, if a Marxist expects the first socialist revolutions to occur in England and Germany (the first heavily industrialized nations), but it instead happens in Russia (one of the least industrialized European nations), he might well find *post hoc* reasons that would explain the facts and save Marxism as a theory, albeit not a scientific one.

Falsifiability is of course not categorical, some theories are more falsifiable than others, and theories that are more falsifiable should be preferred for exactly this reason. So a theory, which generates unlikely, surprising, or just plain weird consequences, is to be preferred over theories that do not. Taken to the extreme, the more patently absurd your theory is, the easier is to falsify it, and therefore the more scientific it is (but the act of falsification removes it from science).

Another important corollary of falsificationism is that you cannot depend on experiments designed to support your theory. Should you design an experiment where the result is interpretable only if it is a confirming instance to the theory, then in Popper's eyes the experiment is categorically not worth doing.

For Popper, the falsificationist approach offered at least three advantages:

1. it solved Hume's problem concerning induction;
2. it allowed theories to be objectively evaluated even before they were put to the test; and
3. it formulated a *critical* method for science, which proceeded through trial and the correction of error.

Popper's philosophy of science gave rise to at least three noteworthy applications outside its subject matter. Firstly, there is his criticism of historicism (the concept of predictability of the wider course of history) and his closely linked political philosophy. These can be found in the little book "The Poverty of Historicism" (1944) and in the very large "Open Society and Its Enemies" (1945). This hefty tome was enormously influential throughout the Cold War and is probably still his best-known work. According to Popper, historicism cannot work because the inherent unpredictability of the scientific process and of the consequences of human actions makes the future fundamentally unpredictable. Therefore, unlike physicists, social scientists cannot make predictions, and never will have that ability.

The laws of physics, which are used to make predictions, are unchanging; the mores of society change continuously. In society we see trends, and these cannot be used as laws for making predictions. Because we cannot recognize truth when we see it, social change advocates should not put their faith in revolution. Instead, it is piecemeal tinkering that leads through trial and error to a better future.

As in science, criticism is the driving force in society, and it is the liberal "open societies," which are conductive to criticism. This leads to a liberal world view. In contrast, authoritarian "close societies" depend on the will of the ruler, which often translates into large-scale projects of social improvements, or to oppressive stasis. The philosophical forefather of closed societies is Plato, whose utopia involves a centralized state, ruled by benevolent philosophers.

Popper's strategy is to show that totalitarianism is based upon historicist assumptions, which are simply false.

Secondly, Popper developed a propensity theory of probability (see Section 4.3) whose aim was to use probability as an objective measure of the truth of scientific theories.

And thirdly, from the structure and ontology of scientific theories he developed a metaphysics of his own. The argument is simple: Each general law or theory can generate an infinite number of predictions, which are always implicitly present, regardless of our ability to spell them out. This is objective knowledge, where "objective" means that it's not dependent on the subject (or the investigator). Since the infinity of predictions emanating from even a few simple mathematical axioms is incompatible with the finite physical structure of a mathematician, we must conclude that objective knowledge belongs to everybody and is not confined to a single head or a book. This makes knowledge formally independent of the knower; it cannot be described by looking at somebody's brain. Therefore, knowledge is in a very real way independent of the physical world; and yet it can influence the events in this world. Thus Popper proposed that in addition to the physical world there are two others, which are as real as the first (physical) world.

World 1 is the world of physical objects, including human artifacts and the slimy substance that is our brain. World 2 contains our thoughts. It is a world of psychology. World 3 is the world of objective knowledge, which is populated with mathematical objects, scientific theories with all their implications, and, among other essences, every character of every piece of fiction ever written.

> I believe ... that even the natural numbers are the work of men, the product of human language and thought. Yet there is an infinity of such numbers, more than will ever be pronounced by men or used by computers. And there is an infinite number of true equations between such numbers, and of false equations; more than we can ever pronounce as true or false. But what is even more interesting, unexpected new problems arise as an unintended by-product of the sequence of natural numbers; for instance the unsolved problems of the theory of prime numbers (Goldbach's conjecture, say). These problems are clearly autonomous. They are in no sense made by us; rather they are discovered by us; and in this sense they exist, undiscovered, before their discovery.
>
> (Popper, 1972)

Thus, the world according to Popper entails human minds directly grasping the denizens of the third world with the same level of reality as physical objects of the first world. Meanwhile, world 3 is not completely independent of us, which differentiates it from traditional Platonism.

World 2 is the creation of world 1 and can directly influence it, while the influence of world 3 on world 1 is felt indirectly through world 2. The interactionist theory of the body–mind connection, which comes from the need of avoiding direct contact between worlds 3 and 1, is described at length in Popper and Eccles (1977). One implication is that because we live in a different world than our theories, we should not be overly hasty in falsifying them. Instead, we should spend time and effort in studying the structure and implications of our theories rather than giving them up on the first whiff of falsification. Therefore, some degree of dogmatism is justified in holding on to one's favorite theory until the dust settles.

2.6   WHY DEDUCTION IS POOR INDUCTION

There have been many critics of Popper's philosophy and in his lifetime Popper took care to answer them as best he could, including by modifying his theories (Popper, 1972; Popper and Bartley, 1993). He also spent more than an ordinary lifetime trying, with little success, to develop an objective probabilistic version of the logic of scientific discovery. Here we look at some of the more interesting objections to Popper's narrowly deductive view of scientific progress.

In the early twentieth century the French physicist and philosopher of science Pierre Duhem recognized that every scientific hypothesis must come with a host of auxiliary hypotheses that are in themselves fallible conjectures, which are based on inconclusive evidence. These concern the apparatus of measurement, laws of nature, and everything else implicitly contained in the proposed hypothesis. This led the great American logician Williard van Orman Quine (1951) to insist that confirmation is holistic, that is, that evidence never confirms or disconfirms any hypothesis in isolation. Confirmation is thus restricted to "total theories," which can be confirmed by their empirical consequences, while their individual hypotheses cannot. This means that there is no rational way to differentially distribute praise or blame over individual hypotheses. It is hard to see how one can specifically falsify the one hypothesis under test, since logically there is nothing to distinguish it from the myriad auxiliary hypotheses. Quine further uses this to cast doubt on the distinction between synthetic and analytic knowledge. When our experience is in contradiction with the vast web of theories, then, as has been a tempting solution in quantum mechanics, it is equally okay to change the concepts of logic as of empirical theories.[3]

Popper's answer was to acknowledge the problem, while insisting that the practical solution lies in provisionally taking the background knowledge or auxiliary theories as granted. Rationality of science is saved by the word "provisionally": any aspect of background knowledge provisionally assumed as safe can be opened up for criticism in any time we like. The practical implication of this is that every scientist has a lot of latitude in deciding how to use the evidence he collects in relation to the hypotheses he tests. The conclusions drawn from the evidence cannot be decided for him and will to a great extent depend on his motivations.

---

3. For example, by eliminating the law of excluded middle so that a statement can be both true and false.

Holistic confirmation can be both a good thing and bad. We absolutely need enough auxiliary knowledge about the workings of our experimental system, to properly interpret its result. For instance, the late eighteenth-century to early nineteenth-century electrolysis experiments of water showed that hydrogen gas rises at one electrode and oxygen gas at the other electrode. The decomposition of water into hydrogen and oxygen should have been an excellent confirmation of the already prevalent Lavoisierian theory that water is a compound consisting of hydrogen and oxygen (rather than being an element)—but wasn't, because the electrodes were separated by several inches and nobody could really understand how a decomposing water molecule could yield its two products in two different places (Chang, 2014). Due to lack of a theory of ions, this mystery of separation went unsolved over several decades, although there was no lack of trying, and kept on hindering the proper interpretation of electrolysis experiments.

As a speculation, accumulation of auxiliary knowledge could lead to an arc of progress over the history of a science. In the early stages, when holistic context is sorely lacking, good science is extraordinarily difficult to do: its theories are by necessity speculative, controversy is rife, and a pluralistic approach, where several paradigms operate in parallel, might be the scientific community's best hope while there is no hope of rationally choosing one over the others. As science grows, then arrives a sweet spot where we have accumulated enough good auxiliary information to effectively test our main theories, after which the progress becomes quicker, more linear, and beautiful to behold. However, as our knowledge base grows and the low-hanging fruit has increasingly been picked, the sheer volume of auxiliary information grows exponentially, while its quality begins to decline (error frequencies go up). This will usher in a world where (i) the auxiliary information becomes unmanageable and (ii) the scientist can always pick and choose a suitably sexy context for his discovery. As it will be difficult to undo the auxiliary information already present, the result could be the ceasing of true progress in science.

A closely related problem for deductive logic, as a harbinger of truth into science, is the underdetermination of theory by data (technically called "contrastive underdetermination" to distinguish it from the "holistic underdetermination," discussed in the previous paragraph). We already saw a glimpse of this in Nelson Goodman's "New Riddle of Induction" where the same set of observations can be equally well used to justify radically different inductive generalizations. Because logically there are an infinite number of self-consistent possible theories; for each theory, and for any amount of evidence supporting it, there will also be an infinite number of theories, which are equally well confirmed by the evidence.

This can be illustrated by the infinite number of mathematical functions, which can be drawn through any conceivable sample of points. With each added data point we eliminate an infinite number of functions, which no longer fit the data, but are still left with an infinite number of possible functions that fit the updated data equally well.

A more physical example comes from *The Scientific Image* (1980) by Bas van Fraassen, who in turn got it from Newton (van Fraassen, 1980). As Newton himself realized, exactly the same predictions are made by the theory whether we assume that the entire universe is at rest or assume instead that it is moving with some constant velocity in any given direction. Since, being a part of the universe, we have no means to detect constant absolute motion of it, which means that we are faced with an infinite number of scientific theories, all making exactly the same empirical predictions.

If we believe in the infinite number of possible theories, Popper's original idea of falsifying theories one-by-one, while trying to devise surprising and unlikely new theories (which are the easiest to test, but for the same reason, far from each other structurally) sounds like an exercise in infinite futility. The magnitude of the problem in actual scientific practice is contentious—although, judging by the history of biology, it is likely to be significant (Stanford, 2006; Mayr, 1982 (Chapter 20)).

Let us return to falsificationism as the criterion for demarcating science from non-science. The critics have pointed out that astrology has often been tested, and refuted. Similarly, the real threats to the scientific status of psychoanalysis come from claims that it has been tested and failed the tests. Clearly, theories may be in principle falsifiable but practitioners may refuse to do so, or not to believe in falsification, or they may change the hypothesis *post factum* to escape falsification. This leaves falsificationism wounded on the field, as it was undoubtedly not pleasant for Popper to contemplate objective knowledge based on the shaky foundation of psychological contingencies of individual scientists.

Next we will consider some criticisms of Popper's falsificationism as a mechanism of generating objective knowledge. Firstly, existential statements are not falsifiable. An existential hypothesis in the shape "there exists x" ranges over an infinite number of individuals, which makes it unfalsifiable in principle. The objection to this is that scientific existential hypotheses are usually circumstantiated. Biochemistry textbooks do not just state that DNA exists—they say where to look for it and what properties to expect from it.

Secondly, some scientific laws are not falsifiable. The first law of thermodynamics states that energy can take different forms, but it cannot be created or destroyed. As a matter of falsification, any counterexample can easily be neutralized by claiming that energy has turned into a new and unknown form. In actuality, every time a real experiment resulted in a counterexample to this principle, physicists rejected other parts of physics, or they proposed existence of new forms of energy.

Thirdly, scientists tend to ignore falsifications. For example, the anomalous perihelion of Mercury, which is incompatible with Newton's mechanics, was known and tolerated for ages. Popper tried to justify this relaxed attitude by arguing that it is rational to stick to a partially successful theory until a better one becomes available.

Fourthly, and most importantly, probabilistic hypotheses are not falsifiable in Popper's scheme. In Modus Tollens we deal with certainties. If we know for sure that a correctly deduced consequence from our hypothesis did not get experimental proof, then we can be sure that the hypothesis is false. And yet, as scientists we can never be entirely sure of our experimental results: there is no such thing as an experimental proof. This leaves us with the next best thing—probabilities. In a probability approach, certainty is substituted by numerical probabilities. There are more than one possible way of making sense of those numbers, some of which try to preserve the objective nature of knowledge, and some of which do not (see Section 4.3). We will see that different ways of calculating probabilities are strictly dependent on particular interpretations of probability, and that the only possible probabilistic version of Modus Tollens depends on a subjective interpretation of probability. Under this interpretation, probability calculation is an act of induction. This is why, taking the objectivity of science as a given and, as a consequence, relying solely on deduction, really left Popper in a lurch (Sober, 2008).

Since it is easier to understand this problem via some simple formalisms, let's start by restating the familiar Modus Tollens. We are falsifying a hypothesis H by its consequence C; therefore,

> If H, then C
> notC
> therefore, notH

So far, so good. Now, let's try a probabilistic version of this:

> P(C | H) is very high
> notC
> therefore, notH

DEF: P(C | H) translates as "probability (P) of the consequence (C), given the hypothesis (H)." This is an example of conditional probability: we are not talking of the global (marginal) probability of C, but the probability of C in the event that the hypothesis is actually true. To see the difference, let's say that probability of rain on a given day is 20%. But probability of rain, given that there is lightning, is very different, being close to 100%. To put it more briefly, P(rain) = 0.2 and P(rain | lightning)≈1. (Note that the reverse, the probability of lightning, given that it is raining P(lightning | rain), is likely to be even less than 20%.)

As we shall shortly see, this inference is invalid.

If the hypothesis states that the consequence is very probably true, and it really is false, should we count our hypothesis as falsified? To answer that, we must understand that equivalently, if H says that some consequence (notC) has a very low probability, and that outcome nonetheless occurs, then we should regard H as false.

P(C | H) is very low
C
therefore, notH

And yet, many a reasonable hypothesis says that its consequences are very improbable (e.g., cell phones cause cancer). Therefore, this form of probabilistic Modus Tollens is useless for a scientist.

Maybe we can save the day by simply modifying the conclusion to the more modest contention that non-occurrence of consequence provides evidence against the hypothesis:

P(C | H) is very high
notC
therefore, notC is evidence against H

As it turns out, even this is not modest enough. To determine if notC is evidence against H, we must know what the alternative hypotheses H_1, H_2, ..., H_n are. This is best clarified by a simple thought experiment (Royall, 1997). In front of me are two large urns containing black and white balls. I want to test the hypothesis that in the first urn there are only 0.2% of white balls. I draw a ball from it, which turns out to be white. Did I just provide evidence against the hypothesis? Actually, I did not. This is because, unknown to me, the first urn does indeed contain 0.2% white balls, and the other urn contains 0.0002% white balls. I just provided evidence for the hypothesis, not against it.

In short, even if evidence supports hypothesis A over hypothesis B, when we introduce an additional hypothesis C, then the same evidence could well make A less likely to be true. This unexpected result leads us to a version of probabilistic Modus Tollens that actually works. This version takes into account our prior knowledge about the system. Prior knowledge is (roughly) about our beliefs about the system *before* we do the test and it is quantified as prior probability (ranging from 0 to 1). For example, if your prior probability for a hypothesis is 0.01, this means that you are 99% certain that the hypothesis is false. Or, equivalently, that rationally you would be willing to place up to 100 to 1 bet for the hypothesis being true.

$P_{prior}(C | H)$ is very high
$P_{prior}(C | notH)$ is very low
$P_{prior}(H) \approx P(notH)$
notC
therefore, P(H) is very low

Finally we are confronted by a valid inference (Sober, 2008). However, this version of Modus Tollens would have been unlikely to make Karl Popper a happy man. The devil is in the prior probabilities. Most inconveniently, we must have both an estimate of the prior probability of our hypothesis, and our probabilistic version of Modus Tollens works *only* if this happens to be close to 50%! To find a prior probability of an hypothesis being true is necessarily an exercise in subjectivity (otherwise, we would already possess knowledge and there would be no need for us to use Modus Tollens in order to test the hypothesis), and even in the instances where we can do this with reasonable assurance, there is no guarantee that it would approximate 50%.

The above leads to an interesting conclusion about the role of propositional logic in science. We can even write it out as an instance of Modus Tollens:

MT: If A, then B
notB
therefore, notA

MT: If we can be certain of our interpretation of scientific observation A, then theory B is falsified
in empirical science we can never be certain of such interpretations
therefore, MT is not a valid form of inference in empirical science

Propositional logic, which is so vital in mathematics, computing, philosophy, and rational discourse, turns out to be worse than useless in experimental science! In experimental science we instead need a method of inference that allows us to operate with uncertain knowledge in a logically valid way, that is, to reason probabilistically. We need a probabilistic logic.

Fortunately, there exists such a logic, which allows to calculate the posterior probability of the hypothesis being true and, to do so, uses any value of prior probability between 0 and 1 (see Chapter 5). Posterior probability quantifies our subjective but rational belief about the hypothesis being true, after learning the results of the experiment. Posterior probability is calculated by the Bayes theorem, of which Modus Tollens is a special non-probabilistic case (Jaynes, 2003).

We still have a story to tell about how it is possible to switch from the falsificationist setting of Modus Tollens to confirmatory mode of hypothesis testing and save the progress of science even in the absence of successful falsification. Although this approach was proposed by Popper himself (Popper, 1963), it appears to be inductive. In the falsificationist mode we have a general theory from which we deduce a testable prediction that, if true, will guarantee the presence of evidence e. The reason, why we wish for the absence of e to falsify the theory is that e may be present not only if our theory is true, but also under many alternate theories. Thus the presence of e cannot unequivocally differentiate between our theory and all of its alternatives (of which there is likely to be an infinite number). Still, what e can do is to provide a numerical measure of how much the observation increases the evidence in favor of our theory. This depends on the severity of test and can be written out as

$S(e, H) = P(e \mid H)/P(e)$

where $P(e \mid H)$ is the probability of encountering e given that our hypothesis is true and $P(e)$ is the probability of encountering e under all possible alternative hypotheses. Note that in the confirmatory setting, if H is true, e does not have to always occur; it is enough if it pops up considerably more often under H than under all the alternate hypotheses put together. Mathematically, severity of test equals the Bayesian measure of degree of confirmation $P(H \mid e)/P(H)$, where $P(H)$ is the prior probability of our hypothesis being true, independently of observing e (Poletiek, 2009). Thus, if we could be sure that we could see e exclusively under our hypothesis, we could actually prove the hypothesis. As this seems impossible, we have to be contented with conducting as severe tests as possible while keeping in mind that the more severe the test, the greater the likelihood that it fails. But it is also true that the more severity a test has, the more its success will increase the level of confirmation after observing e. Even if we cannot give a numerical value to $P(e)$, as must usually be the case, we should still keep in mind the severity of test as a useful heuristic that helps us to avoid conducting meaningless tests and misleading ourselves that observing the results will bolster our favorite theories.

So far we have shown that classical induction is not truth-preserving from particular observations to general scientific theories—and therefore cannot be used as justification for certainty of scientific knowledge. We have also seen that deductive inferences of propositional logic cannot easily be transformed for use in situations, where their inputs contain uncertainty. Neither Modus Ponens nor Modus Tollens is a useful basis for dealing with uncertainty inherent in scientific observations.

We will now proceed by examining how causal hypotheses are raised and tested in scientific practice. To get a hang of this thorny subject we begin by examining a controversy from the history of epidemiology. Namely: does smoking cause lung cancer?

2.7   DOES LUNG CANCER CAUSE SMOKING?

Lung cancer was a very rare disease once. In 1878 it represented only 1% of all cancers seen at autopsy in the Institute of Pathology of the University of Dresden. By 1918 this percentage had risen to almost 10% and by 1927 to more than 14% (Witschi, 2001). Similar exponential growth was observed in Britain and the United States. What could be the cause of this sudden epidemic? Since the turn of the century was rife with many societal and economic changes, the possible causes that were seen as viable options at the time ranged from increased industrial emissions, asphalt fumes, food, poisons, viruses (influenza), and lifestyle changes to the habit of smoking. To solve this epidemiological puzzle, Richard Doll and Bradford Hill did a retrospective study where they questioned patients, who had just received a preliminary diagnosis of lung cancer, about different components of their lifestyles. It did not take them long to notice that in nearly all patients with the initial diagnosis of lung cancer, who happened to be smokers, the initial diagnosis was later confirmed, while being

FIGURE 2.4   Roland Aylmer Fisher with pipe (1946). *Source: Photo from University of Adelaide digital library (http://hdl.handle.net/2440/81670).*

a non-smoker turned out to be a very good predictor for subsequent disconfirmation of the diagnosis. In other words, there was a strong correlation between
smoking and lung cancer. They published in 1950 and the conclusion of their
study, and the follow-up studies, was that smoking increases mortality due to
lung cancer by about 22- to 24-fold, and mortality due to cardiovascular problems by about two fold (Hill et al., 2003).[4]

However, the interpretation of this result as showing a causal link between
smoking and cancer quickly found its critics, the most eminent of whom being
the pipe-smoking geneticist Roland Fisher (Figure 2.4), who as one of the main
architects of neo-Darwinian synthesis has been called "the greatest biologist
since Darwin" (by R. Dawkins) and whose attempts to create statistical methods
suitable for his own analytic needs in agricultural science and population genetics
earned him the accolade of being "the genius who almost single-handedly created
the foundations of modern statistics" (by A. Hald). Fisher pointed out that the fact
that smoking and cancer are correlated does not necessarily mean that one causes
the other. Instead, both could be caused by a common cause: "The suggestion,
among others that might be made on the present evidence, that without any direct
causation being involved, both characteristics might be largely influenced by a
common cause, in this case the individual genotype..." (Fisher, 1958). He then
pointed out an experimental direction, which could help to clarify matters:

> it was well within the capacity of human genetics to examine whether the smoking
> classes, to which human beings assign themselves, were in fact genotypical dif-
> ferentiated, to a demonstrable extent, or whether, on the contrary, they appeared

---

4. The epidemiologic link between smoking and cancer was discovered in Germany already in
   1929, was not especially questioned on scientific grounds, and led to important public policy
   changes supervised by Adolf Hitler himself (Witschi, 2001).

to be genotypical homogeneous, for only on the latter view could causation, either
of the disease by the influence of the products of combustion, or of the smoking
habit by the subconscious irritation of the postulated pre-cancerous condition, be
confidently inferred from the association observed.

Furthermore, even if we did manage to prove that lung cancer does not have an appreciable genetic component, we still would not know if smoking causes cancer, or if vice versa, a precancerous condition causes people to smoke more. Indeed, for all we knew, smoking could be a folk-cure for cancer, or a palliative medicine:

If, for example, it were possible to infer that inhaling cigarette smoke was a prac-
tice of considerable prophylactic value in preventing the disease, for the practice
of inhaling is rarer among patients with cancer of the lung than with others. Such
results suggest that an error has been made of an old kind, in arguing from cor-
relation to causation.

Keeping in mind that while most lung cancer patients are smokers, most smokers are not lung cancer patients (nor will they ever be, as smoking causes more deaths via heart disease than via lung disease), and that smoking is also closely correlated with mental disease (more than 80% prevalence in schizophrenia patients (Flint and Munafo, 2014)): which causal direction sounds more plausible to you?

1. smoking and cancer are correlated and smoking in one's teens causes cancer—with rather low probability—several decades later; or
2. smoking and cancer are correlated and precancerous irritation of the lungs can be alleviated by smoking, thus causing most cancer patients to smoke.

Fisher further bolstered his argument by a twin study, according to which "the smoking habits of monozygotic twins were clearly more alike than those of twins derived from two eggs. The monozygotic twins are identical in genotype and the clear difference in these data gave *prima facie* evidence that among the many causes which may influence the smoking habit, the genotype is not unimportant." More specifically, 80% of the 104 pairs of monozygotic twins were concordant with respect to smoking, compared with only 41% of 49 pairs of dizygotic twins, providing strong evidence that smoking is a fairly strongly genetically determined trait.

In a separate argument Joseph Berkson (1899–1982) noted that the habit of smoking was also linked to several very different ailments, such as heart disease, emphysema, and cancers of the bladder and pancreas (Hill et al., 2003). This violates Koch's postulate of specificity of association for showing epidemiological causation. In Koch's and Pasteur's late nineteenth-century germ theory of disease the goal is to find a specific pathogen, which causes a given disease. In this theory causation is strict (you expect tuberculosis always to be associated with specific bacilli, not just in 90% of the cases), and tends to be relatively straightforward (a reasonable time between infection and disease). So the rule is: one exposure—one disease. Clearly the association of lung cancer with smoking falls outside this paradigm.

Berkson also saw that if smoking happens to be bad for you, but not through causing lung cancer, then it stands to reason that people, who do get lung cancer and who are already harmed by smoking, would have been more likely to end up in a hospital (and in the sample of Doll and Hill) than "mere" lung cancer patients. This would bias the statistical analysis and lead to a false claim of association.

Another great statistician, Jerzy Neyman (1894–1981), remarked on another form of bias: if smoking helps survival in lung cancer patients, these survivors would, by the virtue of being alive, be more likely to see the questionnaires of epidemiologists, and this could lead to spurious association, even if the incidence of lung cancer was equal between smokers and non-smokers.

In presenting these arguments, it was not my intention to make you take up smoking. In all likelihood you know full well that smoking is bad for you; but this would not have been the case in 1960. Imagining ourselves as contemporary spectators of the fray, the question arises: How can we actually prove that smoking causes lung cancer in people?

We could take a large number of teenagers, divide them randomly into two groups, give one group 20 years' worth of cartons of cigarettes with instructions of how to use them and to the other groups, say, a supply of cigarette-shaped pens, and see what happens over the next 30–40 years. But then, the ethics committee wouldn't be thrilled. What was in fact done was to follow a large number of British doctors over many years to see how their smoking habits correlate with health outcomes. The result was the same as always, but the study design still could not exclude that the smokers had something in common, besides smoking. Since only a randomized study design could help with this sort of criticism and for ethical reasons such a study cannot be done, we have to convince ourselves by force of argument, rather than by experiment.

The winning argument was provided by Jerome Cornfield in 1959. In response to Fisher's genetic theory of lung cancer causation, he pointed out that "if cigarette smokers have 9 times the risk of nonsmokers for developing lung cancer, and this is not because cigarette smoke is a causal agent, but only because cigarette smokers produce hormone X, then the proportion of hormone-X-producers must be at least 9 times greater than that of nonsmokers" (Cornfield et al., 1959). This argument disposes of the identical twin results Fisher used to bolster his theory of common cause of lung cancer and smoking. Fisher never replied to Cornfield in print.

What about the criticism that smoking is associated with multiple diseases, making the associations non-specific (and therefore non-causal)? To this argument, the response was that cigarette smoke itself is a complex mixture, and like the Great Fog of London in 1952, could increase deaths from multiple causes. In essence this means that cigarette smoke contains multiple separate causes for multiple separate effects.

This long-running debate stimulated Bradford Hill to propose nine criteria for claiming that a statistical correlation is causal. These are (i) the strength of the association, (ii) its consistency, (iii) its specificity (iv) that cause precede the effect, (v) dose-responsiveness, (vi) biological plausibility, (vii) coherence with the natural history of disease, (viii) consistency with experimental results, and (ix) analogy with related established causal associations. Looking at this panoply of conditions for determining causality, it becomes clear that none of the above can be seen as necessary and sufficient for accepting a causal hypothesis. This means that, in the end, inferring causation from correlation is always probabilistic. Even in the best-case scenario, assuming all nine criteria have been satisfied beyond reasonable doubt, there must always remain room for doubt as to the causal nature of the connection. In the case of smoking causing lung cancer, this room for doubt is certainly not great. Interestingly, the large measure of certainty that we have about the causal link between smoking and lung cancer comes nearly exclusively from correlational studies:

> *it is likely that … the smoking of cigarettes … would have [never] been recognized as a cancer causing agent had it not been for the fact that a previously very rare disease increased in parallel with increased consumption of a widely distributed and highly addictive agent … It is an interesting thought that experimental toxicology has little contributed to our understanding of the disease. There are very few—some might say none at all—studies in which it has been unequivocally demonstrated that tobacco smoke can cause lung cancer in experimental animals.*

(Witschi, 2001)

The apparent ease of inferring causation from correlation in conjunction with the simplicity of calculating correlations has led to a huge popularity of correlative studies in some branches of science (especially in economics, social sciences, and epidemiology, where doing controlled experiments is often impossible). Nevertheless, the use of these methods can be highly problematic. Here comes an example of a hypothesis-driven correlative study of causality (Matthews, 2000). European mothers have for centuries taught their children that babies are bought by storks (an alternative hypothesis being that they are found under cabbage leaves—but who would believe a silly thing like that?). To scientifically test this venerable theory we could start by finding out whether there is a correlation between birth rates and the number of storks in different European countries. Exactly this was done with the result that there exists a correlation: $r = 0.62$, $p = 0.008$. What do these numbers mean and what can we conclude from them is the subject of the remainder of this chapter. Anyhow, the title of the paper, in which these numbers were presented, is "Storks Deliver Babies ($p = 0.008$)."

2.8   CORRELATION, CONCORDANCE, AND REGRESSION

Before we delve into the various ways whereby biological scientists derive causal connections from correlation, we should briefly look at different methods of measuring statistical associations. Correlation, regression, and concordance are three separate statistical methods for looking at associations between variables. Although at first glance they may seem similar, they are not interchangeable: every method has its own uses and will fail if misused. Correlation analysis measures covariance, that is, it looks to what extent two variables are linked together. The main output of correlation analysis is a single number, the correlation coefficient. Regression analysis gives a formula that describes the relationship between related variables: its output is a mathematical model that is supposed to (partially) describe reality and to make predictions about future data. Concordance measures the level of agreement between separate measures of the same thing. If you want to know how well different measurements (say, by different researchers) of the same phenomenon concord with each other, you should do a concordance analysis.

2.8.1   Correlation

Let's suppose that you want to test the hypothesis that Shetland sheepdogs are small because of the small size of sheep in Shetland. So, as a first step, you organize a scientific expedition to measure the sheep and sheepdogs in several Shetland and mainland Scotland locations. As a result, you have pairs of measurements (sheep—sheepdog), for which you ask the question: are the sizes of variable x (the sheep) linked with variable y (dog)? To help you answer that question, you will calculate the correlation coefficient, which is a summary measure of the covariance between the two variables. It does nothing more than indicate the extent that one variable changes as the other variable gets bigger or smaller.

Correlation analysis applies when two variables are independently measured but it is important to assess the range of values covered by both variables in order to avoid situations where a very large change is correlated with a variable covering a much smaller scale.

The correlation coefficient r can range from −1 to 1. If $r > 0$, the two variables tend to change in the same direction, if $r < 0$, they tend to change in opposite directions. For the biologist, it is the r squared (r^2), which is the relevant measure. r^2 is the fraction of the variance, which is shared between the two variables. For example, if $r^2 = 0.65$, then the variation in variable x can explain 65% of the variation in variable y (and vice versa). Unfortunately, correlation is symmetrical: if x is linked to y, then the link of y to x is exactly as strong. This means that on its own correlation is useless for determining if the smallness of sheep caused the small size of sheepdogs or, indeed, if there is any kind of causal connection at play.

There are many ways of calculating correlation with their different sensitivities and underlying assumptions. We will mention two. Firstly, the formula for the parametric Pearson correlation coefficient is:

$$r = \frac{1}{n-1} \sum_{i=1}^{n} \left(\frac{X_i - X_{mean}}{SD_x} \right) \left(\frac{Y_i - Y_{mean}}{SD_y} \right)$$

where X_{mean} is the sample arithmetic mean of X values and SD_x is the standard deviation (see Section 4.1) for the X values

$$SD_x = \sqrt{\frac{1}{n-1}\sum(X_i - X_{mean})}$$

This measure of correlation assumes that not only are both variables x and y normally distributed, but also that the joint probability distribution of x and y is bivariate normal. If either of the variables has a non-normal distribution, then their joint distribution cannot be bivariate normal and any relationship between the two variables might not be linear. Nonlinear relationships can even arise if both variables have normal distributions. The parametric Pearson correlation coefficient measures linear relationships only!

The second alternative is nonparametric Spearman rank correlation coefficient (r_s), which is about 10% less efficient as the parametric version, but does not assume normality of probability distributions or linearity of the relationship. "Nonparametric" means that you don't need to know the probability distribution of the data. It is computationally very similar to the parametric version; only it transforms the pairs of the two variables to ranks, while retaining the pairing. The null hypothesis, which is being tested, is that there is no monotonic relationship between the two variables. This means that nonparametric correlation analyses do not detect all nonlinear associations between variables, just the monotonic ones (a monotonically increasing function does not have parts, which are decreasing, and vice versa). Because of the ranking of the data, nonparametric tests do not allow presenting the coefficient of determination (r_s^2)—thus only the r_s should be presented (Marino, 2014). There exist more modern methods for measuring dependence that are not burdened by the assumptions of normality, linearity or monotonicity (Székely and Rizzo, 2009), none of which are in widespread use in biology.

Regardless whether you use a parametric or nonparametric version of correlation analysis, all r values should be accompanied by a P value or a confidence interval (CI) (Chapter 4). P values and CIs are needed because any observed effect may result either from a real biological difference between sheep and sheepdogs or, alternatively, from random sampling effects arising from intragroup variance plus measurement error. P values (P stands for "probability") are used to provide a single-number summary for the amount of trust that a reported non-zero r value represents a real effect, as opposed to random noise. If the P value is very small, you could infer that the null hypothesis that your effect is due to random noise is probably incorrect. A CI points to a range of r values that are predicted to contain the true r value (but a 95% CI does not imply that the true value resides in the interval with 95% probability; see Section 4.10 for details).

The assumptions behind correlation analysis (Motulsky, 2010) are:

1. random sample (all sheep and all sheepdogs have an equal chance of being measured by you);

2. paired samples (measure x and y from each pair);
3. sampling from one population (your sample determines what statistical population you are really studying. If the sample is biased, you will misrepresent the object of your study);
4. independent observations (your measurement on Monday does not influence which specimen and how you'll measure on Tuesday);
5. X values were not used to calculate Y values (or vice versa);
6. X values were not experimentally controlled (e.g., using time or concentration as a covariate would necessitate using regression analysis instead of correlation analysis);
7. both variables are Gaussian (for parametric analysis only);
8. all covariation is linear (for parametric analysis only).

An important point about correlation is its non-transitivity: if A is correlated with B and B is correlated with C, this does not mean that A is correlated with C (or not correlated with C) (Ellenberg, 2014). Thus it would not be particularly surprising—regardless that cholesterol levels are correlated with stroke—when a compound that lowers cholesterol did not affect mortality.

The lack of correlation in no way implies lack of a relationship between variables and identical r values can indicate very different underlying patterns. Therefore it is pointless to perform correlation analyses when there is already an established link between the variables. In addition, correlation analysis is useless when different measurements are made of the same thing or when computation of one variable contains the other variable. A pointless exercise in correlation analysis would be, for instance, to calculate the correlation between protein concentrations and its rate of accumulation in the cell.

It is important to differentiate between true correlation value present in a statistical population and the sample correlation value (statistical population is the group of units of measurements to which we draw conclusions from our sample, see Section 3.2 for details). The observed correlation between two samples (A and B) is always lower than the true correlation $r_{A,B}$ because of variation in measuring both the instances of A and B. The observed correlation between samples A and B

$$r_{obsA,obsB} = r_{A,B}\sqrt{(r_A \times r_B)}$$

where r_A is the correlation of different members of sample A versus members of sample A in different measurements (e.g., the reliability of A). For example, it has been estimated that even if true correlation is 1, the observed correlation of real brain imaging studies by functional magnet resonance imaging and various personality/emotional traits is unlikely to exceed 0.74 (Vul et al., 2009). It was further discovered that a good portion of actual studies in this field exhibited higher measured correlations, implying that either the true correlation value is larger than 1 (which is impossible), that the reliability of measuring personality traits (A) and/or blood oxygenation in the brain is much greater than usually thought (which is unlikely), or that calculation of r from experimental data is often erroneous (which seems to be the case).

The authors further ascertained that about half of studies on correlation between brain activity and behavioral measures of personality, emotion, and social cognition computed the correlation by first computing a separate correlation across people for each voxel in the brain image, then identifying highly correlated voxels, and then reporting the mean correlation for just these correlation "hot spots." This is equivalent to looking for random correlations between a weather station data and thousands of different financial instruments of the stock market over a time period, picking out the pairing with the highest correlation coefficient, and then using this model to predict future movements in the stock market (if it gets warmer in Nova Scotia, sell your stocks in New York!). This also happens to be essentially how correlation analysis is commonly used in science; often all you need to add to the mix to publish is a discussion on the plausibility of variation in A causing the variation in B.

Another downside of correlation analysis is that it is very sensitive to outliers in the data. If outliers are likely to be present, you have every reason to doubt the outcome of the analysis. If the underlying structure of the data is nonlinear, the value of r, calculated from linear correlation, becomes meaningless. Even highly structured data can result in $r=0$! Importantly, the size of r is directly related to the range of the sample data values. A larger range will produce a larger r value. Therefore, inclusion of extreme values (outliers) can erroneously increase an r value.

2.8.2   Concordance

Suppose that you have done a proteome measurement by mass spectrometry. Now, having determined thousands of protein concentrations you are interested in knowing how much of the variability between different proteins is real and how much is due to inter-measurement error. So you repeat the experiment and plot the results of both experiments. A common mistake is to next do a correlation analysis and to present the resulting high r^2 as evidence that experimental replications are in good concordance with each other. Unfortunately, your high correlation just indicates that the replications are of related experiments, which is something that you already knew.

What is required instead is a measure of agreement, which correlation analysis cannot give. It gets worse. The value of the correlation coefficient depends on the range of the data compared, not just on the association between the variables. If, for example, you happen to compare the inter-experiment variation in protein copy numbers and mRNA copy numbers by correlation analysis, you are very likely to get a higher r for the protein experiments, simply because the range of protein copy numbers is at least three orders of magnitude larger than is the range of mRNA copy numbers (Schwanh a usser et al., 2011). The result will give you no useful information on the relative quality of the protein and mRNA measurements.

So what we need is not a measure of correlation but a measure of concordance (the level of agreement between different measurements of the same thing), which would indicate whether one experiment gives on average higher values than the other, and whether the discrepancies change with the magnitude of the value. While correlation is all about how one variable changes with the other (plus a random variation component), concordance analysis is also interested in directional bias, which correlation indices ignore. If one experiment consistently gives higher values than its replication, correlation between them will be excellent while concordance is poor.

Analysis of concordance in dimensional data can be achieved by directly analyzing the increments (increment = measurement 1 − measurement 2) of every value measured in the two experiments (Feinstein, 2001). If the sum of all the calculated increments is close to zero, there is no overall bias between the two measurements. If the sum is negative, then measurement 2 has a tendency of giving higher values than experiment 1.

The next question we might ask is: what is the mean or median of absolute increments (all negative increments are turned into positive values) as an absolute value and as a ratio to the mean value of each repeated measurement? To address directional bias, we might ask: do relative increments directionally change with the measurement value (does the magnitude of the measured value influence the accuracy of measurement)? We can calculate a correlation of increments versus measurement values to find an answer.

If we add average deviation or interquartile range (recommended) to the mean or median increment, we immediately have a measure for distribution of bias between the experiments. The same operation conducted with mean or median absolute increments gives us not a measure of bias, but of the spread of error in measurement.[5]

If we want to look for concordance between more than two experimental replications, the best thing to do is to arrange the experiments into pairs, to do the calculations for each pair, and then to determine an average result.

For categorical or binary data, if we wish to assess two independent observations of the same thing, the most popular measure of concordance is Cohen's kappa statistic (κ).

Cohen's kappa gives us the proportion of observed agreements, normalized to the proportion of chance agreements.

$K = (\text{Observed agreement} - \text{Chance agreement})/(1 - \text{Chance agreement})$

If the two measurements give negative and positive results in absolute numbers as follows:

---

5. Should you want to continue with inferential statistical analysis of increments, then you should instead calculate squared increments, standard deviations (SDs), and use means instead of medians.

Observation 1	Observation 2	
	Negative	Positive
Negative	A	B
Positive	C	D

Then the proportion of observed agreement is

$$\frac{A + D}{A + B + C + D}$$

And the proportion of chance agreement is

$$\frac{(A + C)(A + B) + (B + C)(C + D)}{(A + B + C + D)^2}$$

Kappa can range from −1 to 1 and concordance is considered substantial if kappa is greater than 0.6 and fair if it exceeds 0.2–0.4.

A problem with kappa is that as a single-number index it cannot differentiate on how close is the agreement on positive versus negative assessments, which means that kappa cannot separate the sensitivity and specificity of a diagnostic test. Also, kappa underestimates the true concordance when the proportion of chance agreement is extreme (very far from 0.5).

2.8.3   Regression

The term "regression" is an historical accident that happened to Francis Galton, who in 1885 compared the height of children to their parents. He saw that the children of both taller-than-average and shorter-than-average parents tended to "regress to the mean," that is, be on average closer to the mean height of all children in their age cohort, than were their parents. This is due to the fact of logic that if a father, for example, happens to be very tall simply due to random variation, his children are likely to be shorter than him. Non-random variation in height behaves quite differently and does not regress to the mean. Regression to the mean is a pseudo-process, which has very real potential to confound actual effects in statistical analysis (see Section 3.3).

In contrast, regression is a type of statistical analysis that tries to build a predictive model from the actual data, with all its assorted random variability. The goal of the model is to accurately predict new data. In other words, regression analysis tries to iron out random variation from the data, resulting in a mathematical model, which describes the underlying non-random properties of the dataset. The simplest regression model is a straight line described as $y = a + bx$, where "a" is the intercept on the Y-axis and "b" is the regression coefficient. Linear regression simply finds the line that best predicts y from x. Regression line is in effect a one-dimensional version of the mean. The squared multiple correlation coefficient r^2 (also known as the "coefficient of determination") assesses how well the regression line fits the data.[6] The value of r^2 can range from 0, when regression explains none of the variance in data to 1, where regression explains all of the variance in the dataset.

Comparing r^2 values to assess the fit of different models can be dangerous. Firstly, r^2 is not suitable for comparing models with different numbers of parameters. And secondly, r^2 should not be used to compare models based on different transformations of their Y values, like $\log(Y)$.

As with correlation, the result of linear regression should be presented with P values or CIs. CIs of the regression line are curves that are wider at the ends of the regression line than in the middle, reflecting greater uncertainties at extreme values.

Unlike in correlation analysis, the connection between x and y is not symmetrical. To regress the data points to the line, regression analysis only considers vertical distances of the points from the line; it minimizes the sum of the squares of those distances. Therefore, before you use this method, you must be sure that there is no significant random variation associated with the x variable (technically known as the predictor variable). The x can be time, concentration, or anything else that you measure precisely. The y variable (dependent variable) is free to vary. This is what you manipulate with treatments and measure in your experimental system. All the biological non-random variation is associated with the y variable. The value of y depends on x, but the value of x does *not* depend on y. Unlike for correlation, regression can be thus said to model a temporal sequence, or even cause and effect.

The assumptions behind linear regression are (Motulsky, 2010):

1. the model is correct (linear in the range we use for prediction);
2. both the x and y values must be continuous and may not be restricted to integers, truncated or categorized;
3. scatter of the data around the line is Gaussian;
4. variability is the same everywhere (neither biological variation nor accuracy of measurement changes over the x values);
5. independence of errors;
6. x and y values were not used to calculate each other;
7. x values are known precisely (all the variation is in the y-direction).

---

6. Technically, r^2 is not an absolute measure of how well a linear model fits the data, it is only a measure of how much a model with a slope parameter fits better than one without (Quinn and Keough, 2002).

The last assumption is safe in many chemical applications as the y variable will often be some experimental quantity, whereas the x variables (concentration, time, etc.) will be measured with good precision.

Regression analysis is a form of mathematical modeling and there is no *ipso facto* reason why you should use a linear model. In fact, in some modern data analysis packages you can automatically try out many different types of models on your data and pick the best-fitting one. This is, however, a dangerous thing to do. The danger here is that, even if your sample data perfectly fit into the model, the fit may not extend to the population that you are really trying to study. Overfitting the data means that chance effects, which are based on the random variation present in your sample, are incorporated to the model and are then used to predict what will happen in the real world.

The thing to do instead is to limit yourself to the types of models for which you have some independent reasons to believe that they might describe your data. For example, to model the degradation of a protein, one might want to try exponential models, as decay processes are often found to be exponential. Or alternatively, one could just try the mathematically simplest model (the linear one) to see if it happens to fit the data reasonably well. A simpler model may have inbuilt advantages. There are less assumptions, the model is easier to understand, and to test.

Also, avoid overdiagnosis of your data. The greater the number of associations, which are tried in a regression model, the greater the likelihood that a fit will be found by chance alone. A chance correlation has exactly the same properties as a true correlation and will appear to give just as good a fit in the regression analysis.

If you have built several mathematical models to fit your data, which one should you prefer? This is a question of model comparison.

A statistical model is not a scientific hypothesis: its goal is not to be true, but to enable predictions about data, which will be collected after building the model. Therefore, the goal of model selection is to find the model, which can best predict new data. It is not to find the model that is true, nor is it to find a model that best fits the old data, which were used to build the model.

There are several methods of model selection, but currently the most popular is Akaike's information criterion (AIC). This measures information that is lost when fitting the individual data points to the model and it also penalizes the complexity of the model (the more complex model is more likely to over-fit the data). AIC allows us to make a list of the competing models from the smallest AIC value (comparatively the best model) to the largest value (the worst model). It also gives us information about the predictive accuracy of the models. The individual AIC value becomes interesting only when compared to AIC values of a series of models specified a priori. If none of the compared models are very good, the AIC approach will try to identify the least bad of them, which might not be enough to make a tired researcher happy. Therefore it is paramount that the candidate models are based on thoroughly researched previous knowledge of the system.

$$AIC = 2k - 2\ln(L)$$

where k is the number of parameters in the model that can vary independently and L is the maximum value of the likelihood function for the estimated model. For small sample sizes an AIC with a correction, called AICc, has been proposed.

$$AICc = AIC + 2k(k + 1)/n - k - 1$$

As the sample size grows, AICc values approach the corresponding AIC values. AIC depends on the usual normality assumptions of most statistics (Sober, 2008). This means that repeated estimates of each of the model parameters must form a normal distribution. The second assumption is that old and new datasets are drawn from the same statistical population. This means that your new data, from which predictions will be made, will be drawn from the same distribution as the old data that were used in model building. So, both the relationship of X and Y and the distribution of X values must be the same across the datasets. Together these assume Humean uniformity of nature (Forster and Sober, 2004). Therefore AIC does not apply when you want to extrapolate from one range of values to another.

As with correlation, regression analysis is sensitive to outliers. Other sources of errors include extrapolation of the model beyond the observed data range and retrospective demarcation of data into zones with their own regression equations.

How reliable are data modeling and model selection in terms of identifying true causal links from the data? An answer is provided by a study whose authors recruited 24 early-career ecologists to analyze the same relatively simple dataset, consisting of a single response (acorn count), three effects (species, site, and year), and seven environmental variables (Stanton-Geddes et al., 2014). In the 20 models built by the participants, no two had the same set of predictors. Of the ten potential predictor variables, not one was included in every model. At the same time, most final models had similar predictive power. This result emphasizes that regression analysis is good for prediction, not explanation. When the authors further looked for the predictive power of the presented models, they concluded that "model selection improves prediction only when few parameters are included in the model. With four or more parameters there was no difference between randomly selecting parameters and using model selection."

2.9   FROM CORRELATION TO CAUSATION

In real life (and in real science) causal hypotheses are often published on the strength of correlations, and this holds for the best of journals. It falls therefore on readers' shoulders to differentiate between serious scientific work and attempts at generating fame and fortune for the authors and journals that publish them.

Fortunately, this is often not hard to do. If you read a causal explanation, based on a measured correlation, you should first try to reverse the direction of the causal link and ask yourself if it makes more sense that way. For example, when a newspaper refers to a discovery of a positive correlation between the number of times people have sex and their life expectancy, the conclusion is invariably that having a rich sex life prolongs life. Yet, on the strength of evidence it could equally well be that healthier people have more opportunity for sex. If confronted with a trivial and (in this case, literally) a sexy alternative explanation for the same phenomenon, a journalist chooses the latter every time. Unfortunately, life is mostly made of the boring stuff.

Secondly, you should ask whether it is likely that both sides of the presented correlation have a common cause—both being correlated with a third factor. If this is so, then we are likely to have been presented with a pseudo-correlation. For example, in our above-mentioned example of storks bringing babies (Matthews, 2000), how likely is it that both the number of babies (the birth rate) and the number of storks are correlated with the areas of the countries taken into the study sample?

Thus it is extremely important to pay attention to what exactly is correlated—in our stork example, had the author used data normalized to areas (storks and babies per square kilometer), the results would have been very different.

Another danger to look out for is the "King Kong effect" whereby a few extreme outliers can easily produce spurious correlations. In our stork example, if only two populous countries with large stork populations are excluded, the correlation falls from 0.62 to 0.16 (Wirth, 2002). Therefore our original stork results lack robustness.

Scientific literature is strife with false leads. For example, electric razors, refrigerators, fluorescent lights, breeding reindeer, owning a pet bird, being a waiter, eating hot dogs, breaking an arm (if you are a woman), being short, and being tall have been all identified as cancer risks on the basis of correlation analysis (Motulsky, 2010). Schoenfeld and Ioannidis have published a study with a rather self-explanatory title: "Is Everything We Eat Associated with Cancer? A Systematic Cookbook Review" (Schoenfeld and Ioannidis, 2013). They randomly selected 50 ingredients from random recipes in a cookbook and trawled the scientific literature for studies associating these with cancer. They found that 40 out of 50 randomly chosen ingredients had articles reporting on their cancer risk. A total of 72% of the studies concluded that the tested food was associated with an increased or a decreased risk, but 75% of the risk estimates had at best weak statistical significance. The median risk ratios (RRs) for individual studies that concluded an increased or a decreased risk were at the very respectable levels of 2.20 and 0.52, respectively. (RR = 2 means that the treatment confers a twofold risk increase.) On the other hand, the RRs from the meta-analyses, which were recalculated from the combined results of the original smaller studies, were on average close to one (median: 0.96) (Figure 2.5). All this does not necessarily mean that cancer has no connection with what people eat, but it does strongly suggest that a lot of individual studies, which measure correlations of this and that with cancer, mainly serve to obfuscate the reader.

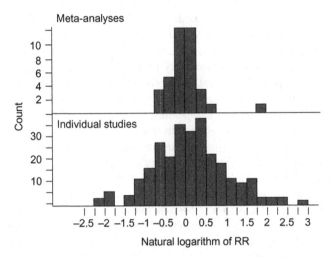

FIGURE 2.5   Comparison of effect sizes from 36 meta-analyses and 255 individual studies, which measure RRs of individual foods on cancer. *Source: Reproduced with permission from Schoenfeld and Ioannidis (2013).*

As a practical matter, newspapers are a notoriously fickle source of information on what is healthy and what is not because they tend to depend on the fast cycle of publishing of individual studies and tend in principle to trust everything that is published in peer review scientific journals. Thus they select their science stories by the news-value of their conclusions and every result, showing that a common food is associated with cancer, certainly has that much to its credit.

One reason for this state of affairs is the ubiquity of statistical data and personal computers, which, if combined, can result in an orgiastic approach to correlations (http://www.google.com/trends/correlate). When very many attempts are made at finding correlations, the laws of chance dictate that a number of spurious "positive" instances will be found. Logically, any correlation thus found could either be not surprising, in which case it is publishable as further proof for existing theories and can be construed as another sign of the steady progress of science. Or then, the correlation can be surprising and therefore the more interesting for the popular press.

For an example of how correlations are turned into science we will look at the evolution of human brain size. Karin Isler and colleagues tested a hypothesis, according to which there exists a trade-off between the size of the brain and the digestive tract; that is, that for energetic reasons, to develop a bigger brain, one must first reduce the mass of his gut (in evolutionary terms, of course). The researchers measured fat-free masses of various organ systems in 100 species of mammals, not to find a correlation between brain mass and fat-free digestive system mass

(Navarrete et al., 2011). From there they concluded that they had thus refuted the expensive-tissue hypothesis. But, they also measured a good number of correlations between different organ pairs and found a smallish negative correlation between brain mass and fat deposits ($R^2=0.26$, $P = 0.006$), but only if the analysis was restricted to a sub-sample of "wild-caught female mammals." This correlation did not hold in the 23 species of primates present in their sample. The scientific conclusion dawn from these results is as follows: "... we find that the size of brains and adipose depots are negatively correlated in mammals, indicating that encephalization and fat storage are compensatory strategies to buffer against starvation." It seems that from the fact that they measured a negative correlation between fat deposits (which may indeed buffer against starvation) and brain size, they concluded that a big brain functions as an anti-starvation buffer as well. Unfortunately, not everything that co-varies is causally connected.

Not only does correlation not imply causality, but also lack of correlation doesn't imply lack of causality! An example of this is found from ecology, where populations of predators and their prey can move up and down in synchrony, then change into anti-correlated behavior, then into a mode where no apparent link can be established. For a researcher studying the interdependence of populations of rabbits and foxes, for example, there may be no long-term statistical correlation between the two, despite a clear causal link (Sugihara et al., 2012).

Such considerations have led to proposal of more refined models to statistically infer causality. One was developed in 1969 as an application of linear regression modeling by the economist Clive Granger, who won the 2003 Nobel Prize in Economics. According to Granger Causality (also known as Wiener-Granger Causality), if X causally influences Y, then including X in a predictive scheme should improve predictions of Y. Conversely, excluding X should make predictions worse. Causal factors therefore reduce predictive accuracy, when excluded from the model. In the words of Prof. Granger:

> The definition has been widely cited and applied because it is pragmatic, easy to understand, and to apply. It is generally agreed that it does not capture all aspects of causality, but enough to be worth considering in an empirical test (http://www.scholarpedia.org/article/Granger_causality).

Unfortunately, Granger Causality applies only in purely stochastic cases, especially in linear systems. In other words, the theory is not relevant for non-stochastic variables, which is unfortunate as nonlinear, deterministic models better describe most things in biology. In addition, intertwining between the parts of a nonlinear system can make "excluding" a variable very difficult (this is called "non-separability"). Another problem is that the Granger test is designed to handle pairs of variables and can readily produce misleading results when the true relationship involves three or more variables. For example, if both rabbits and foxes are driven by a common third process (like weather), with different lags, one might erroneously accept Granger Causality. Furthermore, causally relevant variables, not incorporated into the regression model, will for obvious reasons not influence the output. Thus, Granger Causality should not be interpreted as directly reflecting physical causal chains.

Sugihara and colleagues proposed an alternative for Granger Causality, which extends to nonlinear, non-separable dynamic models, where Granger Causality is helpless (Sugihara et al., 2012). They called their method "convergent cross mapping." It is an empirical method that takes data and tests for causal links. Assuming that X drives the values of Y, the idea is to measure the extent to which the historical record of Y values can estimate the values of X. Presumably this happens if X is causally influencing Y. When causation is unilateral, then it is possible to estimate X from Y, but not Y from X. This runs counter to intuition (and Granger Causality), and suggests that if the foxes drive the rabbit population, we can use rabbits to "predict" the fox population, but not vice versa. Their approach does not involve forecasting *per se*, but predicts contemporaneous or past states of causative variables. Accordingly, it strives to predict the past from the future, the cause from the effect. In this approach causes are not used to predict the effects.

2.10   FROM EXPERIMENT TO CAUSATION

By definition, causes are means for producing effects. This opens the door for exploiting the link between cause and effect. In science such manipulation is called "experiment" (but note that disentangling causal connections is not the only reason why we do experiments; see Section 3.1). The experimenter's job is to intervene in the course of nature to bring about events that would not have otherwise occurred. An experimental intervention is thus an independent cause that overrides or modifies the original causal mechanism, which is under study. If we can switch the original cause of an effect off and on at will, while leaving everything else unchanged, we may conceivably claim that we understand a causal connection. The logic here is similar to that encountered in shop signs: "You break it, you own it!"

By manipulating the cause the experimenter limits herself to the question "what is the effect of manipulating this cause?" and never asks "what causes this effect?"[7]. Since any experimental approach uses manipulation of causes to study their effects, if a cause and its effect (or different contributing causes) happen to be intertwined to a degree that controlled manipulation of one without the other(s) is in practice impossible, the experimental method breaks down (Holland, 1986). Also, evolutionary causes, being historical, seem to be fundamentally different from the everyday proximate causes, which operate in

---

7. Judea Pearl, after turning this quirk of the experimental method into a definition of causality "Y is the cause of Z, if we can change Z by manipulating Y", has worked out an influential formal theory of probabilistic causality (Pearl, 2000).

individual organisms and, unlike the proximate causes, they usually cannot be discerned by experimental manipulation of individuals (Mayr, 1982). This is not to say that evolution cannot be an experimental science: experimental studies of growing organisms under defined conditions to measure their adaptation in real time were already in progress in Darwin's time (Lenski, 2011). By today bacteria have been grown for over 50,000 generations under strictly defined conditions, without the end for increases in fitness in sight (Wiser et al., 2013). However, these experiments still manipulate the cause (growth environment and conditions) to measure changes in effects (specific phenotypic and genotypic adaptations). They are well able to discern the genomic causes for phenotypic novelties emerging in the course of experiment, but cannot be directly used to find the causes for specific preexisting adaptations (phenotypes).

Experimental method is geared to provide inferences from study samples to general biological laws and regularities; it is not suitable for finding the causes for specific historical events.

The experimental approach to eliciting causation can be itself a cause for some rethinking of the concept of cause. In experiment we always compare the experimental treatment with a control or mock treatment. Therefore, causation, as measured experimentally, is always relative to some other cause (terms "cause" and "experimental treatment" are interchangeable under this view). More specifically, the effect of a cause (or treatment) is always relative to another effect of another treatment. In other words, treatment causes the effect with the magnitude of experimental value minus control value. Hence comes the "fundamental problem of causation," which states that it is impossible to observe the experimentally induced value and the control value on the same unit of observation (Holland, 1986).

Of course, you may sometimes do the experimental treatment on a subject and follow this up with the control, but then you must assume that it is still the same subject, that is, you must assume Humean uniformity of nature. This assumption may seem pretty safe in some cases (e.g., when the unit of observation is a thermometer) and unsafe in others (when the unit of observation is a cancer patient), but the point is that one can never know for sure. This approach presupposes both temporal stability (constancy of response over time) and causal transience (effect of treatment 1 does not change the system enough to change its response to treatment 2, which is applied later). Another assumption is that of unit homogeneity. An experimental biologist, who measures reaction rates of wild-type and mutant enzymes in parallel in different test tubes, assumes that the reaction mixes are identical to each other, except for the enzyme itself. In practice, this identity is achieved by preparing the reactions carefully, so that they "look" identical.

Another way around the fundamental problem of causation is statistical. In this approach the researcher will assume that there are many different units of observation, which belong to a statistical population (see Section 3.2). Here the individual units are not identical but a large random sample of them is as a whole representative of the statistical population, from which they were sampled. The crucial assumption here is *independence*, that is, that the variable, whose value we use to infer causality, is independent of all other variables, which we could measure in individual units. If this assumption holds, the researcher can measure the experimental value on some units and control value on others and calculate the average causal effect.

For the average causal effect to be applicable to individual units of measurement, we need the assumption of constancy of effect. If this assumption holds, then causal effect is the same on every unit and we can draw causal inferences at the unit level. If the assumption of constancy of effect does not hold (as must often be the case), then we are left with causal inference at a group level, a good example of which would be a medicine for a disease that has a small effect on the patient population as a whole but a very large effect on a small number of patients. The medicine may be very useful, but the choice of statistical population used in the inference can lead to wrong conclusions about it.

To supplement the above rather abstract discussion with an example on experimental design, suppose that some visiting Martians want to check if the power grid of your town causes the lights to come on in buildings. To do an experiment they pick the largest building in town and cut the power line that enters it. As it happens, they have cut the power to the town's hospital, which means that the back-up generator kicks in, the lights remain on, and the Martians draw all the wrong conclusions.

There are several morals to this story of failed experiment. Firstly, it was a mistake to pick the largest building. Instead, a representative building or a random building would have probably been a better choice. This is because the hypothesis under experimental testing was about all the buildings in town, not just the largest one, and it is entirely in the realm of possibility that the largest building is somehow special or unrepresentative of the buildings in general.

Secondly, one should not be content with picking a single building. Even if it is a random pick, it is possible that there is something unusual about it purely by chance. The more buildings one randomly picks, the smaller the probability that dumb chance will lead to misleading results.

Thirdly, experiments need controls. There are two main types of controls. Positive controls are designed to test if the experimental system really works as we assume that it does in the absence of the experimental manipulation (treatment). In the power grid example a positive control would be to ask for proof that the line was indeed cut properly and that there would be no other power lines entering the building. Or it would be to check if the lights really do go on each evening and off each morning (depending on how you define your "experimental system," which can be tricky in some situations). A positive control should work even if the "real" experiment doesn't. In contrast, negative controls try to exclude the possibility that the experimental perturbation worked, but for the wrong reason. A negative control checks our understanding of the experimental system, not of the experimental manipulation. If the cutting of the power line had left a house in the dark, maybe this outcome would not have been caused by the cutting of the line *per se*, but by some other aspect of the experiment. Perhaps the owner of the house saw approaching his house a Martian wielding a sharp knife, decided he had no need for a visitor just then, and switched off his lights? If the experimenter thinks that this is a real possibility, she should design a control to differentiate between these two possible causes for the shutting off of the lights. A negative control should not lead to the perturbation in nature that the real experiment does, so a passable idea would be sending the Martian with the gleaming knife to the wire and back again without cutting anything, and do it for half of the houses in the study sample. Thus we would see if it was the act of cutting, or the act of approaching, that caused the light to go out.

The sad thing about both the positive and negative controls is that you never can do them all. There are always more confounding factors in your experimental design than you have resources to control for. This means that you should make your controls count. Some people give priority to controls that are most likely to lead to recognition of the invalidity of the experiment (we might call them Popperians), others prefer controls, which are standard in their field of study, and are therefore likely to elicit smiles of recognition from reviewers of their manuscripts. One might also prefer controls that are faster, cheaper to run and easier to interpret. In the end it becomes an optimization exercise, which is just another way of saying that there is no single right answer to the question, which controls, and how many, to do.

And fourthly, you need background knowledge and you need luck. It may so happen that the lights in houses are indeed caused by the electrical power flowing through the power grid, but because every house has a back-up generator, a simple experiment will always give misleading results. If one employs a more complicated experimental scheme and knocks out the back-up systems first, then the result of the main experiment would be very different from what it was before. To generalize, this means that for an experiment to properly shed light on a specific causal connection, we must first have had to more-or-less correctly guess the true structure of the universe. But then, to be able do that is precisely why we run experiments.

This dilemma seems to be especially galling in biology, which contains in abundance stochastic nonlinear systems rich in parallelism, redundancy, and feedback loops (Dougherty and Shmulevich, 2012). The structure of mathematical models needed to describe things like cancer or signaling networks, can be extremely complex and the parameters of the models may not be measurable in the same samples, cells, or organisms due to existing technical limitations of measurement.

Although the experiment is not, and can never be, a foolproof automatic method in determining causality, it seems to be the best method that we have. Most of science (and some of philosophy) vitally depends on experiments (in philosophy's case, thought experiments) and, accordingly, Chapter 3 mostly deals with experiments. But before we sink our teeth into it, let us ask one final question about causality.

2.11 IS CAUSALITY A SCIENTIFIC CONCEPT?

In 1912 Bertrand Russell published a paper, "On the Notion of Cause," with the intention to show that "the word 'cause' is so inextricably bound up with misleading associations as to make its complete extrusion from the philosophical vocabulary desirable" and that "the reason why physics has ceased to look for causes is that, in fact, there are no such things" (Russell, 1912). To do so, Russell first analyses a dictionary definition, where an effect of a cause is defined as anything that is thought to take place in consequence of another process and shows that this definition is both psychological and circular.

Secondly, he looks at a definition of cause as a type of correlation, according to which "Cause and effect ... are correlative terms denoting any two distinguishable things, phases, or aspects of reality, which are so related to each other that whenever the first ceases to exist the second comes into existence immediately after, and whenever the second comes into existence the first has ceased to exist immediately before." Here Russell attacks temporal contiguity of cause and effect, showing that logically both cause and effect must persist for a limited time, after which he further shows that either the cause is a process, involving change within itself; in which case only its latter parts are relevant to the effect and we can "diminish the duration of cause without limit." Or, alternatively the cause does not involve change within itself, in which case "it seems strange that the cause, after existing placidly for some time, should suddenly explode into the effect." Therefore, any cause and effect must be separated in time by a finite time interval, which naturally leads to the famous paragraph:

> The law of causality, I believe, like much that passes muster among philosophers, is a relic of a bygone age, surviving, like the monarchy, only because it is erroneously supposed to do no harm.

This sweeping statement, aimed both at philosophy and natural sciences, is based purely on logical analysis of concepts of causality, purporting to show their inconsistency. A more recent metaphysics puts it like this: "all there is to the world is a vast mosaic of local matters of particular fact, just one little thing and then another" (Lewis, 1986).

On the other hand, anybody who regularly reads scientific literature must recognize the ubiquity of causal terminology in it. For instance, it would be very difficult to imagine biology without recourse to causal explanations involving natural selection, pathogens and disease, or signals and response. The recognition of this fact has led philosophers, including an older Russell, to nevertheless attempt theories of causation (Hitchcock in Psýllos and Curd, 2008). In 1948 Russell proposed his "causal lines" theory, whose reason for existence is to explain the identity of a quality or structure through time. A series of events form a causal line "if, given some of them, we can infer something about the others without having to know anything about the environment." A causal line may contain constancy or gradual change, but not sudden change. According to this theory, a "thing" is not a single entity, but "a string of events having a certain kind of causal connection with each other."

In objection to causal lines Wesley Salmon (1984) noted that causal lines presuppose an inferer. Does this mean that causality does not exist in nature? Moreover, causal lines cannot distinguish between physical processes and pseudo-processes. What all real causal processes must have in common is their adherence to the laws of nature. An example of a pseudo-process is a spot of light moving on a wall. This movement can, without violating special relativity, be faster than the speed of light and it therefore cries for distinction from causal processes.

As an alternative, Salmon proposed the "mark transmission theory" of causation, according to which a process is causal if it is capable of transmitting a local modification in structure (a "mark"). Pseudo-processes do not transmit a mark. So a causal process, according to Salmon, is essentially action by contact, with no spatio-temporal gaps in the causal link (think of billiard balls).

In his criticism of Salmon, Christopher Hitchcock (1995) drew a line between explanatorily relevant marks (as mass or velocity of billiard balls) and irrelevant ones (color of the ball or the chalk mark left by the cue) and showed that the model cannot differentiate one from the other. A feistier example is of a rather ordinary man, Mr. Jones, who for some reason, alongside Mrs. Jones, eats birth control pills and indeed never gets pregnant. The pills dissolve, enter the bloodstream, and are metabolized in both the husband and wife (admittedly in slightly different ways); yet they are irrelevant to Mr. Jones's non-pregnancy and relevant to Mrs. Jones's. Thus, the ability to transmit the "mark" may or may not be causally relevant, and therefore cannot be used as an explanation or definition of causality.

Other influential criticisms of the MT theory include arguments that it excludes many genuinely causal but short-lived effects; that it requires temporal uniformity of causal processes in the absence of interactions, and that marks be transmitted in the absence of *additional* causal interactions (dropping this requirement would reintroduce pseudo-processes, while leaving it in place would exclude even most carefully controlled lab experiments from the realm of causality).

Another popular approach is to view causality as physical transmission of a conserved quantity, like energy, linear momentum, charge, or information. Under these theories the discovery of causes is purely an empirical matter. They have the advantage of excluding the pseudo-processes and the disadvantage that many seemingly legitimate causal claims outside physics would also be excluded. If I killed my potted plants by not watering them or prevented someone preventing an accident, are we talking about causes, pseudo-causes, or fake causes? Maybe the full account of causality must wait for the full reduction of my behavior to the laws of physics?

As none of the abovementioned theories (there are also probabilistic theories of causation relevant to quantum mechanics and epidemiology, and quite a few others) seems very good for explaining and delineating the wider scientific uses of causality, and no theory is widely accepted as the winner, we are left with a dilemma.

If Russell of 1912 is right and we need causality as much as we need the queen, why is everybody still talking about it? For a tentative answer, we have to go back to the nineteenth century to John Stuart Mill, who opinioned that "The Law of Causation, the recognition of which is the main pillar of inductive science, is but the familiar truth, that invariability of succession is found by observation to obtain between every fact in nature and some other fact which has preceded it." In other words, causality is nothing but a convenient simplifying term for a heterogeneous bunch of complex natural phenomena.

To better understand this viewpoint, it behooves us to think of a typical biological explanation of, say, bacterial chemotaxis. A researcher dissolves a salt crystal in the medium, observes the bacteria swimming away from it and comments on their wish to escape from danger. Of course everybody is aware that bacteria do not think or have wishes, but in this context a little anthropocentrism can go a long way in making our talk of signals and receptors and rotating flagella shorter and easier to follow (but less precise). In other words, talking about intentions of bacteria can sometimes be useful in a scientific context, but this does not make it a scientific concept. A similar argument can be made for causal explanations, making them nothing more than verbal shortcuts for referring to wildly different and often very complicated descriptions of reality. If MAPKK causes phosphorylation of the MAPK by direct catalysis and smoking causes cancer by a thousand cuts inside the lungs of a smoker, are we really justified in using the same word to describe these processes, as if it meant something as a scientific term?

Specifically, it may be desirable to purge causal language from epidemiological research by arguing that a statement to the effect that smoking causes lung cancer is either scientifically empty or redundant (Lipton and Ødegaard, 2005), while causal language can be defended as providing a sort of informal guidepost for common sense in medical hypothesis testing (Phillips and Goodman, 2006). Some attempts at normative uses of causality in epidemiology are reviewed in Kaufman and Poole (2000) and Rothman and Greenland (2005).

> The view that "causal thinking" is an umbrella term, suitable for quick and dirty use, is supported by a recent brain-imaging study, which revealed that we have two separate brain circuits for thinking in causal terms: one for mental causation, such as how unfairness makes us angry; and the other for physical causality. The evidence strongly suggests that the two circuits are physiologically mutually antagonistic, so that we cannot use both concepts of causality simultaneously (Jack et al., 2013).

If causation is nothing but a convenient heuristic, why should we regard it as a central concept of scientific thinking and develop all these methods (philosophical and mathematical) to recognize its manifold manifestations?

TABLE 2.1 Truth Table for Implication 'If A Then B'

A	B	A→B
True	True	True
True	False	False
False	True	True
False	False	True

Could causal language be harmful in science? Here two possibilities come to mind. Firstly, when we state that X causes Y, we may inadvertently mask a complex underlying relationship under a pseudo-explanation, which implies knowledge, where there is none. In 1938 John Baker disentangled a single concept of biological causation into ultimate and proximate causes, which enabled him to clearly differentiate between causes responsible for the evolution of a trait and causes responsible for inducing a genetic program for the expression of the trait (Mayr, 1982, p. 68). For example, the proximate cause for a breeding season might be the length of day and the ultimate cause the availability of food. To talk of proximate and ultimate causes interchangeably in this context would pretty much ensure that no useful testable hypotheses would come of it.

And secondly, causal interpretations, when given to logical tools, may confuse us as to the nature of the scientific method itself. In propositional logic implication "if A, then B" (or A→B) is sometimes thought to imply physical causation. This appears to be an instance of assuming that if a term used in logic also has a meaning in the real world, then the two uses of the word must be related. Unfortunately, this is usually not so. While in the material world finding a bloody knife in the butler's cupboard might implicate him in a crime (and probabilistically, at that), the logical implication merely describes a specific set of co-occurrences.

Even worse, in formal logic every true proposition implies every other true proposition. The only situation where A→B is logically false is when A is true and B is false (Table 2.1) (Jaynes, 2003, p. 12).

In probability theory we use conditional probabilities in the form $P(A \mid B)$: probability of A, given the occurrence of B. For example, if A is "rain" and B is "cloudy," then $P(A \mid B)$ is the probability of rain when we are sure that it is cloudy and $P(B \mid A)$ is inversely the probability that it is cloudy when it rains. Clearly $P(\text{rain} \mid \text{cloudy})$ does not equal $P(\text{cloudy} \mid \text{rain})$. Such conditional probabilities are sometimes erroneously given a causal interpretation: "Inverse probability is also called the "probability of causes," because it enables the estimation of the probabilities of the causes underlying an observed event" (Galavotti in Psýllos and Curd, 2008). To test your intuition on the subject, suppose that P(B | A) is high and P(A | B) is low. Knowing this, which is more likely: A causing B, or B causing A? In a probabilistic version of Humean causality, where the cause always precedes the effect, A would likely be the cause B. Now, let's think again about smoking and lung cancer. P(smoking | lung cancer) is very high because smoking increases the risk of lung cancer by about 20-fold and because it is the main risk factor for lung cancer in the general population. At the same time, P(lung cancer | smoking) is much lower since most smokers die of other causes. Hold on, did we just conclude that cancer causes smoking?

Moreover, because, as we will learn in Chapter 5, probability theory demands that conditional probabilities may always be reversed, there is a simple rule, Bayes theorem, to derive P(A | B) from P(B | A), making a mockery of the dictum that causality, like history, is about one damn thing happening after another. While physical causality propagates only forward in time, logical inferences work in both directions. This is very fortunate because otherwise our friend Sherlock Holmes (not to mention every working archeologist and evolutionary biologist) would be at a loss about how to infer from present evidence to a past deed. Thus we may make use of all the relevant information available to us, without asking whether the information is about a period before or after the phenomenon under study.

Not only the rules of logic and probability theory, but also the laws of physics seem to be cause-free. For example, the equation $F = ma$ tells us that force is described by mass and acceleration, not that it is caused by mass and acceleration. Should one accept a causal interpretation, then from the equivalent notation $m = F/a$ comes that not only is force caused by mass but that mass is also caused by force (Pearl, 2000).

As the laws of classical physics are time-reversible, the concept of time is flexible in relativity theory and highly controversial in the physics community as a whole (Smolin, 2013; Barbour, 2001), it seems that a biologist can freely pick and choose the sort of causality (or lack thereof) as he pleases, without hindrance from logicians, statisticians, or physicists.

REFERENCES

Barbour, J., 2001. The End of Time. Oxford University Press, Oxford, UK.

Brown, J.R., 1994. Smoke and Mirrors: How Science Reflects Reality (Philosophical Issues in Science). Routledge, London.

Chakravartty, A., 2007. A Metaphysics for Scientific Realism: Knowing the Unobservable. Cambridge University Press, New York.

Chang, H., 2014. Is Water H_2O? Springer, Heidelberg.

Cornfield, J., et al., 1959. Smoking and lung cancer: recent evidence and a discussion of some questions. J. Natl. Cancer Inst. 22 (1), 173–203.

Dienes, Z., 2008. Understanding Psychology as a Science. Palgrave MacMillan, London.

Dilworth, C., 2006. Metaphysics of Science. Springer, Heidelberg.

Dougherty, E.R., Shmulevich, I., 2012. On the limitations of biological knowledge. Curr. Genomics 13 (7), 574–587.

Ellenberg, J., 2014. How Not to Be Wrong: The Power of Mathematical Thinking. The Penguin Press, New York.

Feinstein, A., 2001. Principles of Medical Statistics. CRC Press, Boca Raton, FL.

Fisher, R.A., 1958. Cancer and smoking. Nature 182 (4635), 596.

Flint, J., Munafo, M., 2014. Schizophrenia: genesis of a complex disease. Nature 511, 412–413.

Forster, M., Sober, E., 2004. Why likelihood? In: Taper, M., Lee, S. (Eds.), The Nature of Scientific Evidence. University of Chicago Press, Chicago, IL.

Frege, G., Geach, P.T., 1977. Logical Investigations. Yale University Press, New Haven, CT.

Goodman, N., 1983. Fact, Fiction, and Forecast. Harvard University Press, Cambridge, MA.

Hacking, I., 1983. Representing and Intervening: Introductory Topics in the Philosophy of Natural Science, Cambridge [Cambridgeshire]. Cambridge University Press, New York, NY.

Hill, G., Millar, W., Connely, J., 2003. "The great debate": smoking, lung cancer, and cancer epidemiology. CBMH 20, 367–386.

Holland, P., 1986. Statistics and causal inference. J. Am. Stat. Assoc. 81 (396), 945–960.

Jack, A.I., et al., 2013. fMRI reveals reciprocal inhibition between social and physical cognitive domains. NeuroImage 66 (C), 385–401.

Jaynes, E.T., 2003. Probability Theory: The Logic of Science. Cambridge University Press, Cambridge, UK.

Kaufman, J.S., Poole, C., 2000. Looking back on "causal thinking in the health sciences." Annu. Rev. Public Health 21, 101–119.

Lenski, R.E., 2011. Evolution in action: a 50,000-generation salute to Charles Darwin. Microbe 6, 30–33.

Lewis, D.K., 1986. On the Plurality of Worlds. Blackwell, Hoboken, NJ.

Lipton, R., Ødegaard, T., 2005. Causal thinking and causal language in epidemiology: it's in the details. Epidemiol. Perspect. Innov. 2, 8.

Marino, M.J., 2014. The use and misuse of statistical methodologies in pharmacology research. Biochem. Pharmacol. 87 (1), 78–92.

Matthews, R., 2000. Storks deliver babies ($p = 0.008$). Teach. Stat. 22 (2), 36–38.

Mayr, E., 1982. The Growth of Biological Thought. Belknap Press, Cambridge, MA.

McGrayne, S.B., 2012. The Theory That Would Not Die: How Bayes' Rule Cracked the Enigma Code, Hunted Down Russian Submarines, and Emerged Triumphant from Two Centuries of Controversy. Yale University Press, New Haven.

Menand, L., 2001. The Metaphysical Club. Farrar, Straus, and Giroux, New York, NY.

Motulsky, H., 2010. Intuitive Biostatistics. Oxford University Press, Oxford, UK.

Navarrete, A., van Schaik, C.P., Isler, K., 2011. Energetics and the evolution of human brain size. Nature 480 (7375), 91–93.

Pearl, J., 2000. Causality: Models, Reasoning, and Inference. Cambridge University Press, Cambridge, UK; New York, NY.

Pettegrew, J., 2000. A Pragmatist's Progress? Richard Rorty and American Intellectual History. Rowman & Littlefield, Lanham, MD.

Phillips, C.V., Goodman, K.J., 2006. Causal criteria and counterfactuals; nothing more (or less) than scientific common sense. Emerg. Themes Epidemiol. 3, 5.

Poletiek, F., 2009. Popper's severity of test as an intuitive probabilistic model of hypothesis testing. Behav. Brain Sci. 32, 99–100.

Popper, K., 1963. Conjectures and Refutations. Routledge, London.

Popper, K.R., 1972. Objective Knowledge: An Evolutionary Approach. Clarendon Press, Oxford.

Popper, K.R., 1992. The Logic of Scientific Discovery. Routledge, London; New York, NY.

Popper, K.R., Bartley, W.W.1., 1993. Realism and the Aim of Science. Routledge, London; New York, NY.

Popper, K.R., Eccles, J.C., 1977. The Self and Its Brain. Springer International, New York, NY.

Psýllos, S., Curd, M.P., 2008. The Routledge Companion to Philosophy of Science [Electronic Resource]. Routledge, London.

Putnam, H., 1981. Reason, Truth and History. Cambridge University Press, Cambridge, MA.

Quinn, G.P., Keough, M.J., 2002. Experimental Design and Data Analysis for Biologists. Cambridge University Press, Cambridge, MA.

Rorty, R., 1989. Contingency, Irony, and Solidarity. Cambridge University Press.

Rothman, K.J., Greenland, S., 2005. Causation and causal inference in epidemiology. Am. J. Public Health 95 (Suppl. 1), S144–S150.

Royall, R., 1997. Statistical Evidence. CRC Press, Boca Raton, FL.

Russell, B., 1912. On the notion of cause. In: Proceedings of the Aristotelian Society. In: Mysticism and Logic. Pelican Books, 171–196 (1953).

Russell, B., 2009. The Basic Writings of Bertrand Russell. Routledge, Abingdon, Oxon [England]; New York, NY.

Schoenfeld, J.D., Ioannidis, J.P., 2013. Is everything we eat associated with cancer? A systematic cookbook review. Am. J. Clin. Nutr. 97 (1), 127–134.

Schwanh a usser, B., et al., 2011. Global quantification of mammalian gene expression control. Nature 473 (7347), 337–342.

Smolin, L., 2013. Time Reborn. Houghton Mifflin Harcourt, Boston, MA.

Sober, E., 2008. Evidence and Evolution: The Logic Behind the Science. Cambridge University Press.

Stanford, P.K., 2006. Exceeding Our Grasp: Science, History, and the Problem of Unconceived Alternatives. Oxford University Press, Oxford; New York, NY.

Stanton-Geddes, J., de Freitas, C.G., de Sales Dambros, C., 2014. In defense of P values: comment on the statistical methods actually used by ecologists. Ecology 95, 637–642.

Sugihara, G., et al., 2012. Detecting causality in complex ecosystems. Science 338 (6106), 496–500.

Székely, G.J., Rizzo, M.L., 2009. Brownian distance covariance. Ann. Appl. Stat. 3 (4), 1236–1265.

van Fraassen, B.C., 1980. The Scientific Image. Clarendon Press, Oxford.

Vul, E., et al., 2009. Reply to comments on puzzlingly high correlations in fMRI studies of emotion, personality, and social cognition. Perspect. Psychol. Sci. 4 (3), 319–324.

Wirth, S., 2002. King Kong, storks and birth rates. Teach. Stud. 25, 29–31.

Wiser, M.J., Ribeck, N., Lenski, R.E., 2013. Long-term dynamics of adaptation in asexual populations. Science 342 (6164), 1364–1367.

Witschi, H., 2001. A short history of lung cancer. Toxicol. Sci. 64, 4–6.

Young, J.O., 2013. The coherence theory of truth. Available from: http://plato.stanford.edu/entries/truth-coherence/.

Part II

The Method

Chapter 3

Study Design

A common approach to teaching experimental design in biomedicine is to start from frequentist statistical methods like null hypothesis testing and then offer some rules of conduct that will help the researcher to meet the assumptions of the statistical test (Marino, 2014; Krzywinski and Altman, 2014; Seltman, 2013; Quinn and Keough, 2002). This brands experimental design squarely as a branch of statistics and it's not the road we take.

Study design is a considerably wider field than statistics and most mistakes made in the design stage of a study cannot be rectified later at the data analysis stage. In fact, standard data analysis methods only have meaning when the study design is good enough to minimize the chances for systematic errors (also known as bias) before any data are collected. In orthodox statistical theory the absence of bias is achieved by the combination of a big enough sample and random and independent sampling. However, we will see that this ideology, while preferred for mathematical reasons, puts the researcher in a position where potentially useful information is lost from analysis and sample sizes need to be inflated.

Study design is not only important, but also the most difficult and least well understood part of scientific methodology. It is there that most of the serious errors, which endanger the validity of results, come in. The main weapon in the arsenal of biologists is experiment and the main reason for careful experimental design is minimizing various errors. Experiments themselves are only means to an end; they are done to test specific predictions derived from scientific theories or hypotheses. In every experiment there is (i) some parameter(s) to be measured, be it length of peacocks' tails or enzyme kinetic constants; (ii) there are the objects or units of measurements (e.g., peacocks, enzyme preparations) that together make up the sample;[1] (iii) there is the experimental manipulation, which hopefully is the only difference in the treatment of the control group and experimental group, and thus defines what is experiment and what is control; and (iv) there is the act of measurement itself, involving a fancy machine, like a ruler or a mass spectrometer. Every measurement is characterized by its accuracy (closeness of its result to the true parameter value), precision (closeness of repeated measurements to each other), robustness (insensitivity to small changes in the measuring procedures), and by its limit and range of detection (Plant et al., 2014).

---

1. A unit of measurement is whatever that is actually measured.

Interpreting Biomedical Science. DOI: http://dx.doi.org/10.1016/B978-0-12-418689-7.00003-X

With each building block of experiment comes its own type of errors: (i) we must strive to measure a scientifically relevant parameter—by doing so we achieve construct validity. For example, when testing a cholesterol-lowering drug, the relevant parameter may not be cholesterol but mortality (LaMattina, 2013). (ii) When deciding how to pick the objects of measurements one is thinking, amongst other things, of the future generalizability of results—of the external validity of the study. (iii) When the user-defined experimental manipulation is indeed the only thing that differentiates the treatment group from experimental controls, then the experiment has internal validity—otherwise it has bias or systematic error. A study has internal validity when its causal conclusions are warranted. (iv) With every act of measurement comes a measure of random error or sampling error. Sampling error is made up of measurement error—which comes from your particular fancy machine and from the way that you use it—and of biological variation present in the sample.

To illustrate the difference between random error and bias we will imagine an experiment where we wish to find the location of a mark (invisible to us) on a wall by the target patterns of two archers shooting at it (Figure 3.1). The first archer's hits are widely spread (imprecise) but they are nicely centered on the mark (accurate or unbiased). This allows us, when having a large enough sample of his shots, to accurately infer the true location of the mark by a simple random error model. In contrast, the second archer gets her hits closer together, but she has a strong tendency of steering her arrows right of the mark (she is precise, yet biased). Thus, under a random error model, the more data we have, the more confidently we will miss the true location of the mark. To make the experiment work we not only need to have the data (the shots) but we also need to know about the amount and direction of bias inherent in the second archer's training, so we can compensate for the bias in our data under a more complex Bayesian model. Bias is directional, random error is not.

1st archer is not very precise
(there is a lot of random error)
but accurate (there is little bias)

2nd archer is more precise but
less accurate (there is more bias)

The amount of bias

FIGURE 3.1 Relationships between precision, accuracy, random error, and bias.

Orthodox statistical analysis concerns itself only with random error, leaving the experimenter free to deal with systematic error or bias as she sees fit. This makes experimental design an art of creating a clearly delimited, fully understood and controlled mini-universe where different types of errors are balanced against each other and against the cost and ethics of doing the experiment. As long as bias is not properly dealt with, it makes little sense to present small P values or confidence intervals (CIs) as evidence for your theory.

The three main principles of experimental design are:

1. Good experiments compare treatment with control. This is because of the need to cancel out sources of bias.
2. You should control for main sources of bias if you can; and randomize if you can't. If direct control is impossible, data analysis can help to deal with nuisance parameters. Systematic pairing of experiments and controls ("blocking") can help reduce bias. Where an effect cannot be controlled, randomization is in order. For example, treatments and controls should be run in random order.
3. You should plan ahead for replication of experiments. Without knowledge of the level of uncertainty in your experiment, your results cannot be interpreted. This is because statistical tests rely on comparison of an effect size with an estimate of uncertainty, which is usually based on the observed variation in experimental replications. Therefore, experiments need to be replicated (Ellison et al., 2009).

3.1   WHY DO EXPERIMENTS?

Most biologists never directly study life. Instead they study narrowly circumscribed and more or less artificial experimental systems. This raises the question: Why volunteer for drawing conclusions which, even if correctly describing the workings of your experimental system, could well be irrelevant to the wider world? What is the upside here?

In Chapter 2 we examined the role of experiments in elucidating causal connections or interactions. The general logic of eliciting causal connections by experiment goes from cause to effect and asks: If we abolish (or change) the cause, will that abolish (or change) the effect? The opposite path—where the scientist asks: What is the cause of this effect?—is asking a why-question (why does this happen?), and is much less often trodden by experimental scientists.[2] While the usual statistical methods of null hypothesis significance testing and statistical estimation are routinely used for disentangling forward causation, the reverse causation is about model checking and hypothesis generation (Gelman and Imbens, 2013).

Eliciting forward causation by experiment can be described as a five-step workflow (Figure 3.2). The experimental approach often starts with asking a clearly delimited question about the existence of a causal mechanism that

---

2. We can approach this question by hypothesizing a list of possible causes and then disrupting them one-by-one by different experiments to see whether only one of them remains associated with the effect.

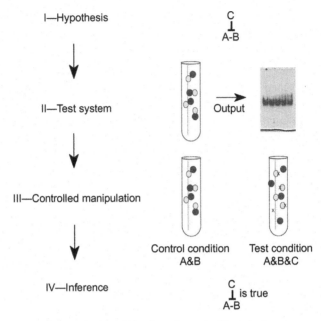

FIGURE 3.2 Components of an experimental test for a causal hypothesis.

connects specific biological objects. The hypothesis could be that an enzyme produces a specific product, that a chemical signal is transmitted from protein A to protein B, that a programmed biological response is responsible for higher organismal fitness under some circumstances, or any number of others. As a rule, the smaller the number of proposed causal interactions, the easier it is to find a workable experimental system and interpret the result of the experiment. And conversely, the more undefined or unknown components there are in the hypothesis, the more chance there is of misinterpretation of the experiment.

Step two is building and validating a defined experimental system, which enables to unequivocally measure the workings of the postulated causal system (there is an output, which we can measure). The experimental system is characterized (at least) by the accuracy, precision, variation, sensitivity, specificity, linear range, limit of detection, and limit of quantification of its output. Ideally your experimental system is an optimal mix in respect of the above characteristics. Because the trustworthiness of the result of your experiment will depend on each of them, as will the required sample size and types of controls, the experimenter should already have an informed opinion about them during the planning stage (McShane et al., 2013). Also, relevant controls that help to exclude possible biases or misinterpretations are specified at this stage.

Each experimental system is also characterized by its structural complexity (and its reverse, simplicity). To understand the importance of simplicity of your experimental system we will need the concept of internal validity. For your experimental system to have internal validity you need to be reasonably sure that it is the experimental manipulation and not some unknown and uncontrollable factor that causes the experimental effect; and this is only possible if your experimental system is simple enough to be tractable. When the groups that are compared differ by more than experimental manipulation, then we have a problem called *confounding*, which is a fatal threat to internal validity of the experiment. More specifically, confounding refers to a difference in the average level of any explanatory variable between experimental and control groups in the absence of experimental manipulation. The factors that cause confounding are called *confounding factors*.

Step three involves manipulating a single parameter in the system to see if breaking a link in the putative causal chain of events will disrupt the chain. The experimenter will look to see whether the manipulation of the experimental system changed the output in the expected direction and range. The experiment needs to be conducted in a controlled way, so that it would be possible to compare the results from at least two conditions (manipulation or not). It is important to exclude confounding factors as causes of effects seen in the experimental group. Confounding can be eliminated by controlling for known confounding factors one-by-one (either in data analysis or in study design, by matching cases and controls by each confounding factor), or more simply by randomization of the experimental and control groups.

Confounding factors that we don't know about and which we therefore cannot specifically control for are called *nuisance variables*. The standard advice of the frequentist school of statistics is that only randomization can exclude nuisance variables. A Bayesian may disagree on principle (Howson and Urbach, 1989; Jaynes, 2003), while supporting randomization on practical grounds as one technique among several to reduce bias (Gelman et al., 2003). Anyhow, to randomize we obviously need to repeat our experiments.

In step four, inferences concerning the presence and sizes of effects are drawn from the output of the experimental system (see Chapter 4). Usually experiments are done more than once to increase the trust of the experimenter that the results are not due to chance and instead reflect real and pertinent aspects of nature. Repeating experiments allows estimating random error (sampling error) present in the system. The influence of systematic error (bias) on the interpretation usually cannot be controlled or estimated simply by repeating the experiment. Bias can be eliminated by randomization or it can be estimated from background knowledge and quantified by Bayesian methods as our level of ignorance concerning the true parameter value (Phillips and LaPole, 2003). Such analysis helps to clarify whether the estimates inferred from experimental results are accurate enough for our specific purposes. It also suggests how likely further research is to produce a substantially different answer to the experimental question.

In step five scientific generalizations from the experimental data-specific inferences are made taking account of the totality of our knowledge on the subject (Chapter 5). The logic behind integrating previous knowledge with our best estimate of the results of the current experiment is, again, Bayesian. It is important to see the difference between steps four and five: step four is all about the factors relevant for the experimental system, while step five is about the factors that are relevant for the underlying scientific hypothesis.

Simple hypotheses are often best served by fully defined *in vitro* models and more complex hypotheses by more versatile *in vivo* systems, like cell lines or model animals, but there are trade-offs either way. Well-defined *in vitro* systems can give misleading results if the tested hypothesis is too simplistic and the actual biological process contains additional functional interactions, which are not included in the experimental system. Also, we still have surprisingly poor understanding of the physico-chemical microenvironments inside living cells. Thus any *in vitro* results could be rendered irrelevant simply by different *in vivo* salt or macromolecule concentrations. On the other hand, when doing the experiment *in vivo*, although the results may be better generalizable, they are also much harder to interpret. A seemingly well-defined manipulation of the experimental system could lead to change in the studied process in some unfathomable roundabout way and in a complex system, like the cell, it is simply not possible to control for everything (or indeed, to even recognize most sources of bias).

Experiment as a controlled manipulation of a clearly defined system is probably a definition that everybody can live with, but there is more to the story. There is more than one type of experimental design, as there are observational study designs (Table 3.1). Both randomized and controlled clinical trials (RCTs) and preclinical lab experiments are clearly experiments by the above definition, but they can be quite different from each other. The traditional gold standard for clinical experiments is threefold: to be seen as trustworthy a clinical trial needs to be controlled, physically randomized, and blinded. The idea is to compare treatment with control (either placebo or another treatment) using randomized samples of patients, who are representative of the patient population in general. Clinical studies are preferably double-blinded, meaning that neither subjects nor researchers know who gets the placebo and who gets the treatment. In pre-clinical lab experiments we cannot draw a sample from a real population (there exists no physical population of experiments in the sense that there is a population of patients), which means that there is no randomization as such. Also, biological experiments are rarely blinded. Instead, we have defined experimental systems and controlled manipulations where, except for the manipulation, we strive to keep the experiments and controls identical both across the experiment–control pairs and over time. This is reminiscent of matching of cases and controls, a method used in a type of observational study to reduce confounding (see below).

TABLE 3.1 Comparison of Preclinical and Clinical Experiments and Observational Study Designs

Pertains to	Experimental Designs		Observational Designs		
	Preclinical Experiment	Clinical Trial (RCT)	Cohort Study	Case-Control Study	Case Series
External validity	Biological replication	Sample is representative to the population of clinical interest	ND[a]	ND	ND
Internal validity, RTM[b]	Defined experimental system[c]	Random and independent sampling	Post-experiment control of confounding	Matching cases and controls	ND
RTM, placebo effect, sensitivity, internal validity	Controlled	Controlled (placebo)	Controlled	Controlled	ND
Researcher and subject biases	ND	Blinding	ND	ND	ND
Time-resolved?	Yes	Yes	Yes	No	Yes

[a]No principled effort is made to control for the factor.
[b]RTM—regression to the mean.
[c]Defined experimental system means that treatment and control groups are identical in every respect except for the experimental manipulation.

Types of Experiments in Clinical Medicine

In the observational study design we may simply observe people who find themselves in different conditions (e.g., we might compare sick people who get a particular treatment with those that don't). In cohort studies, if people who get treatment fare better than controls, we conclude that the treatment may do more good than harm. In case-control designs people are not followed in time but two or more groups (healthy and sick, for example) are compared to find differences between them, like risk factors that predispose to disease. In lieu of full randomization case-controls studies can use matching of cases and controls in respect of a limited number of pre-specified factors (i.e., sex, age). For each case a matching control is selected that has identical values on the pre-determined matching factors. Matched data require more complex statistical analysis than unmatched data but allow using smaller samples, ideally without loss of efficiency in the analysis. Studies can suffer greatly for lack of matching controls, as indicated by case-studies of cancer biomarker failures (Diamandis, 2010).

Observational studies are non-randomized, non-blinded, may use non-independent sampling, but are usually still controlled in the sense that a treatment group is compared with a control group.

All designs depicted in Table 3.1 are used to pry into causal hypotheses, all can start with stating a formal hypothesis, all involve human manipulation of the normal course of nature (although in an observational study the manipulation is often not done by the researcher), but only two of them are experimental. The salient difference between observational and experimental studies seems to be use of more representative samples by experimental study designs, which can be achieved by randomization or by fully defined experimental systems.

To better grasp the power of randomization, let's imagine a typical observational study where we follow patients who either take drug A or drug B for an ailment. Because nobody knows which drug works better, the researcher follows patients' electronic medical records and compares the morbidity and mortality outcomes of the two groups. The weakness here is that patient groups that get drug A and B might be different from each other to begin with. For example, if drug A happens to be more expensive, it is possible that doctors tend to prescribe it to wealthier patients; and we know that wealthier patients tend to be healthier overall. Since we were able to think of this particular type of bias, thus making it a known unknown, it is possible to statistically control for it in the data analysis stage. Unfortunately, the list of unknown unknowns can never be exhausted as it is, well, unknowable. So the question arises, how to turn the observational design into an experiment. The answer can be surprisingly simple. If we are comparing two drugs for the same condition and do not know in advance, which one is better, then all that one needs to turn an observational study into an RCT is a friendly looking green button. Before the doctor touches the prescription pad, she will push the button, which activates a random number generator and produces either "drug A" or "drug B" on a computer screen. Thus the observational study can be turned into a controlled randomized clinical trial in a second. This design is still not blinded, which may pose problems in some contexts and be acceptable in others, but its advantages are excellent external validity (the sample is representative of the patient population), potentially huge samples, and low cost (Staa et al., 2012).

In conclusion, there are three basic reasons why scientists tend to prefer controlled experiments to simple observations. One is that experiments are our best chance to answer the question: what is the effect of this cause? Secondly, experimental study designs help us to fight bias. And thirdly, they help us to differentiate between true effects and random play of chance.

3.2 POPULATION AND SAMPLE

People use statistics not because it is beautiful or powerful or sexy. They do it because there is no better option. The need for statistical inference arises from us being smaller than the world we study. We are limited in both time and resources and when we wish to study an aspect of the world, the chances are that we will not be able to look at every relevant manifestation of it. Therefore we must be content with studying only parts of the whole and from there to draw inferences to the entirety of our scientific problem. This is where inferential statistics comes in. However, to draw inferences from a limited number of observations to the world that allowed these observations is an inductive process. And as we learned from our qualitative discussion in Chapter 2, induction can be a cruel mistress. This is because it tries to accomplish something that is very hard to do. We should keep this in mind when interpreting our statistical calculations. It is always the researcher, not the algorithm, who has to make the real decisions as to what assumptions to accept, which algorithms to use for making inferences, and finally, in which conclusions to put her trust.

In statistical lingo the totality of the phenomena we wish to study is called the "population" (or "collective") and the part of it that we actually study is called the "sample."[3] We start by deciding which population we want to study, then we draw a sample from that population, measure the individuals belonging to this sample and calculate the values of a sample statistic, for instance the sample mean. This part is called descriptive statistics because here you describe the sample, rather than the population (Figure 3.3). In the next step we make inferences from sample values to population values. This activity is called "doing inferential statistics." The inference can of course land closer or farther from the truth and it is the job of the statistician to discover methods that allow to get closer, or at least to predict, how close we are. Due to basic limitations of

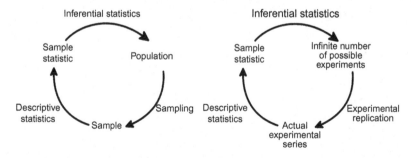

FIGURE 3.3 Statistical inference from sample to population.

---

3. "Sample" can mean different things to scientists and statisticians. For a biologist, a sample is usually a single batch of material subjected to analysis; for a statistician, a sample is a representative collection of items taken from a statistical population. We are using the second meaning.

the inductive process we can never be sure that we have actually hit the truth. From this comes the need for probabilistic descriptions of statistical predictions. In statistics a sample value is called a *statistic* and a population value (population mean, etc.) is called a *parameter*. The population parameters are fixed, that is, they have a single (albeit unknown) value, and sample statistics are left free to vary. A parameter value can be estimated from known values of statistics, but it can never be known.

To be able to grasp what statistics can do for you, it is important to understand the tricky concept of statistical population. Make no mistake, statistical population is very different from biological population. In biology a population is often defined as a collection of actually or potentially interbreeding individuals, or as a collection of individuals of the same species sharing a living space. This means that an individual usually belongs to exactly one biological population and this population is "real" in a very real sense. It is basically a thing (a unique collection of individuals, to be precise). On the other hand, a statistical population is any collection of objects, which we wish to study. A statistical population can sometimes be as real as the biological variety; for example, in clinical studies inferences are drawn for well-defined patient populations, of which we often know a lot. Or then it can be as ephemeral as a collection of all possible results of experiments, which will never be conducted, but could have been.

In statistics books the sample is usually defined through a population: you start by defining a statistical population, which you would like to study, and then draw a representative sample from it. Unfortunately, in practice the opposite also holds: the statistical population, which you actually study, is strictly defined through your sample. Now, the sample is simply a group of individuals (or individual measurements), each member of which had an equal chance of being sampled. This means that each sample necessarily corresponds to a population; but only in very carefully collected samples have we any idea what this population is. Therefore we must always suspect that the population, to which we actually draw our statistical inferences, is not the population that we think we are studying.

The meaning of "representative" in sampling can be elusive. In statistical theory a random sample (defined by the equal chance with which each item of the population can be drawn to the sample) provides unbiased estimates of the population in the long run and it is representative of the population in this very narrow sense. However, long run convergence of many random samples to the true population value says nothing about an individual sample's representativeness. A particular sample can be very different from the population and the larger the sampling variance, the more likely it is. Therefore, one should strive for samples which are not only unbiased, but which also have small variance.

We will generally use the term "representative" in its non-specialist sense denoting the fit of a specific sample to its population. This usage allows us to postulate that every sample is representative of some population (even if we don't know which one).

TABLE 3.2 A Comparison of Biological and Statistical Concepts of Populations

	Biological Population	Statistical Population 1	Statistical Population 2
Individual	A swan	A swan	A single experiment
Sample	NA[a]	A known number of randomly picked swans	The series of experiments, which was actually done
Population	All actually or potentially interbreeding swans	All swans that had an equal chance of ending up in the sample	Infinite number of potential experimental series

[a]NA—Not applicable.

The difference between how the statistician and biologist view the object of statistical inference (the population) is nowhere greater than in experimental biology where a statistical population does not consist of entities encountered in the material world. While in experimental science the sample is defined as the experimental series actually done by researchers, the population is defined as the infinite number of potential experimental series, which were never done, but could have been (Figure 3.3 and Table 3.2). The reason for this fiction is mathematical convenience. Because the classical probability theory was developed from the seventeenth century on by thinking about sampling marbles from well-shaken urns (incidentally, with replacement of marbles), assumptions of the resulting statistical tests include random and independent sampling from infinite populations, which is all very sensible mathematically, but not so much as a theory of the physical world.

The need to generalize from sample to population introduces a weird concept of the world to which statistical inferences are made as consisting of an infinite number of experiments, which were never done. How to get from this strange universe to our own must remain an open question.

Statistical populations have another unusual property. Every reader of this book is a member of a single biological human population. But he or she is also a member of an untold number of statistical populations. For example, you might belong to a population of men, a population of PhD students, a population of green-eyed and left-handed people taller than 195 cm, and so on. When we are interested in a population, we draw a sample from it. If the sample is representative of that population, we can draw valid inferences about the population from the sample. If the sample is not representative of this population, it will surely be representative of some other population, which means that the inferences we draw are still good, just not for the population that we think they are. This means that our conclusions will be biased.

We can try to raise the probability that our sample is representative of the population we have in mind. As it is easier to draw representative samples from populations that exhibit little variation, we need knowledge of the population variation to decide if a sampling procedure is likely to lead to representative samples. Equally obvious is that a larger sample is more likely to be representative than a smaller one. By drawing the sample randomly and independently, so that each individual in the population has an equal chance of ending up in the sample and its making it into the sample is independent of the identities of other individuals in the sample, is often considered the most important advice in ensuring representativeness. By randomization one theoretically forgoes the need to look at the actual structure of the sample (meaning less trouble for the scientist), but this also leads to discarding potentially important information and could necessitate larger sample sizes (and thus more trouble for the scientist) (Dienes, 2008). Here is an obvious need for compromise.

Random sampling is not the only type of sampling that is used by scientists, but it has two very substantial advantages over other methods of sampling. (i) A random sample allows extrapolation from the sample to the population and (ii) it allows estimating how much the conclusions would change if another random sample would be taken from the same population.

Stratified Sampling

Random samples can be a poor choice for monitoring individuals from a continuous sequence, because they can have widely varying intervals between successive items. Furthermore, if the statistical population contains sub-groups, which have different properties, the variance in a simple random sample can swamp any meaningful effects. A solution would be to divide such a structured population into strata (subpopulations), which are then sampled individually.

We could (i) sample equal number of items per stratum, which is sensible when there is no additional information, but it must be understood that combining the results of the individual strata will not give an unbiased estimate of the population, unless all the strata are of equal size.

We could (ii) sample the strata proportionally so that the number of items sampled per stratum is proportional to the fraction of each stratum in the population. Thus the expected mean for the composite of stratified samples is the same as the mean for the population. The variance will be smaller than for simple random sampling.

We could (iii) set the number of samples per stratum in proportion to the size and the sampling standard deviation (SD) of each stratum. This is the optimal strategy, but its use presupposes substantial prior information (Ellison et al., 2009). Of course, each type of stratified sampling requires its own specialized analytic techniques.

As already mentioned, increasing the sample size reduces the probability of it not being representative of the population. The idea behind random sampling is that any factor (besides the experimental manipulation), which could influence the result of your experiment, must be present in equal "concentration" in the study group and control group. If this is the case, their influences on experimental results will be equal for the two groups and will thus cancel out. For this trick to work, we need (i) not too many confounding factors to partition and (ii) not too small a sample size. If the sample size is comparable with the number of confounding factors, some combination of these will be by definition present in one group or the other and randomization cannot in principle remove bias (Dienes, 2008). Unfortunately, we generally have no idea about how many nuisance variables there might be in our experiment and thus cannot be sure that a particular sample is sufficiently representative of the population. This is a strong argument for using the simplest possible experimental designs.

Below we will discuss how statistical power to see true effects in experiments depends on the sample size (among other things), and how the concept of power is commonly used to calculate optimal sample sizes for a given type of experiment. However, as we just saw, the required sample size also depends on the complexity of the experimental system. To my knowledge there is no formal method to take the latter into account in setting the sample size for an experiment. The advice for randomized medical trials seems to be that with a sample size greater than 200, randomization is "an insurance in the long run against substantial accidental bias between treatment groups" (Gore 1981 as cited in Howson and Urbach, 1989). The "long run" in the previous statement means that even with large samples randomization will not guarantee elimination of bias from a particular sample, it will only make it more likely. As long as we cannot enumerate all the potential sources of bias in our experiment, we can only say that after randomization the probability of bias is somewhere between 0 and 1. Moreover, in most biochemical experimentation it is unclear what "random sample" means, because the underlying population is imaginary and conceptually depends on the very same sample that is "drawn" from it. The best that can be done in this situation is to strive for the independence of repeat experiments by controlling as many components of the experimental system (like buffer batches and enzyme preps) as closely as possible and to try to match the samples by keeping the compositions of the experimental and control groups as similar as possible. This naturally leads to paired experimental designs where individual experimental and control groups are treated in parallel.[4]

We can think of representativity as the goal and randomization, independence, and similarity as the methods, which help us to achieve it. Without properly ensuring representativity during planning and conducting your experiment subsequent standard statistical tests are likely to lead to grossly misleading conclusions.

---

4. It is very important that paired experiments are documented as such, as this opens the door for sensitive statistical tests for paired data.

3.3 REGRESSION TO THE MEAN

We start this chapter with four questions to test your intuition.

1. Suppose that you run a school for fighter jet pilots (a nice thought, I'm sure). Occasionally you have cadets returning from their flight lessons who did exceptionally well. So your inclination is to prize them. And then you will no doubt see cadets, who have just made a potentially dangerous mistake and deserve a put-down. Having done these things for a while, you notice a disturbing regularity. The cadets, who got the prize, tend to do noticeably worse the next time they go flying, which means that you will probably want to keep your prize to yourself in the future. And conversely, the people, who got harsh words from you, mostly show clear improvement the next time they go airborne. In fact, the worse the offense and harsher your words, the greater the subsequent improvement! Therefore your natural inclination would be to never say a good word to a student and give them the full force of your scorn whenever they deserve it (Kahneman, 2011). Right or wrong?
2. Suppose that you tested a bunch of children for their mathematical ability. Then you fed fish oil to your sample and tested them again. You found that on average there was no effect of fish oil on math skills. But then you had the idea of looking at the subgroup of children who got the lowest results in the first test. To your delight you found that these disadvantaged children indeed showed clear improvement after eating fish oil. Should you believe that the fish oil treatment really works?
3. Suppose that you wanted to know if eating fish oil helps to cure insomnia. You recruit insomniacs for your study, administer the treatment and find that after eating fish oil the patients indeed sleep better on average. Should you now believe that fish oil cures insomnia?
4. Suppose that you measure the "good" high-density lipoprotein (HDL) cholesterol levels in a sample of healthy individuals, the cut-off for the healthy level is set at 40 mg/ml and an individual called Bob just measured low at 38 mg/ml. Should we conclude that Bob has an unhealthy level of cholesterol and needs medication?

All four are trick questions. If your treatment actually works, you will certainly see improvement. And yet the answer to all four questions is no. This is because, if your treatment does not work, you are still likely to see improvement. And we may assume that most treatments that are tried out do not work. In all four cases the change in outcomes attributed to treatment could have easily arisen by dumb luck. In fact, in all four cases it is very likely that the changes were caused by chance events. In the first three examples the treatment was not administered to a random sample of the population that one wishes to study, but to a sample that had (at least to an extent) selected itself.

In the first example, it was the cadets who were either good or lousy pilots and who happened to have a better-than-average or worse-than-average day; in the second the same explanation applies to children who either sucked in math or just had a bad day; and as for the third example, as everybody knows diseases come and go and their symptoms fluctuate over time. So it is likely that many patients visit their doctor's office when they feel worse than usual and thus are more likely to feel better afterward, even if they get no meaningful help from their doctor.

In the fourth example, we may assume that Bob is an average person (as by definition most people are) and therefore his low HDL cholesterol level is likely to be an unusual result, meaning that it is likely that had we done more measurements over several weeks, then we would have found that his usual HDL cholesterol levels are in the normal range (Barnett, 2004).

All this means that on top of real effect (if any) of the treatment we often see a pseudo-correlation, called regression to the mean (RTM). RTM is a purely statistical phenomenon, masquerading as a causal process, which can make natural variation in repeated data look like real change. Importantly, RTM is not concerned with time; it happens forward in time, it happens backward in time, and it happens across measures collected at the same time. It will happen even with no treatment. It is important to note that RTM is a group phenomenon, so there is no way of knowing which way any individual will move. As a group descriptor, however, RTM tends to go either up or down, dependent if the sample consists of below-the-population-mean or above-the-mean individuals. The more extreme is the sample group, the greater the RTM. The less correlated are the two variables, the greater the RTM. When the two variables are perfectly correlated, there can be no RTM. But this is unlikely to happen in practice as there are always both biological variability and error of measurement present in your samples (Figure 3.4). RTM is in fact the inverse of correlation.

The formula of RTM is:

$$P_{rm} = 100(1 - r)$$

where P_{rm} is the percent of RTM and r is the correlation coefficient between the two measures. If $r = 1$, there is no RTM, if $r = 0.2$, there is 80% RTM and if $r = 0$, there is 100% RTM. Interesting examples of RTM in action can be found in Bland and Altman (1994).

To pry the biological effect of interest loose from RTM there are mainly two things that one should do, both of them in the study design stage. Firstly, if the sample is perfectly balanced in the sense that exactly half of the individuals are below the population mean and half are above it, the RTM as a group

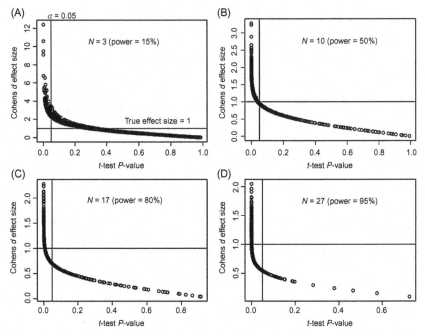

FIGURE 3.4 Setting a quality threshold for data can lead to reporting of seriously inflated effect sizes. Two normally distributed populations were simulated (mean = 0, SD = 1 and mean = 1, SD = 1). From each population were drawn 1,000 samples of indicated sizes (*N*), which were compared with student's *t*-test. (A) *N* = 3, power = 0.15; (B) *N* = 10, power = 0.5; (C) *N* = 17, power = 0.8; (D) *N* = 27, power = 0.95. *P* values and mean effect sixes (EZ) of each comparison are plotted. The true effect size (1.0) and the significance level (0.05) are denoted by lines. Note that when the power of the test is less than 0.5 and the significance level is set at 0.05, it is very difficult not to grossly overestimate the effect size.

phenomenon cannot occur. Therefore a lot of effort is placed in random sampling in order to increase the chances that the sample will be representative of the population.

The second trick is to use the control (placebo) group. If the treatment group and the control group are representative of each other (randomization again), then they should have the same amount of RTM, which means that simply by subtracting the control group mean from the treatment mean should get rid of RTM. It also means that many instances of the seemingly exciting "placebo effect," are really just instances of RTM. The urge to interpret placebo effects as biologically interesting is an example of the universal need to find meaning in the world, even if the original reason behind the placebo-control design was simply to remove statistical noise from the data (there exist data analysis methods that enable to separate RTM from true biological placebo effects in case-control study designs). In addition, analysis of covariance (ANCOVA) is a recommended statistical tool that helps to adjust each subject's follow-up measurement according to their baseline measurement (Barnett, 2004). The more baseline measurements were taken during the study, the easier it is to quantify RTM with ANCOVA.

RTM has a propensity of popping up in unexpected places. An example is the near-mythical phenomenon where many established scientific truths seem to wear off over time. This surprisingly widespread decline in effect sizes reached global consciousness with a spirited discussion in The New Yorker by Jonah Lehrer (13.12.2010) and it is especially serious in medicine, where drugs have been known to lose most of their potency over a decade. The existence of the decline effect has been verified in large metastudies of clinical trials (Gehr et al., 2006; Dwan et al., 2008; Pereira et al., 2012). Over ninety percent of the very large effects described in medical literature became smaller in subsequent trials, on average by about threefold (Pereira et al., 2012).

The most parsimonious explanation of the decline effect is decidedly unglamorous. What happens is simply that when true effects and collected samples are moderately sized, some samples will give reduced estimates of the effect size by chance and some samples will give inflated estimates.[5] Now, inflated effect sizes are more likely to result in statistical significance, which means that the original descriptions of highly significant findings often give out inflated promises (Figure 3.4). Because in the youth of a research program only the positive results tend to get published and the negative ones get buried as boring, the scientific community is presented with a skewed sub-sample of positive results. While initially we need statistically significant data to publish newsworthy research, later on even lesser effect sizes and negative results will have enough news value to be publishable because they can now be compared with the originally published effects. Such downgrading of effects is much more widespread in medicine than it is in experimental biology—by virtue of medical scientists actually repeating each other's work, which is very rare in the biological literature.

As a take-home message to the experimental scientist: when an experimental series results in a surprisingly large effect, the chances are that the true effect is more modest. It is likely that many more effects in the literature are overestimates than underestimates, and the scientists that put them there have no inkling of this. To combat the decline effect we should publish the results of every experiment that we do, which is unlikely to happen anytime soon.[6]

---

5. The smaller the sample size and larger the variation in the statistical population, the larger the fluctuations, which are expected to occur in your sample statistic.

6. If indeed we decided to publish every negative result, we should develop new ways of addressing the biases presumably prevalent in the small lower-quality studies, which today largely vanish unpublished. To publish everything would remove an important type of bias from the literature, but, no doubt, would exacerbate other biases.

Indeed, whenever we set a fixed threshold between non-significant and significant results, we will find that the magnitude of the decline effect (and of RTM) depends on where the threshold line was placed. All that matters is that numerical evidence be converted into a decision to either accept or reject an effect as "real." It doesn't matter whether the selection of real effects from chance effects is achieved by statistical significance testing or intuitively. As long as the selection is lax, the sample of "significant" results will contain many false-positives, but the effect sizes of "real" results are on average close to true values.[7] When the selection is more stringent and there are fewer false-positives, the samples that make it into statistical significance are more prone to inflated effect sizes and RTM. In significance testing the significance level $\alpha = 0.05$ means that researchers' long-run frequency of falsely rejecting the null hypothesis is exactly 5% (provided that the assumptions of the test are met). Setting the significance level at 0.01 will result in fewer false rejections of the null. But it will also result in more false-negative results (which is widely understood) and in more inflated effect sizes for the results that are statistically significant (Friston, 2012; Gelman and Weakliem, 2009; Ioannidis, 2008). Let me say it again: people, who are better at excluding false-positives, are more likely to publish grossly overestimated effect sizes!

The seriousness of this form of RTM to clinical research was shown by plotting of a large number of effect sizes (odds ratios) from clinical metastudies against sample size (a proxy of the false-negative error rate), which revealed a clear negative relationship between the two. While the median odds ratio of large studies ($N > 10,000$) is around 1.5, meaning that experimental condition differs from the control condition by 1.5-fold, for the smaller studies ($N < 300$) the median odds ratio is inflated to about 3.5 (Ioannidis, 2008). Small clinical studies really should be taken with a grain of placebo medicine.

We could of course search for a balance between sample size, specificity, sensitivity, and inflated effect sizes using traditional statistical methods, but to really abolish the problem we would have to give up the habit of making black-and-white decisions about whether a particular effect is real or not, and instead attribute to each effect a probability of lying in some specified interval—or in other words to substitute action with belief.

3.4 WHY REPEAT AN EXPERIMENT?

To consult a statistician after an experiment is finished is often merely to ask him to conduct a post mortem examination. He can perhaps say what the experiment died of.

R. A. Fisher (Presidential Address to the First Indian Statistical Congress, 1938)

---

7. This is because there are few false-negatives that fall under the decision line.

The question "Why repeat an experiment?" may seem simple but as we would like to argue, deceptively so. We can think of at least three different reasons to repeat an experiment.

1. Many molecular biologists see their results dichotomously; either there is a strong effect relative to control, or there isn't. For them repeating an experiment is simply an insurance against the observed effect being caused by some qualitative mistake in running the experiment, like forgetting to pipet the enzyme into the reaction mixture. Thus repeating the experiment once or twice seems generally sufficient to reach a confidence about the result and it is all right to publish just a representative experiment as a graph or table. There is no room for statistics or probabilistic thinking here. A lot of classic molecular biology has been done this way; and rather successfully at that. Over the past couple of decades, however, this approach has become less popular as successive generations of reviewers have learned to expect and demand (at a minimum) means with error bars from prospective authors.[8]

2. Another and very different reason to repeat an experiment is to use statistical hypothesis testing or parameter estimation to try to exclude random effects as a cause for the observed difference between the study group and the control group (Chapter 4 gives an in-depth treatment of these methods). When using conventional statistical methods, the successive experimental replications have to be "random" and independent. The success of this approach depends on the sample size, which should be neither too small nor too large.

3. A further reason to repeat experiments is to use randomization for minimizing RTM and other forms of systematic error by making them cancel themselves out between the control group and the experimental treatment group. Although gaining this goal requires similar basic experimental design as in (2), the sample sizes may be quite different. Unlike for (2), there are no available methods to estimate such sample sizes with any degree of accuracy. Yet, excluding random error as a cause for the observed effect without controlling for systematic error seems a hollow victory.

As a specific example of randomization, it is often sensible to randomize the order in which treatments and controls are measured. This is because the accuracy and precision of the measurement apparatus can drift over time. Randomization here has two effects. Firstly, the treatments and controls now have the same level of drift. Randomization reduces the effect of drift by a factor proportional to $\sqrt{n}$. Assuming a linear drift size D and $n = 6$ in each of two groups, the standard deviation (SD) of the mean drift effect over many experiments is about 0.17D (Ellison et al., 2009). Secondly, although randomization of the measuring order

---

8. These days it is quite usual to come across reviewer comments on the lines of "the results look fine but would the authors please do some statistics anyway."

led to an increased SD for both the treatment and the control measurements, your statistical test is now less likely to lead to a false-positive result. Yes, the power of the test to find true effects went down, but at least the observed effect size is now correctly compared with the full variation in your experiment, including the drift effects.

When fighting drift, an alternative to randomization is to use blocking designs. The idea of blocking is to group the experiments so that in each run of the apparatus a full complement of treatments and controls are measured. For example, if you want to compare a control condition C with treatments A and B and repeat this experiment six times, then you will have six runs, each with conditions C, A, and B. The results of each experimental run are first analyzed together, after which all the runs (blocks) are compared to decide if any of the treatments stands out. The statistical method of choice here is two-way ANOVA (Ellison et al., 2009). Blocking can disappear drift effects completely, and it leads to increased statistical power.

Before we commence our discussion on experimental replication, we must speak a word of caution on biology versus chemistry.

Methodologically, chemistry is quite different from biochemistry. This may sound wrong, but it isn't. The secret ingredient, which forces us to approach biological objects (including macromolecules) differently from chemical objects, is evolution. Evolution really means natural selection and natural selection cannot work without a level of heritable variation in the population. In other words, variation, being the fuel of evolution, is a real and necessary property of every biological system.

In chemistry variation is a property of the measurement. If, for example, we study the molecular weight of ethanol, we can be sure that in reality every ethanol molecule has the same mass (assuming the same isotopic composition), the same reactivity and atomic composition. We can be equally sure that repeated measurements of any of the above will differ from each other somewhat and if this measurement error is random (unbiased), the results will likely fall into a normal distribution, whose peak corresponds to the single real value present in cosmos.

In contrast, even a simple 18 base-pair hairpin ribozyme exhibits intrinsic 50-fold heterogeneity in its folding and unfolding rates when measured at single molecule level (Okumus et al., 2004). Functional heterogeneity has also been well described by single-molecule studies of many enzymes, including the biggest of them, the ribosome (Aitken et al., 2010). No two macromolecules are identical in their structure and functioning. Some of that heterogeneity has regulatory value, some is accidental (random decoding errors and chemical modifications) and some may be the consequence of intrinsic limitations in macromolecular folding.

Methodologically this means that while variation is not a scientifically interesting property of the chemical system and needs to be minimized, in biological systems variation is both unavoidable and potentially interesting, and thus needs to be described. Calculating variation from the biological replicates and technical replicates, and subtracting one from the other, can separate biological variation from measurement variation.

As long as your sample consists of measurements of individual entities, be they peak lengths of finches or reaction rates of individual enzyme molecules, it is important to ask if the biological information that you are interested in, really is likely to reside in the average (mean) of the measurement values or, conversely, in the tails of the data distribution. The structure in the data and how we look at it can crucially depend on whether the individual measurements in our sample are of individual entities or ensembles of many individuals, like in classical biochemistry, where each measurement is of an average score of individual molecules collected from many cells or organisms.

Variation also poses a problem for biological research as its presence makes it harder to distinguish the real between-the-groups differences from random sampling effects, which can result from inside-the-group variation. The larger the inside-the-group variation, the more likely it is to draw a random sample that is not representative of the population of interest. Thus a researcher must have an idea about the size of variation in the experimental system. The larger the intra-sample variation, the larger becomes the sample size, which is needed to reliably recognize true differences between samples. On the other hand, as the true effect size grows over size of variation, we can have progressively smaller samples and still identify true effects.

Sample Size and Statistical Power

The frequency, with which a study is expected to correctly call true effects, is called its sensitivity or power. A major challenge of experimental design is to decide on a suitable sample size so that the experiment would be reasonably well powered. This is often a grueling decision that results from tradeoff between cost, specificity, and sensitivity of the study. Specificity of the study is the frequency of correctly calling the true negatives (see Chapter 4 for more precise definitions). Because scientists are taught to be especially wary of publishing incorrect results, they tend to want to maximize the specificity of their tests. Unfortunately, this comes with the cost of reduced power. To increase the statistical power while keeping the desired specificity level you need to increase the sample size and/or the effect size and/or to reduce variation present in your data. What road to take depends on whether you can increase the effect size by increasing the "intensity" or dose of treatment, whether you can define your study population more precisely thus reducing its variation, whether you can increase the accuracy of measurements, and on the feasibility of increasing the sample size.

If the sample size is too small, the first thing that happens is that most or all of the true and interesting effects are mislabeled as statistical noise. The easiest solution would be, when in doubt, simply to do more experiments until you are satisfied that there really is an effect (or not). However, whether this is a wise course of action depends on the type of statistical analysis implemented. In principle, there is nothing wrong with collecting as much evidence as is needed to make up one's mind, and then to stop. This is a perfectly normal course of action in likelihoodist and Bayesian statistical approaches (Sections 4.11 and 5.4). However, for technical reasons traditional statistical significance testing does not allow stopping the experiment simply when a significant result is achieved. By doing so every series of experiments will be guaranteed to sooner or later yield a statistically significant result. This shortcoming of the statistical procedure leads to the requirement that stopping rules must be decided before the start of experiment. A downside of this is that when you set the significance level to the usual 0.05, decide on a sample size n, run the experiment n times and get a result just short of significance, then it is theoretically impossible to drive your result into statistical significance by doing some more experiments. According to statistical theory you are supposed to declare a non-significant result.[9] Such recalcitrance to additional evidence is not a general property of scientific method, it is just another unpleasantness of frequentist null hypothesis testing, which makes power analysis, and thus setting an optimal sample size, a crucial part of this particular method.

In most statistics textbooks one finds exhortations that the power of a study must be dealt with in the planning stage, not retrospectively in data analysis. The reason is that any realistic power analysis needs realistic estimates of effect sizes and variance as its input; not the observed effects, which are likely to be inflated and variation that is (perhaps) reduced (Perugini and Gallucci, 2014). To calculate the sample size needed for a set power a researcher should have an idea about effect sizes that would likely occur based on external knowledge (not her data) and estimate the variability that is expected to occur in the data (Gelman and Carlin, 2014). If the observed effects turn out to be much larger than the range of reasonable effect sizes predicted from background knowledge (be it a literature search, meta-analysis or preliminary studies), then it would be extremely hard to interpret this in any other light than error. When the relevant parameters are estimated from prior knowledge, as opposed to sample data, it doesn't matter whether the power analysis is done after collecting the data.

Amusingly, it seems that in the current theoretical framework guessing the effect size prior to conducting the study is a major part of "objective" frequentist statistics, while doing exactly the same thing after collecting the data would be seen as "subjective" and Bayesian. In practice one would presumably still want to know, *before* conducting a study, how much time and effort it is expected to cost.

---

9. It is possible to run another series of experiments, calculate the relevant statistics, and then do a meta-analysis of both datasets.

There is another approach to statistical power that absolves you of the need to guess the amount of variation in your future data (Dienes, 2008). One can simply use a preset CI in lieu of a stopping rule. Because CI tend to get narrower with increasing sample size (and with decreasing variation in data), it is enough if, before starting the experiment, it is agreed on at which size of the interval the experiment will be stopped. Confidence intervals are calculated after each experiment and the experiments are stopped as soon as a CI of pre-set width comes up. If the interval does not contain the null value, then the result is statistically significant, otherwise it is not. This method conveniently removes the need for power analysis with its required precognition of effect sizes and variation, but the downside is that one cannot predict how many experiments are needed to arrive at a result.[1]

---

1. It must be noted that sample sizes to achieve short CIs (and thus precise inference) are often much larger than are needed to achieve acceptable power (Maxwell et al., 2008).

Power analysis is routinely done in clinical studies (although mostly suboptimally; Clark et al., 2013), but not so often in laboratory animal studies and apparently almost never in *in vitro* biological experiments. I would hazard a guess that the reason for this omission is poor understanding on the part of many experimental biologists of the statistical nature of experiment. People who think of experiments as truth machines—as opposed to probability-generating machines—will naturally see experimental design in terms of logic, rather than probabilities. They will thus turn a lot of attention to designing controls and not a lot to the question: why do we repeat experiments? They may think that they reach appropriate sensitivities in their studies intuitively, but the available facts do not support this. A retrospective analysis of published neuroscience literature concluded that the median neuroscience paper had a 79% chance of missing an interesting result because of too small sample size (Button et al., 2013). Similarly bleak views have been offered for experimental social psychology (Fiedler et al., 2012).

Since the term "power" designates the frequency of seeing a relevant effect, given that one really exists, a lack of power could simply be interpreted as low efficiency of doing science: inconvenient, but at least not introducing false effects into literature. In reality things are more complicated. When the power of studies is low, it becomes much more likely than the conventional 5% that a statistically significant effect is in the wrong direction—an error with the opposite sign is called "type S error" (Gelman and Tuerlinckx, 2000). The concept of type S errors is Bayesian: you need to come up with a prior probability for your effect size to calculate type S error frequencies.

Another thing that happens when the sample size (and power) is small is that the effects that you actually do see are often greatly inflated—this is called a "type M error" (Gelman and Carlin, 2014) (see also Section 3.3). If the statistical power of a study is 20% (which means that 80% of real effects are not identifiable), one should expect that the effects that are found are inflated by about a third (Button et al., 2013). According to the analysis of Button et al., about 30% of effects, which are published in neuroscience papers, are more than twofold overblown.

3.5   TECHNICAL VERSUS BIOLOGICAL REPLICATION OF EXPERIMENTS

There are widely recognized to be two types of experimental replication (replication simply mean repeating an experiment). The goal of technical replication is to determine the level of variation in measurement and/or sample preparation. Obviously a researcher would prefer this type of variation to be minimal, but since she, despite her best efforts, is unlikely to ever drive it to zero, she will need to measure it to be able to do the power analysis. Power considerations are the main reason why technical replicates are done (the other being that you may be interested in the quality of your apparatus). Technical replicates bring internal validity to the study: they can add confidence that you are really measuring the parameter that you think that you are measuring, that the measurements are accurate, etc.

There are many ways to design technical replicates. For example, you could aliquot the cells you wish to study, lyse these separately and run each sample through your instrument of choice. Or, alternatively, you could divide the sample after lysis and run each aliquot through the instrument more than once. Both are technical replications but they provide different information (so you might consider doing them both).

Biological replicates are conceptually quite different from technical replicates. The goal of biological replicates is threefold: (i) to provide external validity to the study; (ii) to characterize biological variation; and (iii) to increase the power of the study to find real effects. Thus, unlike with technical replicates, the reason behind doing biological replications is not only to increase the power of your study, but also (and perhaps more importantly) to provide an appropriate level of generalizability.

As with technical replicates, the experimenter has a lot of choice in how to design biological replications. However, with biological variation comes a trade-off. You could design your experiment to minimize variability and thus maximize power (= sensitivity), for example by using genetically identical pure-bred mice. This makes it easier to see effects of your experimental manipulation or treatment. The downside is concomitant reduction in the external validity of your results. This means that the better your experimental system is technically (the less variation in measurement), the worse it tends to be biologically; the bigger the chance that results obtained from it are not generalizable to real biological populations. Doing biological experiments can be akin to walking a tightrope between biologically misleading results and no results at all. This tightrope is made of biological replications.

In chemistry or physics, where variability comes mostly from measurement, the experimental replicates tend to be technical. In biology, where variability is usually mostly biological, doing technical replicates leads to serious underestimation of the variability and thus overconfidence in the results. But biological replicates can be a tricky concept. For example, when you do your experiment in ten mice from the same litter, what is your sample size? The answer depends on what level you wish to generalize your results to. If you are happy to generalize to the level of that particular litter (but good luck publishing it!), then your sample size is ten. On the other hand, should you want to generalize to mice in general, your sample size is one (Vaux et al., 2012). If your preferred level of generalization is to mice as such, and you decided to collect data from 40 individuals belonging to four different mouse strains, then in analyzing the data you should, as a first step, average all the data in each strain and then use the resulting four numbers for further statistical calculations, like P values or CIs.

So, what is your sample size when you use 100 mice drawn from different litters of the same inbred laboratory line? When you measure cell division during a single flow-cytometer run, but from a hundred thousand cells, is the sample size 100,000? What if you use 100,000 people from three very different human populations to study genetic association with a disease? Can the sample size be both 100,000 and three? In biochemistry, when you do four repeat measurements using a single enzyme preparation, what is the sample size? Is it four, or just one?

The recurrent question: "What is the sample size?" can be rephrased to: "What constitutes a data point?" An interesting example is provided by two meta-studies of largely the same data. Two groups, one in Newcastle, UK and another in Stanford, USA, took together all the available evidence about whether organically grown fruits and vegetables contain more nutrients than the conventionally produced cheaper (but pesticide-ridden) varieties. The Newcastle group found a statistically significant 12% increase in secondary metabolites in the organically grown produce (Brandt et al., 2011). In contrast, the conclusion of the Stanford group was that "The published literature lacks strong evidence that organic foods are significantly more nutritious than conventional foods" (Smith-Spangler, 2012). The reason why these groups reached very different conclusions from the same data was not that one did a better analysis. What happened was that they did slightly different analyses. If a paper reported data for crops grown in separate years, Newcastle group's meta-analysis used each year's data as a separate data point. This is very reasonable because weather conditions vary a lot from year to year and variations in crops are therefore a given. In contrast, the Stanford group averaged the multiple years into a single data point. This also is very reasonable and is similar to how human clinical drug studies are routinely analyzed. In the clinic the effects of medicines are expected to be consistent and variations can indicate inferiority of the drug. Which study design you prefer, is entirely up to you.[1]

---

1. Whichever design you choose, your knowledge of the different results obtained by the other group should lower your trust on either conclusion.

A real life example about external validity can be found from the field of caloric restriction (CR). Experiments conducted from the 1930s in various organisms from yeast to primates, including a range of laboratory mouse strains, have showed that restricting caloric intake tends to prolong life expectancy. CR is widely seen as the best-proven method for extending life in wide varieties of animals. Currently several trials of CR in humans are under way. But then, there are several examples of species or laboratory rodent strains that do not exhibit any CR-dependent increase in life expectancy. To clarify matters, Harper et al. used wild-caught house mice to study CR, only to find no overall effect (Harper et al., 2006). CR-mice had relatively more early deaths and less late deaths than *ad libitum*-fed controls and had significantly less tumors in their old age. There are more than one possible explanation for the discrepancy between many laboratory lines and wild mice. CR effect might possibly be an artifact of selection of laboratory lines, as researchers may unconsciously prefer less aggressive, more docile (in other words, lazy, stupid, and slightly ill) animals for breeding. Also, it is known that the wild-derived mice eat less than lab mice when fed *ad libitum*, which could mean that it is not the life extension by CR that is measured in lab lines, but rather the shortening of life caused by eating too many mouse equivalents of hamburgers by control groups.

In 2009 Colman et al. published results of a 20-year CR trial in rhesus monkeys (Colman et al., 2009). Their results were promising, to say the least. Besides publishing dramatic pictures of a nearly hairless geriatric-looking control animal and a happily youthful CR animal (Figure 3.5), they were able to show clear and fairly large differences between control group and CR-group mortalities, demonstrating a real reduction in mortality upon CR. Or maybe not: it is also possible that what their study really shows, is abnormally high mortality of the control group, which was fed *ad libitum*. This interpretation is supported by another large rhesus monkey CR study, which found no effect of CR on life extension (Mattison et al., 2012). As both studies had high-quality veterinary support in comparable experimental settings, studied the same species, and tested the same intervention, what could account for the differences in survival? The likeliest answer proposed by Mattison et al. is that while the Colman et al. control animals ate purified food containing 28.5% sucrose *ad libitum*, the Mattison et al. animals ate a natural mix with 3.9% sucrose and control animals themselves were slightly caloric-restricted. Accordingly, at 17 years, the control monkeys of Mattison et al. weighed 12% (males) or 18% (females) less than the controls of Colman et al. It may also prove relevant that the monkeys used by Mattison et al. originated from China and India and had greater genetic diversity than the strictly Indian colony used by Colman et al.

Here are the morals from this story: (i) in the land of graphs there is no preferred direction: if your study group data rose above the control group, either

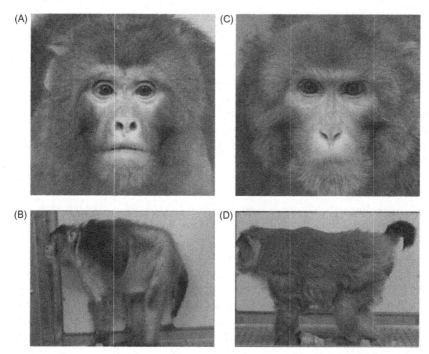

FIGURE 3.5 Caloric-restricted monkeys (panels C, D) versus control animals (panels A, B) at 27.6 years of age. *Source: Reproduced with permission from Colman et al. (2009).*

graph might have exhibited a "treatment" effect and moved relative to the other; (ii) when people aim to generalize from yeast to mouse, from mouse to monkey, and from monkey to man, external validity is an art, not a science; and (iii) starving oneself into old age is not for everyone.

3.6   EXPERIMENTAL CONTROLS

In Section 2.10, we discussed controlled experiments in the context of disentangling causal hypotheses. Here we give a more formal treatment and look at some examples of the uses controls can have in biological experiments.

So, what are controls for? Experimental control is a somewhat unhappy umbrella term as different types of controls are done for quite different reasons and in quite different ways (Table 3.2). In general terms controls are often seen as an insurance that the effect observed after the experimental manipulation is really caused by the manipulation, not by other factors. An analysis of 417 experimental results presented in weekly lab meetings of four top-notch molecular biology labs revealed that in 85% of cases the results were defined by comparison of the experimental condition with other (control) conditions (Dunbar and Fugelsang, 2005). In only 15% of the cases did researchers present their results singly (e.g., without a control). Another use for controls is that surprising results emanating from the control condition allow scientists to generate novel study questions and hypotheses (Dunbar, 1995). In other words, controls structure experiments to enable researchers to take advantage of unexpected findings. Indeed, unexpected results are very common in the "raw data" of molecular biology, making up over half of total results discussed at lab meetings (Dunbar and Fugelsang, 2005).

For the vast majority of unexpected results the reaction is to think of possible methodological errors, repeat experiments, and change experimental protocols, including the controls. In contrast, when several unexpected results occur in a research program, the typical reaction is to come up with new causal models to explain the data and help to design new experiments to fill in the gaps in the models. Thus, experimental controls are used both to weed out spurious effects in testing explicit causal hypotheses and to generate novel serendipitous findings. They both narrow and widen the scope of research.

More specifically, controls can reduce the chances that random effects are improperly designated as biological (specificity), they can increase the chance that the fragile biological effect is rescued from statistical swamp of random noise (sensitivity), or they can be used to exclude, one by one, the known sources of bias. Philosophically, one could argue that controls are there to help us to cope with the Duhem-Quine problem, which roughly states that any test of a particular hypothesis is simultaneously testing many ancillary hypotheses about the study methodology. Controls make this ancillary testing more explicit and increase the chances that the conclusions of the study are about the original scientific hypothesis, not about the methodology of the study. In other words, when designing experiments, we should always ask "Why else?" and then try to show "Why not."

Whether a condition is seen as a control or an experimental manipulation is not a structural property of the experimental system. Controls are defined by the overarching scientific hypothesis, which gives the experimental system its relevance. For example, when the research hypothesis is that a novel enzyme can amplify DNA, the appropriate negative control would be to do a reaction lacking the enzyme and the positive control would be to use a commercially available DNA polymerase in an otherwise identical PCR reaction. On the other hand, if the hypothesis were that a dinosaur bone still contains dinosaur DNA, the negative control would be to omit from reaction material extracted from the bone (template) and maybe substitute it with material scraped from the surface of the same bone (see below). The experimental systems would be very similar, but the different controls would enable to draw very different conclusions.

It is the controls that harness the abstract hypothesis in front of the concrete experimental system and allow experiments to produce output relevant for the particular hypothesis.

The major point of any experiment is to inflict a clearly defined manipulation (treatment) to a well-defined slice of the universe (the experimental system). We do controls to be sure that the resultant change in the system was indeed caused by whatever it was that we thought that we did to it. The reasons why we can mislead ourselves include bias and random effects (RTM, see Table 3.2). As was discussed above, RTM should equally affect the experiment and its controls. Thus a simple subtraction of control value from experimental value can go a long way in reducing RTM.

Things get more complicated when designing controls for detecting bias in the experiment. Each potential manifestation of bias needs to be controlled for separately and thus needs to be envisioned before the start of experiment. For example, when studying how a compound affects growth of a cancer cell line, we might suspect that male and female cells could respond differently. We would then use an equal number of male and female cell lines in both the study group and the control group (and thus "match" the two groups by sex). But then, we could also suspect that the cell lines purchased for our study are misidentified and plan to re-analyze their genetic composition and phenotype. Indeed, several analyses have shown that 15–30% of cell lines routinely used by researchers are misidentified or contaminated, making this type of control highly desirable (Lorsch et al., 2014). In short, the number of potential controls is about as large as the number of potential biases—that is, very large indeed—and ideally we should prioritize our controls by the probability of encountering the relevant biases, and by the severity of these biases.

As the number of potential biases is limitless, experimental controls are often not very efficient in fighting bias. Controls should be adequately matched: when designing an experiment the two experimental groups should not differ on factors that are likely to affect the results (Rubin, 2006). Matching controls is an alternative to randomization: the right balance between these complementary techniques of experimental design depends on our knowledge (or the lack thereof) of our experimental system, its complexity, on the cost of the experiment and on ethical considerations. Typical matching of controls involves sampling equal number of sexes, age groups, etc. to both the control group and the study group (Table 3.3).[1]

---

1. Technically speaking, matching control and experimental groups is a limiting case of the "blocking designs" described above.

TABLE 3.3 Some Types of Controls

Type of Control	Controls for	Mode of Action	Reason for Existence	Comments
Inter-group comparison	RTM, placebo effect	Subtraction	Increases specificity	
Internal normalization	Technical variation	Division	Increases sensitivity	Assumes constancy of internal standard during the experiment
External normalization				Need of spiked-in external standard
Positive control[a]	Bias	Logical analysis	Excludes specific confounding factors	Checks that experimental system works independently of manipulation
Negative control[a]				Checks that manipulation is needed for the outcome

[a]Note that both positive and negative controls involve inter-group comparisons, but all inter-group comparison controls do not fall into the positive or negative control category.

Bias loves ambiguity. Thus, keeping your experimental system simple is the key in excluding bias. If the experimenter has a complete understanding of the experimental system, she can also have complete control over factors that can affect that system. Unfortunately, no useful system in biology is ever likely to be simple enough to guarantee complete human understanding and control over it. Often the best one can do is to take care that the unknown biases would at least be similar between different arms of the study. For example, when comparing a mutant enzyme with its wild-type version, it is a bad idea to purify the two using different protocols, or even to purify them on different days, or even to purify them always in the same order. You can see where this kind of thinking leads: to eternal worrying and anal retention of the tiniest experimental detail.

In doing positive controls our aim is to validate the experimental system (the assay). Failure to do so is likely to lead to negative results and failed experiments, which can be unpleasant for individual researchers but are unlikely to distort the knowledge base of the fields, where non-publishing of negative results is the norm. Things are different with negative controls. In the negative control we often try to perturb the original manipulation while changing the general experimental setup as little as possible, to see if the behavior of the experimental system reverses to the pre-treatment mode. If that is the case, we may conclude that the original treatment was indeed the cause that effected the change in experimental system.

To illustrate the importance of controls in excluding bias, a couple of real-science examples are in order.

3.6.1   Example 1. Negative Controls

What follows is a true story of scientists of good repute trying to amplify dinosaur DNA and getting a nice PCR product. The problem was that the dinosaur DNA turned out to be 100% identical to turkey's (Holden, 2000). Although the closest living relatives of dinosaurs are believed to be birds, the researchers very properly scratched their heads with the question: Why turkey? Thinking of turkey sandwiches, a typical negative control is a PCR reaction without added DNA template. If this gives no reaction, and the actual experiment does, then it stands to reason that the experimenter did not contaminate the reactions with his lunch. Indeed the team checked for turkey DNA in turtle bones, dirt, and burlap from the site where the dinosaur bones originated, and none tested positive. A reasonable conclusion would be that since the associated debris was not contaminated with turkey DNA, the DNA associated with the dinosaur bone must have belonged to the dinosaur. This argument was apparently enough for the group to tentatively present their results in a scientific conference, albeit not enough to publish.

Because the chemical half-life of DNA is measured in tens of centuries even in the best conditions, as opposed to millions of years required for dinosaur bones (Allentoft et al., 2012), and both DNA sequence constancy during 65 million years and convergent molecular evolution between dinos and turkeys seem unlikely, a healthy level of doubt should have been kept. It was still possible that their results were nothing more than a bit of bad luck and turkey contamination simply happened to be in one tube and not the others.

In a situation like this, another control is in order, the usual one being sending the bone fragment to another lab in another city where they prefer sushi and comparing samples from the surface and the interior of the bone. If the result of the experiment still points to turkeys, it is at least potentially publishable. Nevertheless, whether you actually believe that the amplified DNA belongs to the dinosaur should have a lot to do with how likely you think it is that it is possible to amplify DNA from 65 million year old bones. It is not only the work at hand but also the totality of scientific context that is needed to form a rational belief about the true state of the world.

In a famous case of PCR contamination, German police left no stone unturned from 2001 to 2009 in their search for a serial killer the public called the "Phantom of Heilbronn" or the "Woman Without a Face." The only connection between the crimes was DNA evidence, pointing to a Caucasian female, which had been recovered from 40 crime scenes, ranging from 1993 to 2009 and including everything from six murders to petty crime. By January 2009, the reward for clues regarding the "Phantom" was €300,000. When the police finally found their suspect, it turned out to be a worker at a factory that produced the cotton swabs used in collecting DNA from crime scenes.

More recently it has been found that contaminating bacterial DNA is ubiquitous in commercial DNA extraction kits, a state of affairs which apparently has biased many environmental microbiological studies (Salter et al., 2014).

Our next fine example of the importance of fairly standard laboratory controls in preclinical science comes from Kakisi et al. (2013). Briefly, in 2006 US researchers found a new gammaretrovirus called XMRV in 40% of prostate cancers from 86 cancer patients, who carried a mutation in the antiviral gene coding for RNAse L (Urisman et al., 2006). This discovery was done with a novel gene array, the virochip, which scans for viruses with a wide range of virus-specific probes.

XMRV was homologous but not identical to known murine retroviruses. This is the only example of a gammaretrovirus implicated in human infection but gammaretroviruses have been previously associated with tumors in animals. In 2009 XMRV infection was confirmed in tumors not carrying the RNase L mutation and virus-positive tumors were associated with higher malignancy (Schlaberg et al., 2009). The authors used PCR, reverse transcription, immunohistochemistry and electron microscopy to identify XMRV DNA, RNA, proteins, and particles in prostate cancers and to specifically demonstrate viral infection in a prostate cancer cell line. XMRV was fully sequenced and integration sites were found in genomic DNA, as is expected from a retroviral infection. However, several subsequent papers, published from 2008 to 2011, failed to identify XMRV in prostate tumors (Robinson et al., 2011). The integration sites of XMRV in prostate cancer tissues, thought to unequivocally confirm the existence of XMRV in clinical samples, are also likely to be contaminant–derived (Garson et al., 2011). The main reason for the discrepancy seems to be contaminating murine DNA in PCR reactions, which can be controlled by using primers specific for repeating DNA elements present in mouse, but not in human, genomes. Overall, the speed and efficiency with which this problem was spotted and the literature for prostate cancer set right is commendable.

However, it does seem a little worrying, how little the panoply of supporting evidence, purportedly showing the presence of prostrate-specific infection, viral replication, proteins, and RNA, seems to matter, after invalidation of the main experiment for virus detection in cancer patients. In hindsight it begs the question, what exactly was the reason behind all this additional hard work with its own control experiments?

The story does not end here. Some chronic fatigue syndrome (CFS) patients also carry a defective *RNAse L* gene. Accordingly, in 2009 a group of scientists demonstrated in a *Science* paper that XMRV is present in the peripheral blood mononuclear cells of 67% of patients with CFS and in 4% of healthy controls (Lombardi et al., 2009). CSF is a serious debilitating disease and its potential viral origin opened the door for patients for a relatively easy cure of antiretroviral medication. Clinical trials were started and there were also reports of patients self-prescribing antivirals against CFS.

Even more serious was the concern about the detection of XMRV infection in the healthy controls. To anyone, who has heard of AIDS, this would suggest the possibility of another emerging retroviral epidemic, potentially transmissible by blood transfusion. In 2010 several countries banned blood donations from CFS patients. A high-profile publication by scientists from the FDA, NIH, and Harvard discovered XMRV-like sequences in nearly 90% of patients with CFS and in 7% of healthy individuals (Lo et al., 2010).

In the meanwhile, a number of studies during 2010 failed to replicate the findings of Lombardi et al. (2009). The source of discrepancy was initially thought to be differences in cell culture and in extraction and amplification techniques of viruses. It was soon discovered that PCR controls with no added template DNA, as well as samples from empty DNA purification columns and reagents used for PCR, could give reaction products, indicating contamination in PCR reagents and other components used for routine viral detection (Tuke et al., 2011). In December 2010, four papers presented evidence that XMRV was a contaminant from PCR components, laboratory material, and cell culture. Apparently a widely used cell line, which happens to secrete XMRV, may have been present in laboratories involved in the initial "discoveries."

This led to close re-examination of the original Lombardi *Science* paper, which was still defended by its authors. Representativeness of the patient sample was questioned, as was the fact that the experiment was not randomized and not blinded (Lloyd et al., 2010). Demonstration of a single peak in the flow cytometry profile of infected blood cells was deemed strange, as it is unlikely for any virally infected cell population to be 100% infected and expressing viral protein. Lack of antibody responses in a large number of "virus-positive" subjects and possibility of scientific misconduct was also raised. Samples from the original studies were re-tested and found to be negative for the virus.

This led some of the original authors to redo some experiments on the original samples used in the Lombardi et al. (2009) paper, the resulting finding being that some of the CFS peripheral blood mononuclear cell DNA preparations were contaminated with XMRV plasmid DNA (Silverman et al., 2011). Although this result seems to invalidate the main finding that most of the CSF patients and few control individuals harbor the virus, this led only to a "partial retraction" of key data of the original paper, but not the results in their entirety, followed by full retraction of the paper by the editor of *Science*. Separately, a paper by another group, showing similar results to the Lombardi et al. work, was retracted by its authors (Lo et al., 2012). Meanwhile, a multicenter blinded study showed no association between CSF and the virus (Alter et al., 2012).

3.6.2 Example 2. Normalization Controls

Normalization controls aim to increase sensitivity of the experimental system by canceling some of the variability inherent in it. A popular application is to use normalization standards to reduce technical and biological variability inherent in gene expression measurements. For example, when analyzing Northern blot or quantitative PCR experiments it is common to express numerical data as ratios of individual mRNA levels to an internal control, which is usually some fairly abundant mRNA for a housekeeping gene.

Because levels of the normalization mRNA are assumed to be more stable than study mRNAs, this control is expected to smooth out technical variability in the assay. It also gives the experiment its reference system—in principle we could measure changes of mRNA levels against cell volume, cell number, total mRNA concentrations, total protein concentrations, etc. For instance, when normalization is done against an internal housekeeping mRNA, the reference system will be the cell (assuming the constancy of expression of the normalization mRNA); when we normalize against total mRNA, the reference system will be different, which means that interpretation of the results should differ as well.

As the levels of normalization mRNA vary (as do its measurements), its use not only reduces the ambiguity of your system, but also introduces ambiguity of its own (because we do not know whether its variation is correlated, anti-correlated or uncorrelated with the variation of each mRNA, whose expression we try to measure). A solution is to use an average of several mRNA levels, or even to normalize to total mRNA levels. This is routinely done in microchip array studies of mRNA expression and proteomic studies by mass spectrometry.

Alas, each normalization technique comes with its own problems. To better grasp this point we will look at quantitative protein mass-spectroscopy where stable isotope labeling of cell cultures (SILAC) is used to compare two proteomes by pairwise comparing the levels of thousands of different proteins (Cox and Mann, 2011). The two cultures—one containing the "heavy" isotope and another the "light" isotope—are grown in parallel, then the experimental treatment is done to one of them, after which ratios of individual heavy–light protein pairs will be determined. For this to happen you need to mix together the two cultures and detect peptide pairs from the mix. Often they are mixed after cell lysis in a 1:1 ratio by total protein weight. Alternatively, the cultures can be mixed before the lysis by cell count.

The first and often-used method allows controlling for the day-to-day technical variation in the mass spectrometer as the two samples are loaded together. The second method also does that, but in addition controls for the variation in lysis efficiencies. Moreover, these seemingly similar controls depend on very different biological assumptions.

When lysates are mixed by total protein mass, the assumption is that the experimental manipulation did not lead to general change in protein levels, that is, that most proteins retain their levels and are therefore directly comparable to corresponding proteins of control culture. If the treatment causes expression of most genes to go down, this experimental protocol simply compensates by taking lysate from many more treated cells than control cells, and the biological effects will consequently be misinterpreted. The second approach does not fall into this trap, but if during experimental manipulation the study cells are growing differently from the control cells (e.g., by becoming smaller, more or less elongated, or harder to lyse), the interpretation of results nevertheless becomes unclear.

Another example of normalization controls comes from the field of oncogenes. It involves the mammalian transcription factor Myc, which is activated by several independent signaling pathways and which in turn activates the expression of a number of target genes, leading to cell growth, proliferation, and, potentially, cancer (Dang, 2012). Myc oncogene has been vigorously studied from the 1980s and currently provides about 1,500 research papers annually. It has never suffered from lack of attention.

A major 2012 review of Myc target genes claimed that from the 3,000 to 6,000 genes that are bound by Myc, only about 700 alter their mRNA levels in response to Myc (Dang, 2012). Genome-wide mapping of Myc-binding sites and gene expression profiling indicated that most Myc binding genes do not respond to increased expression of Myc, suggesting that Myc binding by itself is not enough to cause changes in target gene expression.

The interpretation, supported by scores of individual research papers, is fairly conventional: myc, in combination with other factors, is a specific activator of transcription that leads to cell proliferation. Nevertheless, the received knowledge was forcefully challenged by two groups contending that expression of the Myc protein is not a specific activator of gene expression at all, but rather leads to higher expression of most genes that were already expressed (but not to the expression of silent genes) (Lin et al., 2012; Nie et al., 2012). This is a novel paradigm of oncogenesis, upending many years of studies, and it was produced thanks to a change in routine controls of the experimental system.

The classical normalization control in microarray assays normalizes the mRNA levels of the experimental culture to total mRNA mass of the control culture. Because this control was done nearly universally, it is only human to ignore its inherent assumption that any change in gene expression is specific and involves a limited number of genes (Figure 3.6). This was ignored to the extent that the seemingly unchanging expression of most proteins, which is an inevitable consequence of the experimental setup, was often used as evidence for the specificity of the effect of Myc on gene expression. All it took to challenge this picture was to substitute the usual control with the technically more demanding spiking of the experimental culture with individual labeled mRNAs (Lovén et al., 2012).

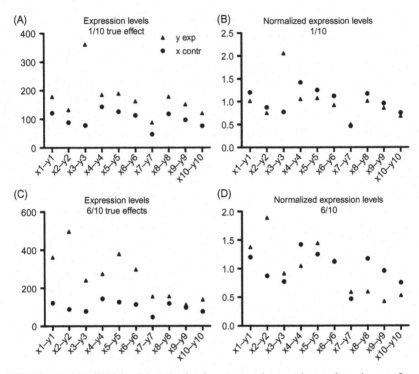

FIGURE 3.6 (A) mRNA levels in a simulated gene expression experiment where nine out of ten "measured" RNAs are similar in control and experimental conditions. $x1$–$x10$ (circles) denote levels of ten different mRNAs from the control condition, simulated from a normal distribution (mean = 100, SD = 30). $y1$–$y10$ (triangles) are levels of the same mRNAs from the experimental condition of no effect (mean = 100, SD = 30) but a bias of 50 units is added to each value. The sole exception is $y3$, whose value is drawn from a normal distribution (mean = 300, SD = 90). (B) Control and experimental values of (A) are normalized against the respective mean value of each series. The $x3$–$y3$ pair can be correctly recognized as exhibiting the single large effect. (C) mRNA levels in a simulated gene expression experiment where six out of ten "measured" RNAs are increased by threefold in the experimental condition. Control values (squares) were drawn from a normal distribution (100, 30), the experimental values $y1$–$y6$ were drawn from a normal distribution (300, 90) plus bias 50 and $y7$–$y10$ from the original normal distribution (100, 30) plus bias 50. (D) Control and experimental values of (C) are normalized against the respective mean value of each series. The normalization now precludes the correct interpretation of most control–experiment pairs.

This minor-sounding change in the experimental protocol resulted in a major re-interpretation of the role of Myc in the physiology of cell, and of cancer. Controls really do matter.

3.6.3 Example 3. Controlling the Controls

A fundamental trouble with controls is that although they are often done to better understand the experimental manipulation, it is as easy to mislead oneself about the controls, as it is about the experiment itself. When we manipulate the experimental system, while leaving the control untouched, the goal is to witness a manipulation-specific effect and we silently hope that the control and experimental conditions are otherwise equivalent. This is, however, impossible to prove and can lead to an infinite regression of controls for controls.

Let us briefly return to CR. The best-studied effector of CR is sirtuin (Sir), which is a NAD-dependent protein deacetylase (Baur et al., 2012). Deletion of Sir2 has been shown to reduce the beneficial effects of CR in yeast, although Sir2-independent lifespan extension in response to CR has also been demonstrated. Overexpression of Sir2 has been shown in many studies to extend lifespan in yeast, worms, and flies (Couzin-Frankel, 2011).

However, the effect of Sir on longevity was soon questioned, beginning with studies on yeast (2004) and mice (2007). A 2011 *Nature* paper further showed that Sir2 overexpression does not increase lifespan in worms and flies (Burnett et al., 2011). This work mainly repeated previous work with the crucial difference that additional controls to the controls were performed by outcrossing the animals to guarantee greater genetic homogeneity and thus greater similarity between the experimental group and the control group. In the worm *Caenorhabditis elegans*, outcrossing of a line with high-level Sir2 overexpression abrogated the longevity increase, but did not abrogate Sir2 overexpression. Longevity co-segregated with a second-site mutation affecting sensory neurons. A *Drosophila* strain that overexpressed Sir2 was long-lived relative to wild-type controls, as previously reported, but was not long-lived relative to the appropriate transgenic controls.

As for Sir2 being the effector of CR, the researchers found that CR increased fly lifespan independently of Sir2. In response the two groups that initially reported the effect of Sir on CR performed additional experiments using improved controls for the genetic background. These results supported their original conclusions that Sir2 overexpression extends lifespan, although the effect in worms was reduced (Bauer et al., 2009; Viswanathan and Guarente, 2011).

The overall result seems to be a notoriety associated with the sirtuin field. In addition to extending lifespan, many different and unrelated functions of Sir have been proposed. These include increased insulin sensitivity, modulating circadian rhythms, improving genome stability, suppressing tumors, reducing inflammation, protecting against neurodegenerative diseases, osteoporosis, frailty and stress, as well as controlling anxiety in mice (Baur et al., 2012). When a factor seems to be involved in every aspect of health and disease, its true function and significance become hard to unravel.

3.7   MULTIPLICITIES

Correcting for multiple testing is what scientists, who do null hypothesis testing in parallel and want to interpret the ensuing P values, have learned to do with reasonable ease and efficiency (see Section 4.8). However, here we set a general context for it, which is a little bit more complicated but infinitely more scary.

But let us start with an innocent question. Suppose that in your capacity as a referee for an experimental paper the authors provide you with an accurate and full description of the experiments included in their manuscript, their results, and statistical inferences. Do you then have sufficient information to estimate the strength of evidence that these results convey to conclusions? You are not asked about the probability of their conclusions being true, just the much easier question of whether the results provide strong evidence in favor of the conclusions.

The peer review process as currently practiced is largely built on the assumption that it is indeed alright to judge the quality of evidence from the quality of presented experiments. This faith is also reflected in the ahistorical nature of scientific prose: scientific papers present the logical structure of the argument (theory–evidence–conclusions), not the tortuous roundabout ways scientists used to reach their conclusions. However, the widespread belief that the quality of scientific conclusions of a paper can be gleaned from the experiments presented is simply not true. It is often not even close to the truth.

To see why, let's start by looking at crime detection, where the strength of DNA evidence can be enough to free innocent people on Death Row, but also to send innocents to prison. In 2008, John Puckett—a California man in his 70s with a history of sexual violence—was accused of a brutal rape and killing that happened in 1972. There were about 20 suspects in the original investigation and Mr. Puckett was not one of them. In the 2003 follow-up investigation based on a search of California's offender database, which at the time contained DNA profiles of 338,000 convicted sex offenders and violent criminals, linked his DNA to crime scene evidence. The expert opinion was that there is a 1 in 1.1 million chance that this DNA match was coincidential. Not surprisingly, John Puckett is now serving a life sentence.

But that 1 in 1.1 million figure is completely misleading, according to two expert committees, one convened by the FBI, the other by the National Research Council. It gives the odds of a DNA match, given that John Puckett is the only suspect, is innocent, and is not related to the real culprit. The question relevant for the crime at hand was, however: how likely it is to find any one person from the DNA database, who has a match? This estimate depends on the size of the database and is not one in a million; it was actually estimated to be one in three.

In the above example the reader was presented with two pairs of odds endorsed by the FBI. If a suspect is first singled out by independent means, the 1 in a 1.1 million figure should be used and if a database trawl is the means of identifying the suspect, the one in three odds are to be used. Unfortunately, things are not as black and white. Since Mr. Puckett had been convicted in 1977 of similar crimes and a witness had described the original assailant as a white medium-height bearded man, maybe we should considerably restrict the size of the relevant database to include only samples from white males of medium height, who are sex offenders? This would have resulted in odds considerably more against the suspect, than one in three (and considerably less than one in a million).

Even worse, the odds calculated from the size of the database depend on a hypothetical: would Mr. Puckett not have been prosecuted if he had presented a match in the part of the database not listing sex offenders? If the answer is "no," then the odds that the observed match in the sex offender part of the database was an error would have been considerably smaller than one in three and there would have been stronger reason to accuse Mr. Puckett. Thus, not only can the statistical competency of a prosecutor substantially influence the weight of evidence before him, but also something that could have happened but never did. It is quite possible that the prosecutor himself cannot say how he would have behaved on the occasion, which never rose.

Similar problems arise in science. The fact that an experiment was done and result secured does not say much about the usefulness of it. Exactly the same data can be very strong evidence in support of a theory or weak evidence in favor of the same theory, depending on the list of alternatives (Section 4.11.).

We illustrate this point by a thought experiment, in which we consider two alternate universes. These are otherwise identical, but in universe 1 lives a scientist called Tom, who is a paragon of health and in universe 2 his doppelgänger suffers from an undiagnosed brain tumor, causing him to hallucinate. Now, Tom 1 decides to take his vacation in a town he has never visited before and he soon notices that the townspeople seem to be exceptionally tall. Measurement and statistical analysis confirm this insight and append a decent P value to it. In the second universe Tom 2 goes through exactly the same motions, only his reason to visit was not vacationing but a hallucination telling him to go there and document the exceptional height of the locals. In fact, Tom 2 sees his mission as a divine commandment.

In both universes we have identical evidence but differing motivations for collecting it. The question is: which Tom can draw stronger conclusions about the exceptional height of the townspeople?

The critical point here is that the evidence strongly supports Tom's conclusions only if he would have been blind to any other surprising feature, which he might have encountered, but didn't (Berry, 2012). The sample of townspeople Tom encountered might have had big red noses, funny gaits, short statures or any number of other things in common just by chance. Therefore, if the town's population was actually of average height, the chance of seeing many unusually tall people in Tom's sample may be small (as reflected by the P value) but the chance of seeing any one of the theoretically possible serendipitous occurrences is quite big. Many small error probabilities sum to a single large probability.

This means that Tom 2, although misguided in his reason to leave home, collected much more valuable evidence in support of the exceptional height theory. And yet, the evidence itself is identical to what Tom 1 got.

It makes a considerable difference whether a result comes from a targeted hypothesis-derived study or is a surprising and serendipitous discovery. When people in their papers present a surprising result, they usually use surprise as a rhetoric device to imply "interesting," but in many cases the truer message could be a warning: "an apparently extraordinary observation becomes quite ordinary when there are many opportunities to observe something unusual. And when we observe something unusual the gods may be sending a message or they may be rolling dice. Researchers think their study is the former. Empirically speaking, it's usually the latter (Berry, 2012)."

This in a nutshell is the problem of multiplicities: if a positive result is selected from many tries that give negative results, then the probability of it being false-positive increases with the number of tries. This is because the tests are never perfect: mistakes are made and errors slip through. The problem is general; it is not specific to any statistical paradigm. Strictly speaking, it is not even a statistical problem. It arises when we test many scientific hypotheses and choose for presentation only the ones which give interesting results. (Or when we test a single hypothesis and "cherry pick" the data that we use in its confirmation.) The result is that only people who have complete information about all the hypotheses that were tested, or would have been tested had the circumstances been different (e.g., had the test results been different), have a chance to give a reasonable interpretation to the results, which are actually presented. Very often even the scientists doing the work cannot claim a full understanding of the extent of potential multiplicities in their studies, because they neither know the full extent of potential results of their experiments, nor can they accurately predict their reactions to such counterfactuals.

Finally, it is time for some good news. Firstly, multiplicities only arise when most of the tested hypotheses are wrong (Rothman, 1990). If this is not the case in your work, then you are in the clear (but perhaps you should ask yourself, what is the point in verifying all these hypotheses that you believe to be true anyway?). And secondly, even where there is a multiplicity problem, but we know the number of tries from which the positive result was selected, there exist data analysis methods that correct for multiple comparisons and keep the frequency of false-positive results at a set level, or at least give an estimate of their frequency (see Section 4.8).

As usual, there is a trade-off. The more comparisons are made, the higher the chance of encountering a false-positive result, and the more stringent criteria we need to accept an effect as real. If the number of comparisons is very large, as in GWA studies where the number of tries to identify a true genetic association can be in the millions, the cost of correcting for the multiplicities becomes a serious issue. Firstly, with increasingly stringent false-positive elimination sensitivity of the study goes down (which is another way of saying that false-negative rates go up); and secondly, the sample sizes needed to see effects often become enormous—from tens of thousands to millions.

Now comes the really bad news, which explains why multiplicities may well be the most serious practical problem for any scientific inference. This is the problem of silent multiplicities (Berry, 2007). Silent multiplicities are the unknown unknowns present in any experimental design and data analysis. While missing data or incomplete information are typical known unknowns, which do not contribute to the problem of multiplicities, the manipulations of existing data, like averaging, removing outliers, doing extra measurements because the existing results look strange or inconclusive, transforming the data, trying out different statistical analyses, doing *ad hoc* subset or interim analyses, putting more weight on data obtained early in the study, burying uninteresting data, and even the results of related studies done in other labs, can sum up as unacknowledged or silent multiplicities and bias the interpretation of your data (Berry, 2007).

> In clinical research a common way to get a date with multiplicities is through subgroup analysis wherein data that fail to exhibit effects in the primary pre-specified outcomes of the study are further subdivided by age, sex, etc., and then re-analyzed. Indeed, most subgroup analyses in published clinical research are *ad hoc* (not pre-specified) and most effects gleaned from them are probably false (Sun et al., 2012, 2014).
>
> Subgroup effects are more likely to be real if the subgroups and directions of predicted effects are pre-specified, if the effects are robust across similar subgroupings, if the subgroups are defined before randomization (not after), and if the results of analyses of different subgroups do not depend on each other (e.g., if the subgroup "men" exhibits an effect, but so does the subgroup "old," and the men in the sample are on average older than women, then the assumption of independence of effects is violated) (Sun et al., 2010). One should be shy in interpreting biological plausibility of a subgroup effect as evidence for the reality of the effect. This is because biology is complex: a great many untrue interactions will still seem plausible for the biologist.

   Importantly, one does not even have to do multiple analyses to incur the risk of silent multiplicities. In the settings of conventional statistical analysis the problem of multiplicities already arises if you do a single and perfectly reasonable statistical test for the data that you obtained, assuming that, had Mother Nature provided you with different data, you would have done a different (and again perfectly reasonable) test (Gelman and Loken, 2014). Simply put, a researcher cannot really know what he would have done, had the data been different and how many different choices he actually has to analyze his *potential* data. So the issue is not only multiple comparisons but also multiple potential comparisons and it is a more serious threat when the effect sizes are relatively small and variation is relatively large (Gelman and Loken, 2014).

As long as silent multiplicities stay silent, we will remain uncertain about the number of tries that we need to take into account in assessing the strength of evidence our experiments provide. A solution would be, before collecting any data, to write down and register your exact methodology, including what parameters are tested and with what tests; and then follow it to the letter. However, this remedy can easily turn into a constriction boa around your throat because you will thereby commit to ignoring all the interesting results, which you were unable to predict before you started the experiment.

A less extreme solution would be to make the silent multiplicities speak by taking them explicitly into account in the experimental design. Although it does not seem realistic to expect that all silent multiplicities can be tamed in this way, here is some advice, modified from Berry (2012), on how to do your best in dealing with them: (i) The study protocol should be made before starting the experiments. It should include the study's goals, a description of experiments and controls, and the full set of data cleaning and analysis methods. It should explicitly address subset and interim analyses of data, the endpoints to be addressed, and the statistical calculations to be made. Ideally, this protocol should be time-stamped and published on the Internet. Every subsequent modification therein should be clearly marked and its reasons explained. (ii) A study report should list what analyses or endpoints or subsets were specified in the protocol, but were not included in the report, and why. Were they not done or were the results uninteresting? (iii) All analyses, which were done, regardless if they are used in the publication, should be described in detail. (iv) Serendipitous discoveries must be separately noted as such. Both supportive and conflicting evidence for the discovery needs to be discussed in the context of the literature. (v) It should be reported and discussed what formal correction, if any, was used to deal with multiplicities. (vi) Publishing your data in several papers does not make the multiplicities disappear, which should then be assessed across the publications.

What is the set of tries that needs to be considered when dealing with multiplicities? For example, when a researcher spends a lifetime searching for antibacterial compounds, does this mean that as results accumulate, the worth of his studies—past, present, and future—diminishes accordingly? This conclusion seems absurd and most statisticians would tell you as much, but I doubt that many people could tell you, why exactly this example wouldn't count as a multiplicity to control for.

Maybe, whether you should think of the totality of your work as a multiplicity problem depends on the overall quality of your science. If your individual experiments test reasonable hypotheses built on solid prior knowledge, are well thought out and executed, then the results are not likely to be saturated by chance findings. And conversely, if your experiments generally test random chance-notions and intrinsically unlikely hypotheses, sooner or later an interesting false-positive will pop up and open the door to high-profile publishing (Begley, 2013).

If it should transpire that real discoveries amenable to our current crop of methods are sparse, then it could well be that most of the papers in top journals are instances of sloppy multiplicity-infested science (while modestly sized true effects cannot successfully compete with the more extreme false effects). A journal, whose reason for existence is to publish a dozen important discoveries each month, will publish the top dozen presented to it regardless of any metaphysical worries about the true state of the world. (If you are in any doubt about this, try to remember the last time your favorite nightly news broadcast was canceled due to lack of news.) One can argue that should good science ever dry up, silent multiplicities will guarantee that, as long as there will be scientists, the flood of seemingly great scientific results will never stop.

3.8   CONCLUSION: HOW TO DESIGN AN EXPERIMENT

Here we take together some advice on what to take into account when designing an experiment. In designing experiments there are three points that need to be stressed.

Firstly, it is paramount to have an explicit study protocol before starting an experiment (Kenakin et al., 2014). This is the best guarantee against sinking into the quagmire of silent multiplicities. Also, it allows you to think through the controls and spot any problems with the experimental design, before time and money have been spent on the actual experiment.

Secondly, already in this stage it is wise to seriously think of the external validity or generalizability of the future results and to find a suitable compromise regarding sensitivity and cost of experiment. When studying the basic working mechanism of a universally important and very conserved enzyme, like the ribosome, it may well be acceptable to collect all the data from a single *Escherichia coli* strain and to generalize the results to bacteria, if not to Life. Such a design conserves resources and lessens variation in the experimental series, thus allowing us to make fewer repeat experiments. On the other hand, when wishing to study a specific mechanism of translational control in bacteria, you should probably want to repeat the experimental series in more than one *E. coli* strain to be able to generalize to the level of *E. coli*, or maybe to Gram-negative bacteria. Prior knowledge about the experimental problem is all-important in any decisions concerning biological replicates and external validity.

Thirdly, no experiment should be planned without some idea of the required number of replications that would give you a reasonable statistical power of seeing interesting and true effects (a good free program for power analysis is GPower (Erdfelder et al., 1996)). Statistical power depends on just four parameters: sample size, significance criterion (α), effect size, and variation of data. Power ($1 - \beta$) is the relative frequency with which the test will correctly reject null hypotheses (Section 4.7). Setting desirable α and β levels (usually $\alpha = 0.05$ and $\beta = 0.2$) and making some educated guesses about expected effect sizes and level of variation in the data will make it possible to get a rough idea about sample size (the number of experimental replicates) and thus the cost and feasibility of the study. There is no point in starting lab work if you cannot reasonably expect to finish in the confines of your budget.

> As a practical warning, many statistics packages allow "retrospective power analysis" from the collected data. Such retrospective analyses are misleading, because they are done using the observed (and thus often inflated) effect size. Also, retrospective power adds nothing to statistical significance tests (as both analyses use the same data, their conclusions necessarily agree). Using the same data twice does not increase confidence in the conclusions. Specifically, low retrospective power does not suggest that the absence of effect was due to low sample size, nor does high retrospective power indicate that seeing no effect makes the actual presence of effect unlikely (Baguley, 2004). Power analysis should be done before the experiment and if that is not possible then in retrospective analysis not the observed effects but expected effect size (typically the smallest scientifically interesting effect size, which you wouldn't like to miss) should be used.

To increase the power, sometimes it is possible, instead of increasing the sample size, to increase the effect sizes in your experiment (usually by increasing the dose of the experimental treatment), or to decrease the variation in the sample by a changing your sampling strategy.

Also, paired designs, where experimental and control treatments are done strictly in parallel and recorded as such, can be very useful in reducing the effects of random variation in the day-to-day operation of the experimental system. Generally, for every statistical test to calculate P values or CIs, there exists a paired version, which will usually give you increased power—if only you were aware of this possibility during experimental design (see Chapter 4). Paired tests assume that there is a dependency between the paired parameter values; that is, that experiments and controls, when done in parallel, have the same sources of error. It is possible to calculate a Pearson correlation coefficient to assess the effect of pairing and to test the assumption of dependence between the paired values.

REFERENCES

Aitken, C.E., Petrov, A., Puglisi, J.D., 2010. Single ribosome dynamics and the mechanism of translation. Annu. Rev. Biophys. 39 (1), 491–513.

Allentoft, M.E., et al., 2012. The half-life of DNA in bone: measuring decay kinetics in 158 dated fossils. Proc. Biol. Sci. 279 (1748), 4724–4733.

Alter, H.J., et al., 2012. A multicenter blinded analysis indicates no association between chronic fatigue syndrome/myalgic encephalomyelitis and either xenotropic murine leukemia virus-related virus or polytropic murine leukemia virus. mBio 3 (5), e00266–e00312.

Baguley, T., 2004. Understanding statistical power in the context of applied research. Appl. Ergon. 35 (2), 73–80.

Barnett, A.G., 2004. Regression to the mean: what it is and how to deal with it. Int. J. Epidemiol. 34 (1), 215–220.

Bauer, J., et al., 2009. dSir2 and Dmp53 interact to mediate aspects of CR-dependent life span extension in *D. melanogaster*. Aging 1 (1), 38–48.

Baur, J.A., et al., 2012. Are sirtuins viable targets for improving healthspan and lifespan? Nat. Rev. Drug Discov. 11 (6), 443–461.

Begley, C.G., 2013. Six red flags for suspect work. Nature 497 (7450), 433–434.

Berry, D., 2012. Multiplicities in cancer research: ubiquitous and necessary evils. J. Natl. Cancer Inst. 104 (15), 1125–1133.

Berry, D.A., 2007. The difficult and ubiquitous problems of multiplicities. Pharm. Stat. 6 (3), 155–160.

Bland, J.M., Altman, D.G., 1994. Some examples of regression towards the mean. BMJ 309 (6957), 780.

Brandt, K., Leifert, C., Sanderson, R., 2011. Agroecosystem management and nutritional quality of plant foods: the case of organic fruits and vegetables. Crit. Rev. Plant 30 (1–2).

Burnett, C., et al., 2011. Absence of effects of Sir2 overexpression on lifespan in *C. elegans* and *Drosophila*. Nature 477 (7365), 482–485.

Button, K.S., et al., 2013. Power failure: why small sample size undermines the reliability of neuroscience. Nat. Rev. Neurosci. 14 (5), 365–376.

Clark, T., Berger, U., Mansmann, U., 2013. Sample size determinations in original research protocols for randomised clinical trials submitted to UK research ethics committees: review. BMJ 346 (1), f1135.

Colman, R., et al., 2009. Caloric restriction delays disease onset and mortality in rhesus monkeys. Science 325 (5937), 201–204.

Couzin-Frankel, J., 2011. Genetics. Aging genes: the sirtuin story unravels. Science 334 (6060), 1194–1198.

Cox, J., Mann, M., 2011. Quantitative, high-resolution proteomics for data-driven systems biology. Annu. Rev. Biochem. 80 (1), 273–299.

Dang, C.V., 2012. MYC on the path to cancer. Cell 149 (1), 22–35.

Diamandis, E.P., 2010. Cancer biomarkers: can we turn recent failures into success? J. Natl. Cancer Inst. 102 (19), 1462–1467.

Dienes, Z., 2008. Understanding Psychology as a Science. Palgrave MacMillan, London.

Dunbar, K., 1995. How scientists really reason: scientific reasoning in real-world laboratories. Nature Insight 18, 365–395.

Dunbar, K., Fugelsang, J., 2005. Scientific thinking and reasoning. In: Holyoak, K.J., Morrison, R. (Eds), The Cambridge Handbook of Thinking and Reasoning. Cambridge University Press, Cambridge, UK, pp. 705–725.

Dwan, K., et al., 2008. Systematic review of the empirical evidence of study publication bias and outcome reporting bias. PLoS One 3 (8), e3081.

Ellison, S., Barwick, V., Farrant, T.J., 2009. Practical Statistics for the Analytical Scientist: A Bench Guide. Royal Society Publishing, London.

Erdfelder, E., Faul, F., Buchner, A., 1996. GPOWER: a general power analysis program. Behav. Res. Methods 28 (1), 1–11.

Fiedler, K., Kutzner, F., Krueger, J.I., 2012. The long way from α-error control to validity proper: problems with a short-sighted false-positive debate. Perspect. Psychol. Sci. 7 (6), 661–669.

Friston, K., 2012. Ten ironic rules for non-statistical reviewers. NeuroImage 61 (4), 1300–1310.

Garson, J.A., Kellam, P., Towers, G.J., 2011. Analysis of XMRV integration sites from human prostate cancer tissues suggests PCR contamination rather than genuine human infection. Retrovirology 8, 13.

Gehr, B.T., Weiss, C., Porzsolt, F., 2006. The fading of reported effectiveness. A meta-analysis of randomised controlled trials. BMC Med. Res. Methodol. 6 (1), 25.

Gelman, A., et al., 2003. Bayesian Data Analysis, Second ed. Chapman and Hall/CRC, Boca Raton, FL.

Gelman, A., Carlin, J., 2014. Beyond power calculations: assessing type S (sign) and type M (magnitude) errors. Perspect. Psychol. Sci. 9 (6), 641–651.

Gelman, A., Imbens, G., 2013. Why ask why? Forward causal inference and reverse causal questions. NBER Working Paper. Available from: <www.nber.org/papers/w19614>.

Gelman, A., Loken, E., 2014. The statistical crisis in science. Am. Sci. 102, 460–467.

Gelman, A., Tuerlinckx, F., 2000. Type S error rates for classical and Bayesian single and multiple comparison procedures. Comput. Stat. 15 (3), 373–390.

Gelman, A., Weakliem, D., 2009. Of beauty, sex, and power: statistical challenges in estimating small effects. Am. Sci 97 (4), 310

Harper, J.M., Leathers, C.W., Austad, S.N., 2006. Does caloric restriction extend life in wild mice? Aging Cell 5 (6), 441–449.

Holden, C., 2000. Paleontology: dinos and turkeys: connected by DNA? Science 288 (5464) 238b–238.

Howson, C., Urbach, P., 1989. Scientific Reasoning: The Bayesian Approach. Open Court, Chicago, IL.

Ioannidis, J.P.A., 2008. Why most discovered true associations are inflated. Epidemiology 19 (5), 640–648.

Jaynes, E.T., 2003. Probability Theory: The Logic of Science. Cambridge University Press, Cambridge, UK.

Kahneman, D., 2011. Thinking, Fast and Slow. Farrar, Straus and Giroux, New York, NY.

Kakisi, O.K., et al., 2013. The rise and fall of XMRV. Transfus. Med. 23 (3), 142–151.

Kenakin, T., et al., 2014. Replicated, replicable and relevant–target engagement and pharmacological experimentation in the 21st century. Biochem. Pharmacol. 87 (1), 64–77.

Krzywinski, M., Altman, N., 2014. Designing comparative experiments. Nat. Drug Discov. 11 (6), 597–598.

LaMattina, J.L., 2013. Devalued and Distrusted: Can the Pharmaceutical Industry Restore Its Broken Image? Wiley, Hoboken, NJ.

Lin, C.Y., et al., 2012. Transcriptional amplification in tumor cells with elevated c-Myc. Cell 151 (1), 56–67.

Lloyd, A., et al., 2010. Comment on "Detection of an infectious retrovirus, XMRV, in blood cells of patients with chronic fatigue syndrome". Science 328 (5980), 825.

Lo, S.-C., et al., 2010. Detection of MLV-related virus gene sequences in blood of patients with chronic fatigue syndrome and healthy blood donors. Proc. Natl. Acad. Sci. USA. 107 (36), 15874–15879.

Lo, S.-C., et al., 2012. Retraction for Lo et al., Detection of MLV-related virus gene sequences in blood of patients with chronic fatigue syndrome and healthy blood donors. Proc. Natl. Acad. Sci. USA. 109 (1), 346.

Lombardi, V.C., et al., 2009. Detection of an infectious retrovirus, XMRV, in blood cells of patients with chronic fatigue syndrome. Science 326 (5952), 585–589.

Lorsch, J.R., Collins, F.S., Lippincott-Schwartz, J., 2014. Fixing problems with cell lines. Science 346 (6216), 1452–1453.

Lovén, J., et al., 2012. Revisiting global gene expression analysis. Cell 151 (3), 476–482.

Marino, M.J., 2014. The use and misuse of statistical methodologies in pharmacology research. Biochem. Pharmacol. 87 (1), 78–92.

Mattison, J.A., et al., 2012. Impact of caloric restriction on health and survival in rhesus monkeys from the NIA study. Nature 489 (7415), 318–321.

Maxwell, S.E., Kelley, K., Rausch, J.R., 2008. Sample size planning for statistical accuracy in parameter estimation. Annu. Rev. Psychol. 59, 537–563.

McShane, L.M., et al., 2013. Criteria for the use of omics-based predictors in clinical trials. Nature 502 (7471), 317–320.

Nie, Z., et al., 2012. c-Myc is a universal amplifier of expressed genes in lymphocytes and embryonic stem cells. Cell 151 (1), 68–79.

Okumus, B., et al., 2004. Vesicle encapsulation studies reveal that single molecule ribozyme heterogeneities are intrinsic. Biophys. J. 87 (4), 2798–2806.

Pereira, T.V., Horwitz, R.I., Ioannidis, J.P.A., 2012. Empirical evaluation of very large treatment effects of medical interventions. JAMA 308 (16), 1676–1684.

Perugini, M., Gallucci, M., 2014. Safeguard power as a protection against imprecise power estimates. Perspect. Psychol. Sci. 9, 319–332.

Phillips, C.V., LaPole, L.M., 2003. Quantifying errors without random sampling. BMC Med. Res. Methodol. 3, 9.

Plant, A.L., et al., 2014. Improved reproducibility by assuring confidence in measurements in biomedical research. Nat. Drug Discov. 11 (9), 895–898.

Quinn, G.P., Keough, M.J., 2002. Experimental Design and Data Analysis for Biologists. Cambridge University Press, Cambridge, UK.

Robinson, M.J., et al., 2011. No evidence of XMRV or MuLV sequences in prostate cancer, diffuse large B-Cell lymphoma, or the UK blood donor population. Adv. Virol. 2011, 1–6.

Rothman, K.J., 1990. No adjustments are needed for multiple comparisons. Epidemiology 1 (1), 43–46.

Rubin, D.B., 2006. Matched Sampling for Causal Effects. Cambridge University Press, Cambridge, UK.

Salter, S.J., et al., 2014. Reagent and laboratory contamination can critically impact sequence-based microbiome analyses. BMC Biol. 12, 87.

Schlaberg, R., et al., 2009. XMRV is present in malignant prostatic epithelium and is associated with prostate cancer, especially high-grade tumors. Proc. Natl. Acad. Sci. USA. 106 (38), 16351–16356.

Seltman, 2013. Experimental Design and Analysis. Carnegie Mellon University, Available from: <http://www.stat.cmu.edu/~hseltman/309/Book/Book.pdf>, pp. 1–428.

Silverman, R.H., et al., 2011. Partial retraction. Detection of an infectious retrovirus, XMRV, in blood cells of patients with chronic fatigue syndrome. Science 334 (6053), 176.

Smith-Spangler, C., 2012. Are organic foods safer or healthier than conventional alternatives? Ann. Intern. Med. 157 (5), 348.

Staa, T.P.V., et al., 2012. Pragmatic randomised trials using routine electronic health records: putting them to the test. BMJ 344 (1), e55.

Sun, X., et al., 2010. Is a subgroup effect believable? Updating criteria to evaluate the credibility of subgroup analyses. Br. Med. J. 340, c117.

Sun, X., et al., 2012. Credibility of claims of subgroup effects in randomised controlled trials: systematic review. BMJ 344 (1), e1553.

Sun, X., et al., 2014. How to use a subgroup analysis. JAMA 311 (4), 405.

Tuke, P.W., et al., 2011. PCR master mixes harbour murine DNA sequences. Caveat emptor! PLoS One 6 (5), e19953.

Urisman, A., et al., 2006. Identification of a novel Gammaretrovirus in prostate tumors of patients homozygous for R462Q RNAseL variant. PLoS Pathog. 2 (3), e25.

Vaux, D.L., Fidler, F., Cumming, G., 2012. Replicates and repeats—what is the difference and is it significant? EMBO Rep. 13 (4), 291–296.

Viswanathan, M., Guarente, L., 2011. Regulation of *Caenorhabditis elegans* lifespan by sir-2.1 transgenes. Nature 477 (7365), E1–E2.

Chapter 4

Data and Evidence

Sure, you can lie with statistics ... but it's a lot easier to lie without them.

Stephen Lagakos

What is the product of experimental science? Why do we write scientific papers? Different scientists and philosophers of science will give different answers to these questions, reflecting the diversity and plurality of science. For some people the reason behind doing science might be just collecting the data or building a database or making a map. Others may wish to present evidence for or against a hypothesis. For still others the reason to write papers is to change the beliefs of the scientific community or society at large, or conversely, to deepen existing beliefs. Then, some people would like to see their results leading to concrete action, like curing cancer or building a bomb. Different people come to science with different hopes, fears, and expectations: science can make you rich and powerful, science can make you wise, science can even make you good (maybe), but it surely cannot do it all at the same time! These different goals require different choices, methodologically and otherwise.

With statistical inference it is possible to do three things (Royall, 1997). Firstly, one can use statistics to keep the long-run error frequencies at a pre-set level. This is basically a question of quality control. Secondly, one may ask how much evidence the data provides for or against a hypothesis. And thirdly, one could ask how strongly he should believe in a hypothesis. Or in other words, how probable it is that the hypothesis is true. (The questions concerning converting one's beliefs into actions fall into the purview of decision theory and will not be addressed in this book.) These three questions are independent and require different statistical approaches. One cannot answer any of them by performing analyses that are intended to answer any of the others. Not even if the correct analysis is difficult to do.

In this chapter we will deal mostly with data analysis by classical frequentist statistics. Our main emphasis here is on how to convert data into evidence. We will first discuss descriptive statistics to make sense of data and then look at inferential statistics, which tries to infer from sample data to the statistical population and, by extension, to reality. Classical frequentist statistics does this

Interpreting Biomedical Science. DOI: http://dx.doi.org/10.1016/B978-0-12-418689-7.00004-1

in a relatively automatic, computationally simple way. The price to pay for simplicity in computation is, however, complexity in the concepts that are needed to make sense of the calculations. In Chapter 5 we will look at Bayesian statistics, whose interpretations are clearer conceptually and which allows direct quantification of beliefs, but which is much harder to calculate.

4.1 LOOKING AT DATA

An approximate answer to the right problem is worth a good deal more than an exact answer to an approximate problem.

John Tukey

Every statistical inference begins with analysis of data. The branch of statistics that analyses data directly and deductively, without making inductive inferences from the data to the statistical population, is somewhat loosely called descriptive statistics. Descriptive statistics includes measures of data averages, like the mean, median, and mode; measures of variation, like standard deviation (SD) and interquartile range; and measures of co-variance between different datasets, like the correlation coefficient.

Descriptive statistics is a collection of methods, which can be used in two fundamentally differing ways. Firstly, it can be used as a statistical summary, which reduces the complexity of data to a few numbers, most popular being the mean and the SD. The function of summary statistics is exactly what the name implies—to look back at what was done and document the results in a concise manner.

The second aim of descriptive statistics is diametrically opposite: to use the data creatively to look forward and to formulate new hypotheses. This approach prefers graphical representations that retain as much of the actual structure of data as possible and it is called *exploratory data analysis* (EDA). It must be noted that the major difference between a statistical summary and EDA lays not so much in the methods used (which overlap) but in the thinking of the researcher. While summary statistics sees data as a given, EDA approaches the data more skeptically and understands that data are always provisional. There is no doubt that the convenience of the summary statistics is universally appreciated. Yet, the EDA approach probably deserves more attention from biologists.

Data analysis should start not with any fancy calculation or modeling, but with simply taking some time to look at the data. Here we see our data at its informative best. With every future calculation we will destroy some of the informational content of our data in order to bring out trends and other salient features. This basic compromise between complexity of data and clarity of exposition leads to the understanding that inferential statistics must use imperfect models to make predictions about reality and that its inferences depend on approximate assumptions about the data and about the underlying reality. For example, a standard t-test takes it for granted that the individual measurements and their errors are independent of each other, that they are randomly drawn from a statistical population, that the two groups being compared are each normally distributed, and that the data at each group have equal variance. When these assumptions are met we may be justified to interpret the results of the test as giving useful information about the probability of observing a certain range of data under a specific hypothesis. However, as we will see in Chapter 5, to widen our inference to the probability that the underlying hypothesis is correct, we must introduce a further strong assumption: that the prior probability of that hypothesis is about 50%. In short, to understand the assumptions behind a statistical procedure is to understand the inferential limits of the procedure.

The researcher and the statistics program think very differently; they have completely different background knowledge, different level and concept of objectivity and different assumptions. Both have things going for them: the researcher has real background knowledge and a feel for the range of useful effect sizes, as well as experience with the specific experimental system and its quality in the context of her lab, while the statistics package is not encumbered by the loves, hopes, and biases of people and can provide output that can be directly compared between different labs and used in principled arguments to resolve disagreements. Statistics can help to bring otherwise disparate arguments to a common footing and to reconcile differing intuitions in a rational way. The statistical methods, however, were not created to make scientific argument obsolete and they should not be used independently of human reason.

The focus of inferential statistics is on the statistical model, while the focus of EDA is on the data (Filliben, 2002; Behrens, 1997). There are two main reasons for EDA. The first is to get an overall picture of the technical quality and inferential limits of the data and the second is to discern patterns in the data. EDA is not dissimilar to detective work, as long as it is understood that the detective keeps an open mind to alternative theories of the crime. Under our parable the detective finally gives the court a single theory of the crime, encapsulated in a statistical model (based on EDA), which the jury then formally tests in order to estimate if the experimental treatment is "guilty" of producing a statistically and scientifically significant effect (Behrens, 1997). EDA itself does not draw conclusions about guilt and innocence but rather generates theories and provides preliminary evidence.

While traditional statistical inference is fairly pedantic and rule-bound, in EDA it all boils down to getting a "feel" of the data. This is achieved mostly through graphical techniques, which tend to preserve the structure of the data much better than single-number summaries, like the mean or SD. In the words of John Tukey, "Numerical quantities focus on expected values, graphical summaries on unexpected values." Because a single mean can easily arise from multiple and very different data distributions, early psychologists routinely accompanied their means with histograms; to the point that a 1935 editorial in the journal *Comparative Psychology* could ask, "Should we not exclude reports in which the group averages of performance are presented without interpretive distributions?" (Behrens, 1997). Needless to say, such considerations seem to be largely absent today.

When, as in most molecular biology, the number of data points is small, at an early stage of analysis there is no need for anything more than a simple one dimensional dot-plot of the individual data points. Just take a deep breath and look: it may give you enough information to decide if there likely is a difference between experimental and control values.

It often pays to start the analysis by drawing a scatterplot of the data. Scatterplot is a version of the dot-plot where we plot the values of variable Y versus the corresponding values of another variable X. For example, Y might be the number of cells in a bacterial culture and X might be time. The scatterplot can be used to compare datasets, to identify trends, to check the variation in the data (and whether variation of Y depends on the value of X), to identify outliers and to check the assumptions of downstream statistical tests (like for a linear versus nonlinear regression model)—and all this in an intuitive way (Marino, 2014; Filliben, 2002).

If there are at least moderate amounts of data ($N > 7$, say) then it may make sense to lose some of the complexity from the data by plotting the medians with interquartile ranges (i.e., to draw a box-plot) to get a general and robust notion of the spread of data, the shape and symmetry of its distribution, its central tendency and outliers (Krzywinski and Altman, 2014b; Streit and Gehlenborg, 2014). For the more discerning customer, there exist versions of the box-plot, which incorporate individual data points, outliers, or confidence intervals (CIs) of the median (Krzywinski and Altman, 2014b).

A popular graphical method for looking at large datasets while reducing some of their complexity is to draw frequency histograms. A histogram is constructed by splitting the range of the data into equal-sized bins, which are usually arbitrarily defined (generally 5–30 bins are used per histogram, depending on the amount of data and the shape of its distribution). For each bin, the number of points from the dataset in that bin is counted and counts plotted so that the x-axis shows the bins and the y-axis shows the number of data points in each bin. The histogram graphically indicates center, spread, and skewness of the data, as well as the presence of outliers and multiple modes in the data (Filliben, 2002). Thus a histogram can help us, among other things, to create a reasonable distributional model for the data.

There exist many more graphical representations that help to gauge the randomness, variation, distribution, etc. of the data, some of which are nicely explained in Filliben (2002).

Are there any obvious outliers that should be removed straight away? If yes, make a note in the experimental protocol that these particular outliers were

Sometimes it is good to transform data so that it can be more conveniently visualized. An example is taking ratios of data values from your experimental condition over the control condition and, perhaps, converting them into percentages. There are a couple of problems with this approach, however. Firstly, ratios can mask the overall low quality of your experimental system. For instance, if you study mutant versus wild-type enzymes and are really bad at purifying your enzymes (by virtue of having a bad protocol), then it stands to reason that although the enzyme activity is very low in all your preparations, your mutants can still be lower than wild-type. Although you might only want to convey the relative activity of mutant over wild-type, your readers expect to have been given all the relevant information on the likelihood that your experimental system accurately reflects the wider world.

And secondly, taking ratios of biological data is likely to result in non-normal distributions for the ratios, even if the ratios were taken from two normally distributed datasets (and the resulting distribution can be relatively inconvenient to back-transform into normal distributions). This means that such ratios should not be used in calculations of SDs, standard errors, CIs, or *P* values. What is a good policy for EDA, can present problems downstream in inferential statistics.

removed and why it was done. This will clean up your data and retain the information necessary for an honest appraisal of the general quality of the experiment. If, for example, you will after 5 years need to compare the old results with some new data obtained from the same experimental system, and you can determine that the quality of experiment has improved significantly over time, this information can be crucial for drawing conclusions. The obvious outliers are data points that are biologically impossible (like human subjects aged 333 years). On the other hand, data points that lie outside the main data spread are not obvious outliers *per se*. For removal of any of those you should have an independent reason (perhaps some glitch in running the experiment, which was noted and documented during the experiment).

The next thing is typically to ask whether the experimental condition and the control condition give different results (whether there is a treatment effect). Here the pivotal question is: what does "different" mean? No two results are exactly alike, so one needs to decide the minimal effect size that would still be scientifically interesting. This is a judgment call for the researcher and to avoid wishful thinking it is preferable to think these things through before running the experiment, not after seeing the data.

The next important question to ask is if one should compare experimental results with controls at the aggregate level of the experimental series or in pairwise comparisons. If the individual experiments in the series were run in parallel with individual controls, then it makes a lot of sense to compare the experiment–control pairs individually. The reason is that any of the myriad ephemeral undocumented day-to-day differences in running of the experiment (maybe the incubator was off by 1°C last Wednesday) are expected to affect individual treatment–control pairs in the same way. The more there are such variations in your experiment, the more the pairwise comparisons will help to increase the sensitivity of the experiment. Of course, if the experiment wasn't run pairwise or wasn't documented fully enough, there is nothing that you can do about it at the data analysis stage.

To be able to better see the differences between experimental and control conditions one needs to take into account both the magnitude of the effect and the variation in the data. Thus it is often good to plot the data as averages and also show a measure of variation. "Average" is a nontechnical umbrella term under which belong several technical measures, including the mean (which usually means the arithmetic mean), the median, the geometric mean, and the mode. When your goal is just to form a general impression of the data, then the median is probably the most useful measure, because it has the most intuitive interpretation. When a classfull of children is ordered by height, the child standing in the middle has the median height. It is natural to talk of him as a child of average height. Median is a robust measure because it is insensitive to outliers. But it is also insensitive to real and relevant structure in the data. Very different data distributions can give identical medians. The prize of robustness is insensitivity, that is, purging of potentially relevant information from the data. Another disadvantage of the median is that it is seldom used in parametric inferential statistics, where the mean rules.

Things are slightly more complicated with the arithmetic mean. The distinction between the median and the mean becomes apparent when thinking of average income. Because a few individuals always earn a lot more than the rest, the mean income of most nations is *ca.* 30% above the median. Using the mean to indicate the average presupposes a symmetrical model of the distribution of data, while using the median does not. While the median is arguably the best way of visualizing averages, when the intention is to do inferential statistics and calculate P values or CIs, the mean is often the better choice because calculations starting from the mean are often more efficient mathematically and lead to more accurate statistical predictions.

Similar dichotomy exists in measures of variation. When the average is given as median, the usual way to show variation is by interquartile ranges. Using interquartile ranges it is possible to show outliers as separate data points, and it gives a rough idea of the distribution of data (symmetrical or not) (Streit and Gehlenborg, 2014). The most used single-number estimator of variation is the SD, which, like the mean (which it commonly accompanies), works best when coupled to the assumption of normal distribution of underlying data (see Section 1.2 for calculation and meaning of SD) (Krzywinski and Altman, 2013b). The popularity of the SD is based on its downstream use in statistical inference, which, assuming that the sample size is small (<30 or so), presupposes normally distributed data.

As an example of EDA we will finally look at some real data, namely the mean January temperatures from 1860 to 2013 of Tartu, Estonia. We are looking at climate data because it is difficult to analyze (it has high variability) and it is associated with a popular scientific theory (global warming). My climatologist friends tell me that there has been global warming in Tartu, concentrated in spring months. So we will now try to figure out if there is evidence for warming in January. Firstly we will plot the temperatures over time and to bolster our intuition will do a linear regression and plot the regression line with 95% CIs for the

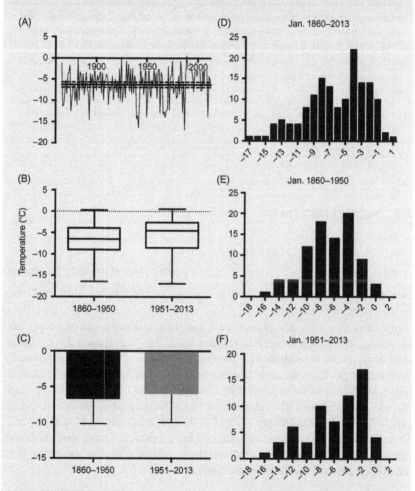

FIGURE 4.1   An example of EDA of the January temperature data (1860–2013) of Tartu, Estonia. (A) Plotting of the time course and linear regression analysis indicate no warming trend. (B) Dividing the dataset into two and doing a box-plot analysis hints at a possible effect. (C) Bar graphs with SDs do not support this. (D–F) Comparison of histograms of the full dataset and older and newer subsets clearly indicate warming.

regression line (Figure 4.1A). Clearly there is no evidence for any warming trend in January. Or is there? Just to be sure, we will next divide our dataset into two: the years 1860–1950 and 1951–2013. Now we can do a box-plot (Figure 4.1B). There seems to be a slight change in the median temperature, although the variation in both dataset seems to be pretty much equal. Also note that the "ordinary" mean ± SD plot (Figure 4.1C) shows no difference. But maybe there is a slight trend after all? My mother always says that when in doubt, do a histogram. Following her advice results in a two-peaked distribution (Figure 4.1D), suggesting that the data do not come from a single normal statistical population. This can be taken as an indication that some change in time may be taking place. Indeed, dividing the dataset into two subsets results in symmetrically distributed data for the 1860–1950 dataset (Figure 4.1E) and non-symmetrical data for the 1951–2013 dataset (Figure 4.1F). Importantly, visual comparison of panels E and F suggests January warming, opening the door for further analysis, like analysis of moving averages, etc.[1]

---

1. Incidentally, neither the *t*-test nor the Mann-Whitney test of the datasets leads to statistical significance.

4.2 MODELING DATA

There is something fascinating about science. One gets such wholesale returns of conjecture out of such a trifling investment of fact.

Mark Twain (Life on the Mississippi)

Life is messy. Yet, science strives to be the opposite. While in reality stuff often just happens in seemingly uncontrolled ways, in science we try to cast these happenings into rules and exceptions—and there is a premium on beauty and simplicity. The simpler the rules and the fewer the exceptions a theory generates, while being consistent with empirical evidence, the better it is generally supposed to be. Leaving aside the thorny question of whether such beautification is justified in terms of generating true descriptions of the world, it is evident that for simple theories to be able to generate accurate predictions of future data, certain assumptions implicitly present in their structures must hold in the real. To avoid confusion in terms, we shall call any constructs, meant to describe the true state of the world, "theories" and those, whose reason of existence is to make predictions about future data, "models":

A mathematical model is neither a hypothesis nor a theory. Unlike scientific hypotheses, a model is not verifiable directly by an experiment. For all models are both true and false ... The validation of a model is not that it is "true" but that it generates good testable hypotheses relevant to important problems.

Levins, 1966

What Is a Probability Model?

A model describes the relationship between two or more variables. The main difference between mathematical and statistical models is that mathematical models are deterministic (one parameter value exactly predicts the other), while statistical models are not (one parameter estimates the other). Probability models are used for making statistical inferences about true parameter values. A probability model is not about your particular data; it is about all possible data that could be generated by the experiment. To build a probability model we must first explicate all possible outcomes or events that might occur. This list of possible outcomes is called the "sample space." Then we need to predict for each individual outcome in the sample space, how likely it is to occur, which leads to a probability distribution of different outcomes over the sample space. This, in effect, is a probability model of our experiment.

How does one assign a probability to each possible outcome of the experiment? In the physical reality different outcomes of our experiments are likely to be fully deterministic, but if we don't know enough about their causes we could make our lives easier by pretending that we are dealing with random processes. When doing so it is good to remember that although some models are useful, all models are false. For simplicity's sake it is desirable to build your probability model as a mathematically easily described probability distribution over the sample space, the most popular of which is the normal distribution.[2] Now we can collect some data and test how well our model predicts our new data (Section 4.5).

2. The simplest probability model (all outcomes are equally likely) presupposes the least knowledge about the causal links that we study. Unfortunately, for exactly this reason this model is also likely to be completely useless for any causally deterministic biological system.

A good model is simple enough, understandable enough, and calculable enough—and, regardless of the aforementioned crimes against Truth, it still leads to good enough predictions about future data. This means that a good model must nevertheless have some semblance to reality. The secret of successful modeling is to find a balance between simplicity and truth-likeness of the model.

In a deterministic universe there is no such thing as a perfect random model to describe it. Thus the assumptions behind a model should be judged in relation to its usefulness: they are not meant to be strictly true. We don't even need to believe that true physical randomness exists to build practically useful mathematical models based on the mathematical concept of randomness. A probability distribution is neither an entity of this world nor a theory about the world. It doesn't matter if the data don't fit the distribution perfectly; all they have to do is to fit well enough for subsequent calculations to be meaningful. And what is meaningful is, of course, a matter of opinion.

Most widely used statistical methods of inference presuppose normally distributed data. The reason is that it is mathematically relatively easy to create accurate and sensitive analysis methods for normal distributions. However, the normal distribution is not a good model for description of many biological datasets. But then, the good news is that for reasonably large samples (for most biological data $N = 30$–40 is thought to be large enough) the Central Limit Theorem (CLT) makes it possible to use non-normally distributed data without a penalty in statistical inferences, which rely on the normality assumption. CLT roughly states that the mean values calculated from many independent samples are normally distributed even if the underlying distribution, from which the samples were drawn, is not normal.[3] In other words, almost any sampling distribution of the mean approaches normality as the sample size increases. This means that if the sample size is large, the inferential methods that rely on normal distribution can be used regardless of whether the population data are actually normally distributed or not. On the other hand, if $N < 30$, the actual distribution of the data matters.

It must be stressed that the distribution of a multitude of simulated sample means is not the same thing as distribution of real experimental data. For one thing, with increasing sample size the sampling distribution approaches normality and the data distribution often does not. Also, the sampling distribution gets narrower as the sample size increases, which is not the case with distributions of actual experimental data. Nevertheless, it seems that many biologists unconsciously see the normal distribution as being "normal" in the biological sense and accordingly try to fit their data into a mold, which is symmetrical, bell-shaped, and falls off steeply from the central peak, that is, it contains a very small fraction of the data far from the center. (The aptly titled book, Excel 2007 Data Analysis for Dummies, puts the biologically highly relevant lognormal distribution into the chapter "Some Really Esoteric Probability Distributions.")

Biology's love affair with the Normal has something to speak for itself as the CLT also applies to the means of multiple different independent random variables, assuming that none strongly dominates the others. This is why real-world variables, whose value depends on many independent factors, tend to have a normal distribution. Unfortunately, many biological variables cannot have negative values and depend on many positive multiplicative underlying factors, which leads to unsymmetrical lognormal distributions that are sharply cut off at zero but go to infinity on the other side (Figure 4.2).

In addition, it should be noted that normal distributions have narrow tail areas: it is extremely unlikely to encounter data that are four or five SDs from the center of a normal distribution. Therefore a normal distribution is likely to be a misleading model for biological distributions that have fat tails. A consequence of slim tails of normal distributions is that even if your data are normally distributed, letting even a few outliers in will wreak havoc with the shape of the

---

3. The meaning of "sample" is statistical: a set of values measured from repeated experiments.

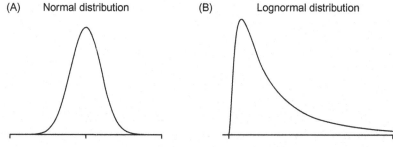

FIGURE 4.2   Normal and lognormal distributions.

modeled distribution. But it is equally true that most biological data are not normally distributed and therefore one should assume that most seeming outliers are in fact biologically meaningful.

A normal distribution is described by just two parameters, the mean and the SD. This is how mean (μ) and SD (σ) define the shape of the normal distribution:

$$f(x) = 1/\left(\sqrt{(2\pi)}\sigma\right)e^{-(x-\mu)^{2/2\sigma^2}}$$

As indicated by the complexity of the mathematical description, the shape that defines normal distribution is very specific. Indeed, most symmetrical distributions are not normal. Also note that a normal distribution goes to infinity in both directions. Biological data, which cannot have negative values, cannot be normally distributed and are often assumed to be lognormal (see below).

SD is a measure of variation in the data that retains the original units of measurement and defines the spread of a normal distribution at approximately 60% of its maximal height. SD cannot provide "error-bars"—it is not a measure of sampling error. It measures variation, which is a real property of any biological phenomenon. SD can be calculated as

$$SD = \sqrt{\frac{\sum(X_i - X)^2}{n - 1}}$$

where X_i is the value from a single measurement, X is the mean value of all the measurements, and n is the sample size (n = how many measurements were conducted).[4] When the data are normally distributed, plus/minus one SD

---

4. In addition to the sample SD shown here, there is at least the population SD, which divides the squared deviations with n, various unbiased SDs, and the SD for small samples, but these are much less used in biological literature and are thus ignored here.

encompasses about 68% of the data, two SDs encompass about 96% of the data, and three SDs 99% of the data. For the reader, who expects normal distribution from the data, this means that when the value of a SD approaches half of the mean value, a good fraction of units of measurements in the statistical population are expected to have negative data values. For example, it has been published in a respectable journal that the alcohol consumption of a sample of Estonians is 29 ± 62 g per week, while the Finns consume 25 ± 125 g (Fischer et al., 2014). To move from statistical to scientific interpretation, it seems that the Peoples of the North have learned how to secrete alcohol—and, if Fischer et al. is to believed, they do this with cigarettes too! Should the reader want to see us Estonians in action, he'd better hurry, as I'm sure that as soon as the tax-man learns of this, such practices will be outlawed.

Importantly, when the data are not normally distributed, two SDs can encompass as little as 75% of the data. For non-normal data the average deviation is arguably a better single-number estimator of variation:

$$\text{Average deviation} = \frac{\sum_{i=1}^{n}\left|X_i - X_\mu\right|}{n}$$

The average deviation, being the arithmetic mean of the absolute deviations, is less sensitive to outliers than the SD, which squares the absolute deviations. The average deviation encourages no assumptions of normality and by its use of arithmetic means of differences it is likely to be more intuitive for most people. Unfortunately, average deviations cannot be used in further statistical inference and meta-analysis, and some journals explicitly prohibit their use.

Whereas many random and independent additive effects lead to the normal distribution (typical example: errors of measurement), many small random multiplicative effects result in log-normally distributed data (Figure 4.2). For right-skewed data the simplest model is the lognormal distribution, which corresponds to a normal distribution for logarithmically transformed data. Indeed, it has been argued that most biological variation is of multiplicative nature and should be described by a lognormal function (Limpert and Stahel, 2011). If the data are normally distributed, they will usually also fit the lognormal distribution reasonably well, so there is usually no great penalty against log-transformation when you are unsure about the distribution of your biological data.

Dealing with lognormal distributions is easy (Limpert and Stahel, 2011). First, take a logarithm of your data (it doesn't matter much if it is $\log_{10}$ or $\log_2$). Then, calculate the mean and SD as usual. Because mathematics does not care about the units of the data there is no penalty for any monotonous transformation of data.

Step three consists of back-transforming the mean and SD from logarithms, which give the geometric mean (x*) and the multiplicative SD (s*). The geometric mean is always smaller than the arithmetic mean. The s* characterizes variation at the original scale of the data, determines an interval containing 68% of the data, and ranges from x* divided by s* to x* times s*. For lognormal data the multiplicative intervals are shorter than the "normal" SD and they never encompass negative values. This not only restores the interpretability of SD but also increases the accuracy for inferring the value of the population mean and thus increases the sensitivity of statistical estimation. Ordinary means and SDs can also be directly converted to x* and s* by a relatively simple transformation, without the need of having access to the underlying data (Limpert and Stahel, 2011).

Here is some general advice on data transformations to achieve greater normality. For positive data values, whose distribution is right-skewed (log-normal), taking a logarithm usually results in a normal-like distribution. Log transforms are particularly recommended if the variance increases with the mean.

If a log transform does not normalize your data, you could try a reciprocal ($1/x$) transformation (assuming again that all values are positive). This is often used for enzyme reaction rate data.

The cube root is a weaker transformation than the logarithm, and it also reduces right-skewness, and it can be applied to zero and negative values.

Square root transform is recommended when the data are counts: it will convert a Poisson distribution to a normal distribution. Cube roots and fourth roots are increasingly effective for seriously right-skewed data (Quinn and Keough, 2002).

The square has a moderate effect on distribution shape and it can be used to reduce left-skewness (assuming positive data values). With left-skewed data it may be a good idea to reflect the data before an ordinary log-transformation. To reflect a variable you first generate a constant by adding one to your largest value. Next you must subtract from this constant each data value. This will result in a right-skewed mirror image of your dataset, which can more easily be transformed into normality.

When we have calculated ratios of positive values, then the ratios lie between 0 and infinity, which is a recipe for skewed distribution (as is transformation of data into percentages). Arcsine transformation can be useful for converting percentages and proportions into normal-like distributions.

4.3   WHAT IS PROBABILITY?

It is probable that improbable things will happen. Granted this, one might argue that what is improbable is probable.

Aristotle (Ellenberg, 2014)

To grasp the methods and concepts of inferential statistics, we must start with a discussion of probabilities. In science we derive generalizations from data and statistical inference is a method that helps us to do that. It can be said that inferential statistics, largely created during the first three decades of the twentieth century, has taken over from non-numerical theories of classical induction, which have been in play since the days of Aristotle.

Because scientific inferences take us from limited amounts of data to more general theories, a level of uncertainty is inseparable from natural science. Thus there is an obvious need to present conclusions in probabilistic terms. It might therefore come as a surprise that we still don't fully agree on probability as a scientific or logical concept. In fact, different statistical methods use widely differing interpretations of probability and this leads to very different methods of inference.

To shake your intuition about the nature of probability and to show that interpretations of probability really matter, here is a problem for you: You are on television (and hopefully not too nervous about it) and have just won a game show. The host—let's call him Monty—presents you with a final choice before you get your reward: three closed doors, behind one of which is a brand new car and behind two others stand somewhat older donkeys. You will point out a door, which you want to be opened, in the understanding that this action gives you 1 in 3 chance of winning the car. Let's suppose that you picked door 1. But, as it happens, Monty, who as you know, knows exactly where the car is, opens door 2 instead, behind which you see a donkey. Now you are given a further choice: are you going to stick with your original plan (door 1) or do you wish to switch to door 3?

When presented with this dilemma, most people retain their original plan. My own reasons for not making a change were as follows: The car hasn't moved, it is still either behind door 1 or not, and knowing that behind door 2 stands a donkey does not change this objective physical fact. So, by changing my opinion I would not change the world and therefore I would expect to gain nothing by doing it. Besides, not changing one's mind without a reason shows character, which is supposed to be a good thing.

In lab experiments about 80–90% of people indeed prove that they have a backbone by sticking with their original choice (Mazur and Kahlbaugh, 2012).

Not changing one's mind when playing this game may sound logical, but it turns out to be the wrong strategy. A simple enumeration of all the possibilities proves that by switching doors one increases the probability of winning from 1/3 to 2/3 (Figure 4.3). Yet, when the Monty Hall (name of the original TV host) problem was originally popularized in *Parade* magazine in 1990, the magazine got over 10,000 letters from indignant readers protesting the solution, nearly a thousand of whom cited their PhDs in giving weight to contrary arguments. Thus it may not come as a surprise that, when allowed repeated tries in the game, pigeons can better humans in switching to the correct strategy (Herbranson and Wang, 2014; Herbranson and Schroeder, 2010, but see Mazur and Kahlbaugh, 2012).

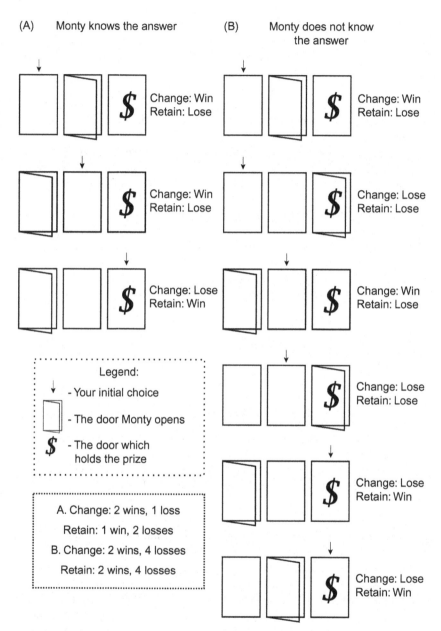

FIGURE 4.3 The Monty Hall problem has different solutions depending on whether the host knows about the placement of the prize.

What Monty Hall did, when opening a door with no car, was not changing the physical facts of the case; instead he changed our knowledge of the facts.

Now, winning 1/3 of the time instead of 2/3 of the time is no mere theoretical construct; it is quite real (and expensive, one might add). Thus, the consequences of getting your probabilities right are real enough—it is just that things themselves do not have probabilities. When we talk about probabilities, we are really talking about our knowledge of the world, not about the world as it really is.

To illustrate this point, let's see what happens when Monty knows as little as you about the placement of the car and opens a door at random. Now the game runs as follows: first you point to a door, after which Monty randomly opens another door. If there is a car behind it, then you obviously lost. If there is a donkey behind the door Monty opened, does it make sense for you to change your initial choice? As is evident from the figure, the answer is "no." Your probability of winning is 1 in 3, regardless if you change your pick or not. When no additional information goes into the game, then no gain can be got from doing the additional work of picking and choosing.

What if you are not sure whether or not Monty knows about the placement of the car? Should you then change your choice of doors when prompted to do so? The answer is obviously "yes" and the probability of your winning the car (residing somewhere between 1/3 and 2/3) depends on how likely you think that Monty knows the truth. This makes it clear that we are talking about your personal and thus subjective probabilities, because presumably Monty knows whether he knows or not. Of course, it is also possible that you don't know for sure if Monty knows and that Monty himself is not quite sure of his knowledge (maybe an assistant, who as Monty knows, likes practical jokes, has moved the car). In any case, your personal probability of winning when switching and Monty's probability of the same will probably be different, unless you decide to share your knowledge.

I dusted out this weird game to suggest that there must be something wrong in how we humans habitually think of probability as a physical property of the material world. Nevertheless, in orthodox statistics probability is commonly seen as a physical property, or to be more exact, as a long-run relative frequency of events.

We will continue our journey into the heart of probability by making a general distinction between two major classes of interpretations of probabilities: probabilities as degrees of beliefs and probabilities as long-run relative frequencies (Table 4.1).

From the table it is evident that both interpretations have their good and bad sides. The probabilities that are easy to calculate are hard to interpret, and vice versa. A common error on the part of scientists is to do the frequentist calculations and then to interpret the results as Bayesian. Typical examples are

TABLE 4.1 Consequences of Adopting a Theory of Probability

Probability as …	Interpretation	Pertains to …	Results in …	Computation	Scientific Interpretation	Applicability
Degrees of belief (Bayesian)	Probability as betting odds (betting on horses)[a]	Our state of knowledge	A degree of belief	Hard	Easy	Wide (single events have probabilities)
Long-run relative frequency (frequentist)	Probability as gambling odds[b] (gambling on roulette)	The state of the world	A decision (rejection of the null hypothesis)	Easy	Hard	Narrow (only long series of random events can have probabilities)

[a]Betting and gambling are quite different occupations. To understand why, imagine a professional better, who spends his days learning everything he can about individual racehorses, so that he can have an edge over the average betting man, and thus in the long run make a profit. Now think of a gambler, who spends his days learning about the roulette. Assuming an unbiased game, however assiduous he is, in the long run he will lose his money with exactly the same frequency as the average tourist.

[b]A more scientific version is that probability is the long-run relative frequency of the occurrence of an experiment's outcome, when repeating the experiment in a random and independent fashion.

interpretations of P values as giving a probability that the null hypothesis is wrong or statements to the effect that 95% CIs mean that the population value is expected to be inside the interval with 95% probability. This approach to statistical inference amounts to substituting a hard problem with an easy one and pretending that solution of the latter also solves the former. At the very least one should not take this road unaware.

4.3.1 Bayesian Probability

Is it likely that it will rain tomorrow? This sounds like an innocent enough question; depending on your location and past weather you will surely find a natural answer. Now, let's specify the question a little: "What is the probability that it will rain tomorrow?" If you think it unlikely that it will rain tomorrow, you may answer offhand that the probability is around 20%, but this will of course beg the question, why not say 30% or any other number.

In order to answer that, we must first ask another question: what does the 20% probability actually quantify? It cannot quantify the actual state of raining or not raining tomorrow, because it either does or doesn't (which means that in hindsight the probability of rain can only be 1 or 0). And yet it seems natural to talk of a 20% chance of rain tomorrow. The natural interpretation is actually not about the weather; it is about you. Probability quantifies your degree of belief in the hypothesis that it will rain tomorrow. And if you think again and still believe that the probability of rain tomorrow is 20%, consistency of your belief requires that you would refuse to bet on it raining unless betting odds are in your favor by at least 1 in 5. Probability as betting odds is not just an illustration; it is a consistent definition of this kind of subjective probability (Hacking, 2001).

Maybe your neighbor Ann is a meteorologist and you will ask for her opinion about the probability of rain tomorrow. She knows things that you don't, so her probability, and betting odds, will likely be different. This illustrates both the subjective and objective nature of this concept of probability: neither you nor Ann knows exactly what the weather will be, but if Ann has more background knowledge, we may assume that her bets on the weather will generally be more successful. It could be argued that things are the same in science: whoever has more relevant information, and knows how to use it, is more likely to bet on a successful theory.

There are downsides to this story, however. The first is that neither you nor Ann has access to all the relevant information. In fact, no one has. Therefore, all Bayesian probabilities are by their nature subjective and measure our ignorance as much as wisdom. The almighty god does not need Bayesian or any other kind of probabilities to say if it will rain tomorrow. He just knows the answer.

It is possible to argue with John Maynard Keynes that to the extent that Bayesian probabilities objectively reflect the state of our knowledge, they are objective in the logical sense: "The theory of belief is logical [...] because it is concerned with the degree of belief, which it is rational to entertain in given conditions, and not merely with the actual beliefs of particular individuals, which may not be rational" (Keynes, 2010). An influential alternative to the logical interpretation of probability as degrees of belief by Keynes was offered in 1923 by his Cambridge contemporary, Frank Ramsey. In this theory degrees of belief are subjective and represent psychological states. Degrees of belief are rational when they are consistent with each other according to the rules of probability calculus (Ramsey, 2011). This is essentially a more sophisticated version of the "probability as betting odds" interpretation, and it is today the most popular interpretation of Bayesian probabilities. Indeed, it is possible to derive all the basic rules of probability theory (Chapter 5) from the Dutch-book coherence principle, which states that when you assign probabilities for various events or combinations thereof you should never place your bets so as to incur a sure loss. Derivation of basic probability theory from this simple principle of gambling is not even difficult—for example see Fuchs and Schack (2013). Because there is no specifically Bayesian or frequentist probability theory, one could say that the so-called objective frequentist statistics is squarely based on the fundament of gambling in a coherent way.

Maybe we should think of Ramsey's interpretation as a fallback position to deflect criticism from the strong Keynesian position of interpersonal rational priors, while retaining the Bayesian logic. As a working compromise, one could imagine Keynesian priors as a worthy goal to aspire to, even without a real hope of ever quite getting there.

4.3.2 Frequentist Probability

To escape from the subjectivity and difficult calculations of Bayesian probability, the original eighteenth-century solution was to define probability objectively, and what is more objective that throwing dice in a game of chance? With a fair dice the limiting long-run relative frequency of each number is $1/6 = 0.1667$, and this is exactly what frequentist probability means. Frequentist probability is meant to be a real and objective property of the world (if sometimes idealized into infinite series).

The confusing of physical models and probability models led to a profusion of "laws of probability" in the eighteenth and first half of the nineteenth centuries, which had an uncanny, if superficial, resemblance to Newton's law of gravitation ($F = g(m_1 m_2/r^2)$. For example, in 1710 John Arbuthnot discovered that the ratio of male to female births is not 1:1, as was expected, but 13:12. From this law of birth he concluded that as the birth ratio is "non-random," god must have a hand in this. Ergo, god exists.

If the law of birth can be formulated and seems to hold as strictly as Newton's laws, why not do a law of death as well? This was the thinking behind J. L. Lamberts' law of mortality of 1765, which modeled then-available mortality data by a highly complex mathematical function. In his thinking "The human species dies in the same way that a prismatic vase, or vertical cylinder, empties itself through a tiny hole in its base" (Hacking, 1990). As the birth ratios of today are very similar to those of 1710 but the death curves are completely different, there is obviously some fodder for thought here.

Classical statistics, which you were taught at school, relies exclusively on the definition of probability as the limiting long-run relative frequency.

The modern version of the frequentist theory was worked out by the physicist Richard von Mises in the late 1920s. His aim was to provide an empirical and fully scientific account of probability (contrasting with the logical approach of Keynes). von Mises' theory is based on two empirical laws: the law of stability of statistical frequencies (the increasing stabilities of frequencies as the number of observations increases) and the law of randomness (the sequence of events that gives arise to the stable frequency is itself random, meaning that a gambling system, which would give a gambler long-run advantage over the odds, is impossible). Although tying together of an axiom of lawfulness (the law of stability of statistical frequencies) with one of lawlessness (the law of randomness) seems a little surprising, if not inconsistent, von Mises took this marriage of order and disorder to be a fact of nature.

For Karl Popper it was a fundamental problem. The long-run limiting frequencies are in von Mises' view nothing but a mathematical idealization of empirical laws of nature. The everyday scientific applicability of von Mises' theory therefore depends on whether the phenomenon one is studying actually is a part of a large sequence of events *and* whether the individual occurrences are random and independent of each other.

Scientific theories are not part of any recognizable series. Therefore, according to frequentist interpretation of probability, they do not have probabilities in any scientifically meaningful sense. This presents an obvious problem: we want science to be objective, but we also want to speak of individual scientific theories as likely to be true or false. The frequentist solution to this quandary is to accept the inevitable and never to speak of probabilities of theories (which would be both useless and non-scientific), but only of probability of data. To be able to talk of probability of data in the frequentist sense, it is necessary to believe in the fiction that your experimental series is part of an infinite series of potential experiments. This is a price that must be paid for the use of frequentist statistical methods in scientific inference. For example, most weather forecasts that show probabilities of rain use frequentist probabilities; for them a 20% chance of rain means that when they run their mathematical model many times (while very slightly changing their input data), then in 20% of the runs the output of the model indicates rain.

The frequentist takes his probability as a conditional probability of observing some kind of data under a specific hypothesis, or P(Data | Hypothesis). This means that if we assume the hypothesis to be true, then we can calculate an objective probability of finding a certain type of data. Of course, assuming that a theory is true is, well, an assumption—not necessarily an objective state of affairs.

For $P(D \mid H)$ to be computable, we need to define the hypothesis H precisely, preferably as a probability distribution. Unfortunately, in experimental science we rarely have a precise expectation of the effect size. And conversely, once the experiment is done we know exactly what the results are and this makes it pointless to assign probabilities to the data. Therefore, we cannot directly study the probability of data under our imprecise working hypothesis (for instance that a drug lowers mortality). Instead, we can define its opposite: a precise hypothesis of no effect (the null hypothesis), and study the probability of finding the data under this hypothesis. Now the hypothesis is fixed and the data are left to vary—and $P(D \mid H_0)$ is the probability of finding the data that you see or more extreme data under the fixed null hypothesis. If P(data | H_0) is very low, we may conclude that the null hypothesis is probably not true and therefore some other hypothesis might be true. If, on the other hand, $P(D \mid H_0)$ is high, then we cannot conclude anything about H_0 being true, because the probability would also be high under many (in fact infinitely many) different alternate hypotheses.

4.3.3   Propensity Theory of Probability

The Bayesian interpretation of probability is applicable to single events, like scientific theories, but it is deemed subjective. The frequentist probabilities are supposed to be objective, but they cannot be applied to theories. Thus there is an obvious need for a synthesis: a theory of probability that would be both objective and applicable to scientific theories.

The most famous attempt along those lines is the propensity theory of probability that was proposed by Karl Popper (Popper, 1959). Popper's propensity interpretation sees probability as belonging to the entirety of the experimental arrangement, rather than to a simple sequence of events. It was intended as a physical (or perhaps metaphysical) hypothesis, not as a hypothesis about logic. Under this theory there is a causal propensity inherent in each experimental system (when we throw a coin to test its fairness, the system consists of the coin, the thrower, local gravity, etc.) that causes a tendency of the result of each individual experiment to be close to the long-run frequentist average. Propensity is essentially a tendency of a possibility to realize itself upon repetition, which is the same as to say that a propensity is a single-case probability. Propensities should be thought of as objective physical properties of the world. The coin's propensity of landing heads is deductively and causally (not probabilistically, as in classical frequentist theory) linked to what would happen in the long run. According to Popper, the propensity interpretation attaches a probability to a singular event as a representative of a virtual or conceivable sequence of events. This interpretation depends on two metaphysical conjectures: (i) that singular propensities exist not only in repeatable conditions, but throughout the physical world and (ii) that despite the absence of frequencies, they satisfy the calculus of probabilities (Gillies, 2000).

On the face of it, propensity as a physical driver or cause of probability seems like an attractive prospect. According to this theory, the long-run relative frequencies are caused by propensities of individual events to occur with certain probabilities, and this means that we could indeed talk of objective probabilities of individual scientific hypotheses.

Indeterministic physical propensities are also promising as a base for interesting metaphysical speculations on the cosmos, free will, and whatnot. One could argue that it is the propensities that let real novelty, creativity, and freedom into the world. Real novelty seems to mean the creation of new propensities, though. For example, what was the propensity of you reading this book in 1990?

There is a serious problem with causal propensities. $P(Y$ at $t_2 \mid X$ at $t_1)$ represents the causal tendency of X at time t_1 to produce Y at a later time t_2. But then, according to Bayes theorem, conditional probabilities can always be reversed as $P(X$ at $t_1 \mid Y$ at $t_2)$, and in scientific inference probabilities are actually sometimes used in reverse as "backwards probabilities," for instance when evolutionary coalescence times are inferred from extant DNA sequences using the "molecular clock" probabilistic model (Sober, 2010). As long as we believe that cause must precede the effect, these considerations seem to make causal propensity theories both logically and practically untenable:

> The abandonment of superstitious beliefs about the existence of Phlogiston, the Cosmic Ether, Absolute Space and Time ..., or Fairies and Witches, was an essential step along the road to scientific thinking. Probability, too, if regarded as something endowed with some kind of objective existence, is no less a misleading misconception, an illusory attempt to exteriorize or materialize our actual probabilistic beliefs.
>
> de Finetti, 1990

When we think about our paradigmatic case for probabilities as frequencies, flipping a coin, we will see that the randomness in the system does not come from the physical nature of the experiment. Had we only followed the experiment closely enough, we would be able to explain the result of every individual flip of the coin without recourse to probability theory. Indeed, the actual construction of a deterministic coin tossing machine, the outcomes of individual tosses of which can be accurately predicted, has been taken on to prove this point for anyone still doubtful of the applicability of Newtonian mechanics (Diaconis et al., 2007). The fact that we need probabilities at all in dealing with the problem indicates our limited knowledge of initial conditions of the deterministic physical processes, which act on the coin. Thus, the physicalist propensity interpretation of probability starts "by throwing away practically all the professional knowledge we have labored for centuries to get" (Jaynes, 2003, p. 327).

Under the Bayesian interpretation frequentist probabilities are just special cases of Bayesian probability that involve long runs of events. The numerical values of long-run limiting frequencies and probabilities as degrees of knowledge are identical only if there is a complete lack of additional information about the experimental system. Seen like that, frequentist probability is a special case of Bayesian probability, assuming that there are observed physical frequencies and there is no additional relevant prior knowledge.

Thus for a Bayesian all forms of probabilities are unified, deal with information—not physical processes—and lead to probability theory as formalized inductive logic, of which deductive logic is a limiting case that operates with certain knowledge. At the end of the day all probability statements quantify the state of our knowledge (Jaynes, 2003, p. 44–52). This means that all probabilities are conditional of prior knowledge.[5] Probabilities as carriers of incomplete information are subjective in this narrow sense. Meanwhile, probabilities can be seen as objective in the sense that they do not reflect the subjective personal aspects of their users. They are meant to incorporate all the available information impartially, regardless of the loves and hates of their user, whose level of knowledge they quantify.

4.4 ASSUMPTIONS BEHIND FREQUENTIST STATISTICAL TESTS

Before we go any further, let me again stress that the reason why we repeat experiments and test for statistical significance (as opposed to simply looking whether the means of comparison groups are different) is to deal with sampling error, which is caused by the level of natural variation in our study system. We do not do these things to fight bias. As a wise man (David Sackett) once said, bias 12 times is still bias. This means that the most important assumption behind statistical tests is that you have already used other means to successfully eliminate bias from the data (see Chapter 3). Also note that it is easy to churn out data that violate any and all assumptions of the statistical test. However bad the input data, the output still looks like a P value and the eventual consumers of your P values, even if statistical sophisticates, usually have to take them at face value. Statistics is a clean business, in here excrement rarely smells.

Typical assumptions behind frequentist statistical tests are:

1. random sample;
2. independent observations;
3. independence of error;

---

5. This is often not evident from Bayesian formulas, where the ubiquitous conditionals in the style $P(A \mid X)$ where X is prior knowledge, are usually omitted for the sake of brevity.

4. accurate data (no outliers);
5. data are normally distributed (depends on the test and sample size);
6. equal variances in samples that are compared (depends on the test);
7. sufficiently large sample size (depends on variation, effect size, data distribution, and the test);
8. inherent plausibility of the scientific hypothesis, in which you are interested in (this will be explained in Chapter 5).

If any of these assumptions is not properly met, statistical testing is likely to do more harm than good and lull the investigator into a false confidence about his results.

I would hazard a guess that the hardest assumption of them to follow in experimental science, both philosophically and practically, is the first. For an experimentalist each random sample is drawn from the imaginary population of experiments waiting to happen. A cost of using the supposedly objective and rational frequentist methods turns out to be a weird metaphysical theory not dissimilar to the Third World of Karl Popper (see Section 2.5).

Randomly sampling individual experiments from the imaginary population means that every potential experiment needs to have an equal chance of materializing in the lab. To realize this requirement, the experimenter must presumably understand not only the range of actual ingredients of her experimental system, but also the range of *potential* ingredients.

The second assumption of independence means that in no aspect of the experimental system, which exhibits variation between repeated experiments, may this variation correlate between experimental replicates. This requirement can be interpreted as demanding a complete reconstruction of the experimental system before each replication, which in biochemistry means new solutions, new enzyme preparations, etc.

The third assumption of independence of errors means that all of the factors contributing to errors of measurement must be independent for each data point (Marino, 2014). This assumption holds for all of null hypothesis testing, parametric and nonparametric, and it is violated, for example, when technical replicates are used instead of biological replicates to generate the data to be analyzed. This violation is caused by the fact that the factors contributing to systematic error for each set of technical replicates are not independent—they are specific to the particular run of replicates (Marino, 2014). This is a very interesting point, so let me state it again: unless you are sure that there is no systematic error in your measurement, technical replicates cannot in good conscience be analyzed by any frequentist test that generates P values or CIs as its outcome.[6]

Another common situation when independence of errors is violated is when measurement values are close to baseline, which cannot be accurately measured and subtracted. Let's suppose that your experimental values are high and control

---

6. It is still okay to use technical replicates to assess your measurement apparatus—here it doesn't matter if your biological parameter is non-representative of an interesting statistical population.

values are close to baseline. Now you have different kinds of systematic errors at play in control versus experiment and comparison by frequentist methods (*P* values, CIs) may not be justified.

The universal assumptions 1–5 and 8–9 can only be effectively addressed in the study planning stage. Therefore the best use an experimental scientist can have for a statistician is not in number crunching, but in experimental design (Berry, 2007).

4.5 THE NULL HYPOTHESIS

The most often used type of frequentist statistical testing is null hypothesis testing. It is actually a collection of separate methods, which are often taught together in statistics courses as the method of testing of hypotheses. Null hypothesis testing is nearly always misunderstood by students and often by scientists, and it has always been highly controversial in specialist circles, both as a statistical procedure and as a part of scientific inference.

Here we will make an attempt to understand the conceptual basis of null hypothesis. We will start our discussion by taking a sympathetic look at the general logic behind the procedure (Chow, 1998; Batanero, 2000).

Let's suppose that we are interested in the active site of an enzyme. We have a catalytic mechanism in mind but we don't know yet if the mechanism applies to our enzyme.

We might call this initial question:

1. A substantive hypothesis.
2. Next we deduce from it a research hypothesis (RH), which is experimentally tractable. For example, from our knowledge of the enzyme we might deduce that the amino acid histidine at position 69 of the protein is a candidate for involvement in the catalytic center (it could be catalytic, be involved in organizing the active site or in controlling the reaction).
3. The next step is to develop an experimental hypothesis (EH), which is actually testable in the confines of our experimental system. In our example, a good bet would be that mutating His 69 to alanine would change the catalytic rate of the enzyme. Therefore, the EH is that the activity of the mutant enzyme will be substantially different from the wild type.
4. From there comes the alternative statistical hypothesis ("alternative" meaning "not the null"). This involves samplings of statistical populations and calculated sample means and can in our case be written out as $H_{alt} \equiv \mu_m \neq \mu_{wt}$, where μ_m is the mean of the repeated experiments with mutant enzymes and μ_{wt} is the same for the wild-type enzyme. Note that from the substantial difference between the mutant and wild type, which was postulated in the EH, all that survives is "any difference in any direction."
5. From the under-defined H_{alt}, which manages to exclude only one value, we will derive its opposite, the null hypothesis (H_0) (Figure 4.4). H_{alt} and H_0 must be mutually exclusive and exhaustive. As H_{alt} entails any size of effect

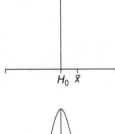

H_0 as point hypothesis. The mean of experimental values is shown as X. The question is: is the difference from the H_0 value likely to be due to chance?

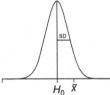

Modeled distribution of individual measurement values under H_0. There is a substantial fraction of individual measurements that would give higher values than X, assuming that H_0 is correct. The spread of this distribution is given by SD.

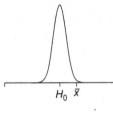

Modeled distribution of many randomly sampled means ($n = 3$) from the previous distribution. It is much less likely that the mean of 3 independent experimental values is larger than X. The spread of this distribution is given by standard error of the mean SEM = SD/$\sqrt{3}$

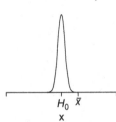

By further increasing the sample size to 10 (SEM = SD/$\sqrt{10}$) we have narrowed our sampling distribution to the point where, if H_0 is correct, it is extremely unlikely that a sample mean exceeds X

FIGURE 4.4 Construction of a statistical null hypothesis.

in any direction, that leaves the H_0 with no effect ($H_0 \equiv \mu_{\mathrm{m}} = \mu_{\mathrm{wt}}$). More generally, a null hypothesis is a precise statement about a value of a population parameter or size of an effect of an experimental treatment. H_0 is frequently but not always set at zero (when at zero, H_0 is sometimes called the zero hypothesis).

Nevertheless, it is not the statistical point hypothesis H_0, which is used in statistical hypothesis testing, but a statistical null model.

6. A statistical null model is a probability distribution that shows the long-run frequency of observing each parameter value in an experiment, given that the statistical hypothesis H_0 is true. Strictly speaking, it assumes that the experiment is repeated (simulated) an infinite number of times. The

usefulness of the statistical model can be ascertained by comparison of the model with outcomes of the real experiment. In the simplest case the need for a probabilistic formulation of H_0 arises from the presence of random error in the measurement. We next assume that the point value of H_0 is the true value and that we know the variation inherent in the measurement. Now we can simulate many independent virtual measurements to get a decreasing concentration of measurement values as we go further from the point value of H_0. We can model this by a continuous probability distribution like the normal distribution. When we add more measurements, the SD of this distribution (which uniquely determines its spread at about half-height) does not change in any systematic fashion.

However, because in our example the distribution merely explicates measurement error, we are really not interested in its spread, but rather in the point value of H_0. Our aim is not to look at the result of a single measurement; it is to compare the mean value of a number of repeated measurements (the sample mean) to the H_0 point value and to figure out if the sample value is far enough from it, to be reasonably seen as inconsistent with the H_0.

7. A way to increase the accuracy of prediction of the population mean from the sample mean is to draw a larger sample. Because now we are looking at the fit of a sample mean, not a single measurement, we have to redefine the null model as the probability distribution of simulated sample means. This is fortunate as long as our aim is to compare point values, because the larger the sample, the smaller will be the SD of the simulated H_0 probability distribution of sample means. More specifically, the spread (at about half height) of H_0 for a given sample size n is defined as

$$\frac{\text{SD}}{\sqrt{n}}$$

which is called the standard error of the mean (SEM). When we substitute the SD value that defines the width of the normal distribution (6) with its SEM, then we have a new and narrower normal distribution, which is the sampling distribution of the mean. The sampling distribution gets narrower as the sample size increases because a larger sample tends to more closely resemble the population from which it came. In the extreme, an infinitely large sample will result in a sampling distribution with zero spread (a point value).

8. The estimation of the population SD from sample data systematically underestimates the population variance at small sample sizes.[7] To make a correction, the tails of the normal distribution (7) are lifted proportionally to the inverse of the sample size (Figure 4.5). The resulting fat-tailed t distribution

---

7. This bias is a separate problem from the increased random variation of small samples.

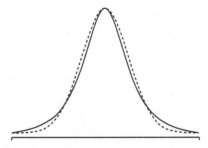

FIGURE 4.5 Standard normal distribution (dotted line) and t distribution (solid line) if the sample size is 3 ($df = 2$).

is the version of H_0 that will be used in the actual test and compared with experimental sample data. To use this correction with small sample sizes presupposes that the data are normally distributed.

It is important to note that the statistical model of the H_0 must be mathematically simple, meaning that the probability function must be fully explicated. If the probability model is a normal distribution, a mathematically simple model must explicate both its mean and SD.

The story as presented above is somewhat simplistic. In statistical theory constructing a null distribution involves a defined outcome space (the range of possible observations), an understanding of the population variance and the use of a mathematically convenient statistic, whose sampling distribution is modeled. Popular test statistics include the t statistic, F statistic and chi square statistic. For example, should we wish to compare the means of two samples while keeping our eye on the question whether these samples originate from different statistical populations, the null distribution can be seen as the distribution of infinitely many values of the t statistic calculated from an infinite number of imaginary samples, where $t = [\mu_1 - \mu_2 / SE(diff)]$ and SE(diff) is the standard error of the difference and μ_1 and μ_2 are the sample means. The SE(diff) between two sample means is the SD of the sampling distribution of the difference between the sample means. But, because knowledge of the full population is nearly always beyond our reach, in practice we estimate the SE(diff) on the two samples:

$$\text{Estimated SE(diff)} = \sqrt{[[SD_1^2(n_1 - 1) + SD_2^2(n_2 - 1)/(n_1 - 1) + (n_2 - 1)] \times (1/n_1 + 1/n_2)]}$$

where SD_1 is the SD of the first sample and n_1 is its size.

If the sampling distribution of the test statistic is incompatible with experimental data, the H_0 will be rejected and the corollary of that action is often thought to be that it is unlikely that the data come from H_0. Simply put: H_0 exists to be killed by experiment. As a larger sample leads to a narrower sampling distribution, such H_0 is more exclusive of actual data. Thus the chance of success of the test in rejecting H_0 depends both on the extent of random error (SD), and on the sample size n used to construct the H_0. Since the H_0 sampling distribution is really about sampling error, by rejecting H_0 we rather modestly reject sampling error as the cause of the difference between our data and the point version of H_0 (5). This only corroborates the hypothesis (4) by excluding a single value, which obviously does not constitute verification for any of the scientific hypotheses (1)–(3).

A problem with using test statistics for generating the H_0 distribution is that by picking a different test statistic it is possible (and, indeed, easy) to change the result of the test. Moreover, if the H_0 is a continuous distribution, there exist pairs of statistics that are related by a one-to-one transformation (and thus carry exactly the same information), such that only one of them leads to the rejection of the H_0 at a specified significance level (Howson and Urbach, 1989).

Testing experimental data against the sampling distribution of test statistic seems to be a sensible thing to do as long as we are not especially interested in the variation around the mean of H_0. Who cares about variation caused by measurement error when we can simply ignore it at proper sample sizes? (answer: the instrument maker). The real question is, however: Who cares about biological variation? (answer: the biologist).

With biological variation it is not necessarily the peak of its distribution or the average performance of a molecular ensemble that we are interested in. If our study population consists of cancer patients that are fully cured by a drug candidate, patients that are harmed by it, and patients that get no benefit or side effects—then are we still interested in the mean response and its deviance from the null model in the form of a normal distribution?

Seen like that, it makes no sense to lose much of the information present in the probability distribution of the data (6) by converting it into a sampling distribution, which focuses attention to its mean while losing information about any structure present in the biological population.

There are two fundamentally different accounts as to what happens when H_0 turns out to be inconsistent with experimental results. The first view is that because it is the H_0, not H_{alt}, which tends to be precisely defined, it is far easier to falsify H_0, and this is what the null hypothesis test strives to do. The natural way to treat a rejected H_0 is to conclude that H_{alt} must be true and climb all the way up the inferential ladder again, from the statistical alternative hypothesis to the substantive hypothesis (Batanero, 2000). The conclusion from the test is then scientific: the substantive hypothesis is supported by the data.

However, the scientific interpretation of the rejection of H_0 is problematic. To see why, it is instructive to compare the logic of falsification of H_0 with Poppers logical falsificationist approach to theory testing (Table 4.2).

At first glance Popper's falsificationism and null hypothesis testing may seem quite similar. Both deduce specific testable hypotheses from a GT and both are geared to experimentally overthrow a specific and restrictive EH. In the falsificationist view only severe tests are worth doing and the severest test is concerned with the narrowest EH. In contrast, in null hypothesis testing the

TABLE 4.2 Comparison of Logical Falsificationist and Statistical Null Hypothesis Testing

Falsificationist Hypothesis Testing	Null Hypothesis Testing	Comments
General theory (GT)	Substantive hypothesis (Histidine 69 is in enzymes' active site)	
RH	RH (Mutation of His 69 will change enzyme kinetics)	RH is a deductive prediction from GT
EH	EH (mutant kinetics $\neq$ wild-type kinetics)	EH states what will be measured in the experiment and what range of results would lead to rejection of EH
NA	Alternative statistical hypothesis H_{alt} ($\mu_m \neq \mu_{wt}$)	H_{alt} is often much less restrictive than the EH
NA	Null hypothesis H_0 ($\mu_m = \mu_{wt}$)	H_{alt} and H_0 are mutually exclusive and exhaustive
Test of EH	Test of H_0	H_0 is usually much more restrictive than EH
Conclusion: If not EH, then GT falsified	If not H_0, then data support the substantive hypothesis	
Deductive logic: If GT, then EH notEH therefore, notGT	Inductive logic: If GT, then EH if EH, then H_{alt} H_{alt} xor[a] H_0 notH_0 therefore, H_{alt} therefore, EH is supported therefore, GT is supported	In deductive logic, affirming the consequent (EH) does not mean that the antecedent (GT) must be true The falsificationist scheme of rejecting the consequent (Modus Tollens) is valid

[a]*xor is the "exclusive or," which means that either H_{alt} or H_0 can be true, but not both.*
NA—not applicable.

statistical derivate of the EH, H_{alt}, is often extremely wide-ranging, encompassing all possible parameter values, except the null. The H_0 is correspondingly more restrictive and thus easy to falsify. However, because it is uninteresting scientifically, its falsification in itself does little that is useful. Because H_{alt} (if not EH) would be supported by almost any conceivable evidence, experimental support of it will not tell us much about the veracity of the substantive hypothesis, from which it was derived via several rather informal steps. Seen like this, the scientific interpretation of null hypothesis testing is hardly defensible.

The second interpretation is that null hypothesis is not concerned with actual scientific populations at all; it is about statistical populations only (Chow, 1998). The null hypothesis merely concerns the data collection procedure and by rejecting H_0 we are merely stating our willingness to exclude the possibility that the observed effect was due to sampling error. Under this interpretation, assuming random and independent sampling, H_0 can be true on most occasions (while under the scientific interpretation H_0 is literally never true). H_0 is thus seen as a statement about how well the data are collected, and nothing else. Under this view null hypothesis testing does not constitute inductive theory corroboration. Instead it provides objective means to exclude chance as an explanation of research data, but is not informative as to what are the non-chance factors that make the experiment work. Rejecting H_0 says nothing about the scientific importance of the data.

Fortunately there are statistical "leftovers" from null hypothesis testing, like effect size ($\mu_{wt} - \mu_m$ in our example) and the concept of statistical power, which, while not directly used in rejecting sampling effect as a cause for the observed effect, can be used in a scientifically meaningful way in evaluating the scientific worth of the data.

The eight-point scheme of null hypothesis testing presented above illustrates a fundamental difficulty in science. The scientifically interesting theories, which address some general feature of the world, tend to be relatively vague and are separated from reality by the conceptual gap between a particular experimental system and the need to generalize to whatever phenomenon the scientist has posited to occur outside the controlled environment of the lab. Technical aspects of designing an experiment and scientific theories that inspired this experiment are separated by a gap, whose bridging *always* requires a leap of faith on the part of the scientist.

It is easy to move down the hierarchy of ever-narrower hypotheses that ends with the H_0 without losing deductive rigor but nigh impossible, after rejecting the H_0, to unequivocally move up again. This difficulty in upward mobility is not specific to statistical hypothesis testing and it also holds after confirming an EH. What, when going down, looks like a single continuously narrowing funnel, will reemerge as a labyrinth of forking paths, when going up. This gap between theory and experiment makes it easy for a scientist to independently manipulate both her experimental system and her scientific theories, until interpretation of one matches the other. Whether this is a good thing, or bad, is a matter of contention (Ioannidis, 2013):

> *The freedom we have in the design of our experiments is so enormous that when an*
> *experiment does not give us what we are looking for, we blame the experiment, not*
> *our theory. (At least, that is the way I work). Is this problematic? No.*

(Former Professor Dietrik Stapel in his inaugural speech, as cited in Ioannidis, 2013)[8].

Accordingly, if you think that you are an experimental scientist, and yet you do not have control over the scientific questions that spawned your experiments, then in all likelihood someone else has. And whoever controls the scientific question, controls the interpretation of experiments, controls the science, and therefore is the scientist.

4.6 THE P VALUE

The workhorse of statistical hypothesis testing is the P value, which can be used in three different ways. (i) P values can be used to control the long–run relative frequency of erroneously rejecting null hypotheses (the Neyman-Pearson procedure). (ii) P values can be used as an informal measure of the weight of evidence against a particular null hypothesis (Fisher's procedure) and (iii) P values can be transformed into CIs.

In the following chapters these three uses will be explained at some length, but for now it suffices to say that P values can be used to ask quite different questions and that they have different functions in their different roles. In the Neyman-Pearson procedure (i) the question is: will I reject the null hypothesis? In Fisher's procedure (ii) the question is: is it likely that the null hypothesis is false? (iii) When bundled into CIs, the P values are used to assess uncertainty about the true value of the population mean, as inferred from the sample value.

To understand the meaning of the P value, we start with the null distribution (8) of the previous chapter. Now, we run a real series of experiments (the sample size equals with what was used to calculate the SEM of the H_0 distribution), record the results and calculate the mean. Next we find the value of the mean on the x-axis of the H_0 probability distribution and draw a vertical line from the axis to the curve of the null function. Now we can calculate the area under the curve delineated by the vertical line, away from the peak of the curve (Figure 4.6). The fraction of the area under the curve, which is delineated by your data, from the total area under the curve, is the probability of encountering these data or more extreme data, under the null. This fraction is the P value. The P value can vary between 0 and 1 and its use lays in comparing the actual results of experiment with the prediction made by the H_0. If the results at hand are unlikely under the H_0, then maybe H_0 does not hold.

There is one more thing to notice. Assuming that we cannot predict to which side from the mean of H_0 our sample mean will fall, we must assume that it

---

8. See Section 7.2 for a very similar approach on the part of the philosopher of science, Paul Feyerabend.

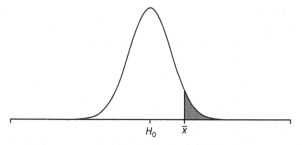

FIGURE 4.6   Graphical representation of the P value. The one-sided P value is the ratio of the shaded area below the graph over the total area; $\bar{x}$ is the sample mean; H_0 is the point value of the null hypothesis.

could fall to either side with equal probability. Therefore, the area under the curve, which is delineated by the P value, must be exactly duplicated on the other side of the curve. This is called "two-sided P value" and it is two-fold larger than the one-sided P value. If you believe that the treatment effect can cause your data to fall to one side of H_0, but not to the other, then it is good to use one-sided P values. But be sure to remember that should your data fall to the wrong side of the curve, you may only conclude that there is no statistically significant effect (even if the two-sided P value would be tiny). You are not allowed to use two-sided P values after you have committed yourself to the one-sided test. If this seems like a bad deal, then stick to two-sided P values.

The P value essentially puts a number on the expectation of seeing similar data to your data, if the null hypothesis is true. Here is a more exact definition. Let's suppose that we have calculated the sample mean and that the two-sided P value is 0.05.

P value 0.05 predicts that (1) if the H_0 is true and (2) if you will draw an infinite number of random samples from the population, then in exactly 5% of the samples will you encounter mean values as far or further from the mean value of H_0.

A shorter version is: P value is the probability of the observed data or of more extreme data, given that the null hypothesis is true, and an even shorter version is: P value $\sim P(\text{data} \mid H_0)$.

P value is a random quantity. Should we repeat our experiment, we would get a different P value. When H_0 is true then the P value will be less than 0.05 in exactly 5% of the experiments. Under a true H_0 all P values are equally likely and thus have a flat distribution between 0 and 1. In significance testing, which scientist happens to have a significant P value under true H_0 on a given day has nothing to do with the quality of the research; it is just luck.

There are many different tests of statistical null hypotheses, all calculating their own P values. They have somewhat different assumptions and they can test different null hypotheses. And yet, each P value gives us a long-run relative frequency that a H_0 is true—meaning that the different P values are not only calculated differently, but also have subtly different interpretations. In Table 4.3 we will list some of the more widely used tests for comparing two or more groups of data (Marino, 2014; Motulsky, 2010).

TABLE 4.3 A Selection of Statistical Tests for Comparing Two or More Groups

Test Compares	Assumes Normality	Unpaired Experimental Design	Paired Experimental Design
2 independent groups	Yes	*Student's t-test* H_0: identical means *Assumes* independence, equal variances between groups	*Paired t-test* H_0: identical means *Assumes* that the difference values between paired measurements are normally distributed
	No	*Mann–Whitney test* H_0: identical distribution of values *Assumes* that the distributions have the same basic shape	*Wilcoxon signed-rank test* H_0: zero median difference between the pairs *Assumes* that the distribution of difference values between paired measurements is symmetrical about the median
3+ independent groups	Yes	*Analysis of variance (ANOVA)* H_0: identical means *Assumes* equal variances *Posttest*[a] for all pairwise comparisons of means: the Tukey test or Holm-Sidak test. If comparing all means with the control mean: the Dunnett test	*Repeated measures ANOVA* H_0: identical means *Assumes* sphericity[b]
	No	*Kruskal–Wallis test* H_0: identical sums of ranks of different groups *Assumes* the same as Mann–Whitney	*Friedman test* H_0: identical medians *Assumes* the same as Wilcoxon signed-rank test, plus sphericity

[a]*The reason for the posttest (multiple comparisons test) when comparing 3+ groups is to pinpoint the statistically significant difference(s) to a specific pair(s) of groups. See Section 4.8 for a discussion of ANOVA.*
[b]*Sphericity means that the variance of the differences between all combinations of related groups is equal. A random factor causing error in one experiment does not influence the error in other experiments—the errors are truly random.*

Because of the universal assumption of independence of errors, none of the above methods can be used to test for the presence of trends in data (over time, space, strength of treatment, etc.). Especially, appending standard errors or CIs to each time point in a series is a classic exercise in futility. What we need instead is a test for trend (see Alexandrov et al., 2012 for a technical review). In statistical terms a trend is characterized by a changing probability distribution. Therefore the H_0 is that there is no change, or that (time) series values are independent and identically distributed. The parametric test for trend is regression analysis (see Section 2.8). For nonparametric analysis of monotonic trends the Mann-Kendall test is recommended. This test is not dependent upon the magnitude of data, missing data or irregular spacing of data on the x-axis. It requires at least four measuring (time) points to determine a trend. There exist versions of the test that are able to take into account seasonal variations from the monotonic trend.

4.6.1   What the P Value Is Not

The above definitions of the P value speak about hypothetical and actual data, given that the H_0 is true. For a scientist this takes the insult of a wrong question (about data rather than about the hypothesis) and adds the injury of two separate hypotheticals (about data and about H_0). Thus it comes as no surprise that more intuitive alternative interpretations of P values abound in scientific practice.

What follows is a list of common mischaracterizations of a low P value. Be aware that these are not only wrong logically. It must be stressed that in most scientifically relevant situations the following interpretations of P values cannot be seen as useful approximations of the correct interpretation.

1. Low P value means that you have disproved the H_0 ($P = $ not H_0). This is false because the H_0 (if a normal distribution) goes to infinity on both sides, meaning that it is theoretically possible to get any sample value, if the H_0 is true.
2. You have found the probability of H_0 being true or $P = P(H_0)$. This is false because whatever the point value of your H_0, it is always possible to think of alternative H_0 point values that have associated probability distributions, which will result in exactly the same P value.
3. You have proved the alternative statistical hypothesis or $P = (H_{alt})$. This is false for more reasons that I can count. For example, even if the H_0 is true, you would get a $P < 0.05$ in 5% of your experimental series.
4. You get the probability of the H_{alt} being true or $P = P(H_{alt} \mid data)$. This would mean that $P(H_{alt} \mid data) = P(data \mid H_0)$. There is simply no way in doing this conversion without very substantial assumptions being met. The information contained in the P values does not come even close to addressing this conversion.
5. You know the probability that you are making the wrong decision if you reject the H_0 or $P = P(not\ H_{alt} \mid data)$. This is logically equivalent to (4) and as wrongheaded.

6. You have a reliable experimental finding in the sense that if, hypothetically, the experiment were repeated a great number of times, you would obtain a significant result on the vast majority of occasions or $P = 1 - P(\text{data})$. This is false because it ignores the assumption that the H_0 is true.

4.7 NEYMAN-PEARSON HYPOTHESIS TESTING

The most used application of the P value in science is statistical hypothesis testing or Neyman-Pearson hypothesis testing (NPHT). This method was created in the 1930s by Jerzy Neyman (1894–1981) and Egon Pearson (1895–1980). NPHT largely automates statistical testing, but its outcome is very specific. It does two things very well. Firstly, it is an action rule that keeps the long-run relative frequency of incorrectly rejecting the null hypothesis (H_0) at a pre-set level. This is the type I error frequency, denoted by α. And secondly, it enables to keep the long–run frequency of incorrectly not rejecting the H_0 at a pre-set level. This is the type II error frequency, denoted by β.

NPHT is at its best in a quality control setting. It is about making decisions. Every test results either in rejecting a null hypothesis, or accepting it. Rejection of the H_0 does not imply truth—it does not imply that the H_0 is false or that the H_{alt} is true. The goal is not the truth but the long-term quality of the product.

In the confines of the Neyman-Pearson procedure it is possible to commit two types of errors, to reject a null hypothesis in vain (type I error) or not to reject a false H_0 (type II error) (Table 4.4). Quality control thus has two conflicting goals: to be stringent enough so that effects are not called falsely and to be relaxed enough so that true effects will not be ignored.

This is how NPHT controls this dilemma. First we must decide on the H_0 probability distribution. Then we will decide on a parameter value that defines the mean of the H_1 distribution. Now we can draw a vertical decision line through both distributions at the place of our choosing and the areas under the H_0 and H_1 curves delineated by this line will define the long-run error probabilities α and β (Figure 4.7).

Where to draw the line depends on a further decision: which is the greater sin, committing type I or type II errors? If you find publishing false effects horrifying,

TABLE 4.4 Quality Control by NPHT

	Reality: H_0 Is True	Reality: H_0 Is False
Decision: keep H_0	Outcome: correct action (1 − α = size, specificity)	Outcome: type II error (β)
Decision: reject H_0	Outcome: type I error (α)	Outcome: correct action (1 − β = power, sensitivity)

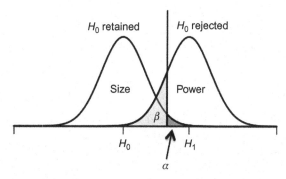

FIGURE 4.7   Graphical representation of the NPHT. The power (sensitivity) of the one-tailed *t*-test is 80% and the size (specificity) of the test is 95%. The significance line is set at 0.05.

then you will want to draw the line far from the mean of H_0. By convention, the position of the line is called "statistical significance level" (α). By definition, α is the long-run relative frequency of erroneously rejecting H_0 or committing a "type I error." $1 - \alpha$ is called "specificity" (in medicine) or "size" (in statistics) of the test. Here is the difference between statistical and biological significance:

Statistical significance is the long-run frequency of committing type I errors. Biological significance is the scientific relevance of an effect of a given size.

The first assumes that H_0 is true and the second assumes that H_0 is false. One is a long-run relative frequency (a number); the other is essentially scientific context (not a number). When you think of it, it is obvious that there is no direct path from statistical significance to biological significance. And yet, in the biological literature the meaning of "statistical significance" is routinely confused with "significance."

If we always act according to the rule—if $P < \alpha$, then reject the H_0—then the long-term frequency of type I errors converges at the α level. Assuming that α is set at 0.05, this means that by following the rules of NPHT, exactly 5% of rejected H_0s are true in the long run.

Unfortunately, drawing the decision line far from the mean of H_0 distribution entails drawing it that much closer to the mean of H_1, thus increasing the frequency of type II errors, which is denoted by β. $1 - \beta$ is called "sensitivity" (in medicine) or "power" (in statistics) of the test.

This is how NPHT works in science:

1. Define two statistical point hypotheses. In biology H_0 usually exemplifies the null effect and H_1 is your expectation for the effect of the experimental treatment. If, as is likely, you cannot predict the effect size precisely, it is a good idea to use as H_1 the smallest scientifically relevant effect size that you would be interested in finding.
2. Model the sampling distributions of H_0 and H_1.

3. In the sampling distribution of H_0 set the significance level α. This will define the frequency of type I errors that the procedure allows.

4. Using both H_0 and H_1 distributions we can do the power analysis to find the sample size that yields the desired type II error frequency (β). We can use the interdependence of the standard error and sample size (SEM = SD/$\sqrt{N}$) to find a minimum sample size that is consistent with the power we chose for our experiment. Power roughly equals with $\dfrac{EZ \times \alpha \times \sqrt{n}}{SD}$ where EZ is the effect size and SD standard deviation.

5. Now it is time to do the experiments and calculate the P value. If $P < \alpha$, then you will reject the H_0. If $P > \alpha$, then you will keep the H_0. If the Neyman–Pearson logic is taken to be that H_0 and H_1 are the only alternatives, then the non-rejection of H_0 implies the acceptance of H_0. Accepting the H_0, however, does not imply its truth. A non-significant result means that we should suspend judgment. If the power of the study is high, then we can conclude the true effect is probably less than the effect size, which was used in the power analysis. Some believe that when the power of the study is high ($\beta = 0.05$, say) and the P value is large, then we could accept the H_0 as likely to be literally true. However, this conclusion presupposes that H_0 and H_1 are the only possible hypotheses that could explain the facts. This assumption does not look promising. However, what we can say is that when the power is high, the prior probability of no substantial effect is fairly high too, and we accept the H_0, then it is unlikely that a true effect larger than the H_1 exists.

By following the prescribed route of NPHT over a lifetime, a scientist will erroneously reject the fraction α of all true H_0s that she tested. But this is not how most scientists use NPHT. Most scientific users are interested in how much evidence their particular data provide against a particular H_0. Accordingly, in biology a single statistically significant result is often interpreted as implying that an explanation of the observed effect in terms of sampling error (chance) can be excluded with the level of strictness provided by the chosen significance level (Chow, 1998). Under this interpretation the contrast between H_0 and H_1 is not a contrast between two different scientific hypotheses, it is a contrast between chance and no chance.

Richard Royall has shown that the evidentiary interpretation of NPHT in favor of H_1 over the H_0 is often unwarranted (Royall, 1997). If α is set at 0.05 and β is set at 0.20 (as is usual in scientific settings), then in about 1/3 of the statistically significant cases the evidence provided by the data for H_1 will be weak or misleading. To make the evidentiary interpretation meaningful at a single sample level, it is necessary to increase the power (usually by increasing the sample size). For example, to reach 95% probability that a statistically significant result produces at least moderately strong evidence for H_1, the power analysis should be done at the $\alpha = 0.05$; $\beta = 0.004$ or $\alpha = \beta = 0.016$ level (Royall, 1997).

Presenting exact P values makes no contribution to keeping the type I error levels at α. In the long run it does not matter if a particular P value was 0.049 or 0.0001. As long as every $P < \alpha$ leads to rejection of H_0 and every $P > \alpha$ to keeping of the H_0, the long-run type I error frequency will be exactly α. In this context publishing of exact P values works rather as a rhetorical device, implying high psychological confidence in rejecting an individual H_0. The common practice in biomedical literature to differentiate between significant and highly significant P values seems to take for granted that a single number can serve two very different functions in the same argument: P value as a long-run relative frequency of type I errors, which assumes that the H_0 is true, and P value as a measure for the strength of "short-run" evidence that the same H_0 is false. It must take a particular type of imagination, sadly lacking in the author, to simultaneously cogitate the truth and falsity of the H_0 with the long-run decision rule and short-run weight of evidence, all emanating from a single P value.

Forgetting for a moment the intricacies of statistical hypothesis testing, ask yourself, how would you personally define the false-positive result rate in your experiments. I'm guessing that your answer would be on the lines of the false-positive rate being the fraction of false-positives from all positives. In the statistical parlance this is called the *false discovery rate* (FDR). The significance level α does not give you the FDR, except when the universal null is true and there exist no real effects (Motulsky, 2010).

Significance level merely constrains type I errors, which assume that H_0 is true. If, for example, 20% of the tested H_0s are in fact not true (there are actual effects) and we do NPHT at $\alpha = 0.05$; $\beta = 0.5$ then, when a H_0 happens to be true, the probability that we incorrectly reject it is 5%. On the other hand, when we look at the frequency of false-positives from the cases where we have actually rejected the H_0, it is not 5% but 29%.

Using $\alpha = 0.05$ does not guarantee that 5% of statistically significant results are false-positive. Table 4.5 enumerates the false-positive rates for different

TABLE 4.5 FDRs at Different Fractions of Real Effects (False H_0s) from All H_0s, $\alpha = 0.05$

Fraction of Real Effects (False H_0s)	$\beta = 0.2$	$\beta = 0.3$	$\beta = 0.4$	$\beta = 0.5$
0.5	0.06	0.07	0.08	0.09
0.4	0.09	0.10	0.11	0.13
0.3	0.13	0.14	0.16	0.19
0.2	0.20	0.22	0.25	0.29
0.1	0.36	0.39	0.43	0.47
0.01	0.86	0.88	0.89	0.91

TABLE 4.6 Type I Errors and the FDR

Reality	H_0 Is True	H_0 Is False
Decision: keep H_0	A	B
Decision: reject H_0	C	D

$\alpha = C/A + C$
$FDR = C/C + D$

fractions of true H_0s. We can see that although the significance level α is set at 0.05, in the high-risk high-impact research situation, where most null hypotheses tested are true, it is easier to obtain FDRs closer to 50% than to 5%.

In the light of this information, would you say that the average FDR in the papers published in *Nature* or *Cell* would be similar to the ones published in ordinary run-of-the-mill journals, like *PLoS One*? Now, did I just change your perception of what it means to be a successful scientist? (Table 4.6).

If a scientist over his lifetime exclusively tests false theories (the universal H_0 is true) and uses NPHT with α set at 0.05, then he should find that 5% of all of his tests will result in the rejection of H_0. However, people, who think that at least some of the effects they will describe over their lifetimes are real, cannot use α as the false-positive rate or FDR (Sorić, 1989).[9] In order to determine what fraction of all rejections of H_0s is erroneous, that is, the FDR, we must first guess how many of the tested H_0s were in actual reality false and how many were true. On the face of it, it may seem that calculations, which require knowledge of the fraction of true rejections of H_0s, cannot be a part of the objective frequentist approach, although they might be suitable for a more subjective Bayesian analysis. We will shortly see, however, that given enough data, the frequentist approach can yield surprising benefits under the multiple hypotheses testing framework in directly estimating the true proportion of all rejections of H_0. The concept of FDR is increasingly being used as an alternative to NPHT when dealing with massively parallel null hypothesis testing.

4.8 MULTIPLE TESTING IN THE CONTEXT OF NPHT

In following the procedure of NPHT at the 0.05 significance level the researcher makes a commitment of erroneously rejecting 5% of all null hypotheses she tests. Thus she can assume that when she tests only one H_0 at a time and the result is statistically significant, then her rejection of this specific H_0 is either correct in the sense that the H_0 is actually false, or an instance of bad luck; but

---

9. In statistical usage only FDR should be used because some statisticians insist that "false-positive rate" indeed means "type I error rate."

in either case a rational choice. (She also tacitly assumes that a good fraction of the H_0s that she tests is false.) If, on the other hand, she tests many different H_0s in parallel and publishes as effects those that were rejected because statistically significant, then the probability of at least one statistically significant effect being a type I error is a lot higher than 5%.

This is the multiple testing problem in NPHT: how to treat results of multiple comparisons? The probability of making one or more type I errors in a set (or family) of tests is called the family-wise type I error rate. The problem of increasing family-wise type I error rate can occur everywhere there are multiple significance tests that are considered simultaneously. If the tests are independent of each other, the family-wise type I error rate can be calculated exactly as $1 - (1 - \alpha)^c$, where c is the number of tests. This means that when the number of parallel tests is 13, there is a 50% chance that at least one of the statistically significant results obtained is a type I error.

As was discussed in Chapter 3, there are two questions pertaining to multiple testing that are as much philosophical as statistical and scientific. The first is: what comprises a family of tests? And the second is: what to do about it? One extreme would be to define the family of tests as every test that has ever been done with a publishable result. This number is both unknowable and very large. The next possibility would be to define the family as the tests that a researcher does over a lifetime, the main corollary of which could be that we should put more trust to the results of people who die young. A more workable definition is that "a family is defined as a collection of simultaneous tests, where a number of hypotheses are tested simultaneously using a single dataset from a single experiment or sampling program" (Motulsky, 2010).

Here is the philosophical question: should we prefer the workable definition and search for lost keys under the lamp? Whatever your answer is, a relatively safe advice is that the decision of how many tests to put into a family should be taken at the planning stage of experiments.

Now to the second question: what to do about the multiple testing problem? There isn't a single answer. The simplest thing to do, when the number of comparisons is very large, is to do nothing. When the number of independent comparisons is known and duly discussed in the paper, we know how many type I errors to expect. If we see a lot more statistically significant results than expected under the universal null hypothesis, then we can be pretty sure that most of these are not type I errors. We will not be able to tell for sure that a given H_0 that was rejected is false, but then, no amount of calculation can change that.

Another piece of advice is only to plan tests that are justifiable by previous knowledge, that is, only to test specific scientific hypotheses. Then the number of tests in each family can be thought of as being one and the problem vanishes. Even if your experimental system measures in parallel the expression levels of 4,000 proteins—if you started the experiment with interest in only 4, the size of the family of tests is 4, not 4,000. But this means that you are committed to a different treatment of any effects seen from the rest of the 3,996 proteins. You may still want to check them out and use the results as an inspiration for future research, but you should not publish the initial results with much confidence in their correctness.

You are also in the clear when the comparisons in the family are complementary. For instance, when in a large drug trial there are positive overall results, it is alright to check patient subgroups for consistency of results. If the subgroup results are consistent, this serves as a good control for the robustness of the effect and the multiple testing problem does not arise. In this case you have a number of statistical hypotheses that pertain to a single scientific hypothesis, which means that in the final analysis a single well-defined hypothesis is tested.

If you do have a multiple testing problem and would like to keep the family-wise type I error rate at the same level as your significance level α, there is a simple procedure available for that. The Bonferroni correction divides your significance level by the number of independent tests (α/c) to arrive at a new significance level. In genome-wide association studies (GWAS) it is not unusual to correct $\alpha = 0.05$ to $\alpha = 10^{-8}$!

The Bonferroni correction is very effective in controlling type I errors but it causes type II error frequencies to get out of hand quickly with increasing number of comparisons. And when the type II error rate is close to 100%, the probability of finding a true effect is close to 0%. (The euphemism here is that Bonferroni correction is very conservative.) In order to get the type II errors back in line, a remedy is to increase the sample size. This is one reason why in genetical studies, like GWAS, sample sizes needed to find effects at acceptable power are often huge (up to hundreds of thousands and even millions).

Another and much cheaper remedy is to use a different correction that compromises on the level of type I errors but is less harsh on type II errors. A popular option is the sequential Bonferroni or Holm-Bonferroni method. Let's suppose that you have five tests in your family of independent tests. You will start by listing the five P values obtained from the smallest to the largest. Then you will construct a different threshold value for each P value by the following rule: for the smallest P value, the threshold is $\alpha/5$, for the next it is $\alpha/4$ and for the largest P value the threshold is $\alpha/1 = \alpha$. Now you start checking from the largest P value to find a P value that is smaller than the threshold. When you have found the first such P value, then all the P values which are smaller than it will be deemed statistically significant. Sequential Bonferroni can save your day if the number of comparisons is relatively low (perhaps in tens) but it still tends to be too destructive of power in modern biological experiments, where the number of parallel comparisons is in the hundreds or more.

Yet another option is to use another type of frequentist test, called ANOVA (one-way ANOVA) that compares all groups in the family at once and gives you just one *P* value. If the *P* value is small, all you can say is that at least one group is different from the others. Should you want to know which one(s), you will be back at doing multiple comparisons. If your goal is to compare every mean with every other mean and you wish to see CIs for the differences between the means, then Tukey test should be preferred. If CIs can be dispensed with, then Holm-Sidak test presents a more powerful alternative. If you wish to compare every mean with a single control mean, then Dunnett's is the post-test to use. ANOVA assumes that samples are normally distributed and that SDs of all compared populations are identical (but this assumption is not very strict when samples are large and equal between groups).

The general idea behind ANOVA is to partition the variability into two components: the variability measured among the group means and the variability within the groups. These are expressed as mean square values, the ratio of which is the *F* ratio. From the *F* ratio a *P* value is obtained. The null hypothesis is that there is no difference in the group means that cannot be explained by intragroup variance, and an *F* ratio close to 1 indicates that the null hypothesis should be retained (Marino, 2014). A small *P* value from an ANOVA does not mean that any of the individual means will necessarily be significant in *post hoc* multiple tests. When the overall ANOVA is significant, but the post-test is not, the interpretation is that there is a statistically significant difference of the means, which however cannot be pinpointed to any specific pair of means. This means that better-designed follow-up studies are indicated.

The currently most popular solution to the multiple comparison problem, when the number of simultaneous comparisons is large, is to go against the logic of null hypothesis testing and ask a different question. You will simply refuse to care about the family-wise type I error rates and everything they stand for, and will instead invest in the FDR (Benjamini and Hochberg, 1995). By doing so you will ask: what fraction of all the discoveries is false?

How can you tell if a H_0 was rejected correctly or not? Unless you are willing to start all over and run another experimental series, there is no principled way to say whether an individual H_0 was rejected because there is a real effect in nature or because of sampling effects, which is chance at its purest. Luckily there is a way of estimating the fraction of all the false rejections of H_0s. This works only if we have a large number of tests of different H_0s, and it works because we can predict the distribution of *P* values under the universal null hypothesis (under which there never are any effects). We also know that if some H_0s are not true, then the distribution of *P* values will be different. Thus, by looking for the discrepancy between the predicted distribution of *P* values under the universal null hypothesis and the actual distribution, we can estimate the fraction of statistically significant results, which correspond to true effects (true $P < \alpha$/all $P < \alpha$). The size of this fraction depends on the significance level, set by the researcher: the stricter the significance criterion, the larger the fraction of true rejections of H_0 in the significance region.

TABLE 4.7 A Comparison of Bonferroni, Sequential Bonferroni, and FDR

Tests Ranked by P Value from Smallest to Largest	Bonferroni	Sequential Bonferroni	FDR
1	α/k[a]	α/k	α/k
2	α/k	$\alpha/(k-1)$	$2\alpha/k$
3	α/k	$\alpha/(k-2)$	$3\alpha/k$
i	α/k	$\alpha/(k-i+1)$	$i\alpha/k$

[a]k is the number of comparisons.

Finding the fraction of false rejections of H_0 from all rejections is what FDR tries to do and the concept of FDR is a great success story of objective methods in scientific inference. Having said that, since the scope of null hypothesis testing is rather limited—it is to fight sampling error, not bias—knowing the fraction of false rejections of H_0 in itself will not tell us much about the fraction of true scientific conclusions.

There are both frequentist and Bayesian interpretations of FDR (Storey, 2003). The original and simplest frequentist calculation of FDR looks somewhat similar to sequential Bonferroni correction, although its interpretation is different (Table 4.7).

The testing in FDR classically proceeds from the largest to the smallest P until a P value is found that falls below the threshold (Benjamini and Hochberg, 1995). When a P value is under the threshold, then this and all the smaller P values are termed "significant." Thus you can estimate the proportion of truly significant comparisons from all significant comparisons.

The accuracy of FDR relies on the P-values being uniformly distributed when the H_0 is true. This brings in a further assumption that calculation of the P values was appropriate to the data, which should always be visually checked by creating a histogram of the calculated P values (Pike, 2010; Krzywinski and Altman, 2014a). The expectation is that the majority of calculated P values correspond to true null hypotheses and therefore the distribution of P values over 0.5, say, should be flat. In contrast, the P values that correspond to non-null cases are expected to be small and thus congregate to the low end, which part of the histogram then displays a logarithmic distribution of P values (Figure 4.8, P value distribution).

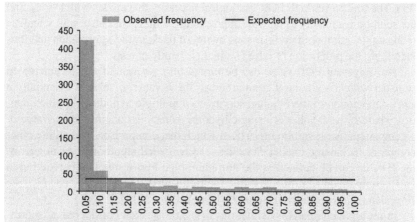

FIGURE 4.8   Frequency distribution of *P* values from a massively parallel proteomics experiment where about half of H_0s are false. The frequencies were presented with a script provided by Krzywinski and Altman (2014a).

The Bayesian interpretation of FDR assumes that there is at least one positive finding, that tests are independent, and follow a mixture distribution (a mixture of true nulls and true effects). It can be defined as

$$\text{FDR} = \frac{\pi_0 \times \alpha}{\pi_0 \times \alpha + \pi_1 \times (1 - \beta)}$$

where π_0 is the prior probability of the H_0, π_1 is the prior probability of the alternative and H_0 and H_{alt} are mutually exclusive and exhaustive, meaning that together they cover the whole hypothesis space ($\pi_0 = 1 - \pi_1$). π_0 corresponds to the fraction of the *P* values below the optimal boundary that splits it into uniform (H_0 true) and skewed components (H_0 false) (Krzywinski and Altman, 2014a). Computationally, π_0 can be estimated as $2/N$ times the number of *P* values greater than 0.5 (Storey, 2003). This Bayesian version of FDR usually gives higher power and better control of the false-positive rate than the original Benjamini-Hochberg method (Krzywinski and Altman, 2014a).

FDR directly connects the frequentist type I and type II errors with Bayesian prior probabilities and it is interpretable in both frameworks (Storey, 2003). FDR increases with increasing type I errors and decreases with increasing power (power equals $1 - \beta$). If only a small proportion, say 10%, of tested H_0s are false, then the π_0 will be low, the FDR will be high and the power of your study can be as low as 10–20% (Krzywinski and Altman, 2014a). In this situation, the higher the number of tests in the family, the lower the power (even if π_0 is held constant at

0.1). The low power/high FDR is a logical necessity that comes with low π_0, and the solution is unlikely to lie with improved algorithms but rather with scientists reducing the number of comparisons by use of background information and thus increasing the proportion of false H_0s in their family of tests.

The Bayesian FDR value can be turned into a q value for each calculated statistic, which is a natural counterpart to the P value. q value is essentially a Bayesian version of the P value, serving as a multiple hypothesis testing quantity, whereas the P value is a single hypothesis testing quantity. The q value of a comparison is the minimum FDR at which that comparison, and all the other comparisons having smaller P values, can be called significant. Analogously, the P value can be defined as the minimum type I error rate that is required to call a comparison significant. While the frequentist P value is $P(\text{data} \mid H_0)$, the Bayesian q value is the reverse: $P(H_0 \mid \text{data})$ (Pike, 2010).[10]

In multiple comparisons situations $q < 0.05$ is often considered a "significant" result, but $q = 0.05$ means something very different than $P = 0.05$. For example, when we have a family of 1,000 tests, then $P = 0.05$ means that we should expect $1000 \times 0.05 = 20$ false effects with this or smaller P value. P values are interpreted over all the tests in the family. In contrast, a q value is interpreted only over the number of tests which have this or smaller q value. If we have, say, 20 q values equal or smaller than 0.05, then it is likely that from those 20 discoveries we have on average $20 \times 0.05 = 1$ false effect.

All this boils down to realization that in situations, where we conduct a hundred plus independent and parallel tests of data from a single experiment, a P value is little more than an intermediate step in the calculation of q values. It is the q value, not the P value, which should be interpreted by the scientist. A convenient q value calculator can be found in Pike (2010).

4.9 P VALUE AS A MEASURE OF EVIDENCE

The aim of the Neyman-Pearson procedure is to keep type I and type II errors at a set level in the long run. This has very little to do with assessing the truth of individual H_0s. A different method of null hypothesis testing was proposed by Roland Fisher in the 1920s. In here the aim is not long-term quality control, but to reach a conclusion about the falsity of a single H_0. The aim is truth, not action. In Neyman-Pearson testing a P value has no meaning apart from being larger or smaller than α. In Fisher's approach the actual value of P is interpreted in an informal way as a strand of evidence among others relevant for the corroboration of a scientific hypothesis, while α and β are not set. Also, there is no H_1. There are good reasons why Fisher's exact P value approach should not be combined with NPHT, which boil down to inflating the weight of the data by using the same P values twice to bolster the same argument (Cohen, 1994).

---

10. For both P and q values "data" really means "data as far or further from the center of H_0 distribution."

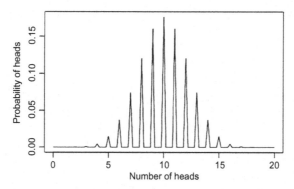

FIGURE 4.9   Binomial distribution of heads in a series of 20 coin tosses when the frequency of heads is 0.5.

To see how Fisher's idea works, imagine testing a theory that the coin that you just picked up from the curb is bent. Your EH is that the frequency of it landing heads is different from it landing tails. You decide to flip the coin 20 times and record the results. The H_{alt} is thus that in the long run the frequency of heads will not be 0.5 and the H_0 is that it will be exactly 0.5. However, because of the sampling error, there will be a predictable frequency or probability of seeing each number of heads from 0 to 20 in the actual sample, given that the H_0 is true.

This will enable us to derive the probability distribution of the H_0 (a binomial distribution in this case), which is the version of H_0 that will actually be tested by the experiment (Figure 4.9). After flipping the coin, the result was six heads and 14 tails and the two-sided P value is 0.115. This means that if the H_0 is true, then the long-run relative frequency for many series of 20 throws to give these results or more extreme is 11.5%. Now the question becomes: what to do with this number?

In Fisher's interpretation a low P value can mean two things: (i) it can mean that you had a run of bad luck with your data (sampling error), which is a fact about blind chance or (ii) it could mean that H_0 is not true, which is a fact about the world. It is up to the scientist, not the statistician, to decide which interpretation to choose. To make a principled choice, it is necessary to incorporate background knowledge into the decision process. In our example, the background knowledge would include information about the curb from which the coin was found. Was it just anywhere in the city or was the coin found in front of a circus where they use bent coins in magic tricks? Depending on the answer it would be rational to argue either that the coin is almost certainly fair or that maybe it isn't. If the experimental result had been different and the P value high, would that suggest that we have good evidence that H_0 is true and the coin fair. The answer is "no." According to Fisher, the null hypothesis can never be proved or established, it can only be disproved in the course of experimentation. Experiments exist to give the facts a chance of disproving the null hypothesis.

The logic of Fisher's approach seems to require a "law of improbability," which can be succinctly formulated as "the smaller the P value (probability under H_0), the stronger the evidence against H_0." However, the law of improbability does not hold. It is not low probability under the H_0 (the P value) that gives us evidence against the H_0. Instead, it is low probability under the H_0 *relative* to a higher probability under a different hypothesis H_1 (see below). When using Fisher's scheme it is entirely possible that an observation, which is very improbable under H_0, is even more improbable under any scientifically meaningful alternative. Because Fisher's approach does not make the H_1 explicit, his inference from P values to strength of evidence is unfounded.

The statistical literature is replete with criticisms of P values as indicators of strength of evidence (Cohen, 1994; Royall, 1997; Goodman, 1999; Dienes, 2011; Gelman, 2013). The first thing to know is that the value of P depends not only on variation in the data and the true effect size, but also on the sample size. The larger the sample size, the smaller the P value. When the sample size is huge, the P value is likely to be small even if the H_0 is true (and H_0s as scientific hypotheses are never exactly true).

To illustrate this, let's test a H_0 that the frequency of heads when tossing a coin is 40% ($H_0 = 0.4$) at significance level 0.05. To reject the H_0 after 10 tosses you need to observe at least 8 out of 10 (80%) heads, at 20 tosses, you need at least 13 heads, after 100 tosses you need at least 49 heads and with 1,000 tosses just 43% of heads (Sober, 2008). A P value by itself tells you nothing about the effect size. This is true for both the Neyman-Pearson and Fisher approaches and has led to humorous statements along the lines that calculating P values "can involve a tautological logic in which tired researchers, having collected data from hundreds of subjects, then conduct a statistical test to evaluate whether there were a lot of subjects" (Thompson, 1992).

If we go into the Neyman-Pearson paradigm and to the above $H_0 = 0.4$ add $H_1 = 0.7$, things get even stranger (Figure 4.10). If $N = 10$ you need 70% of heads to reject H_0 in favor of $H_1 = 0.7$ at $\alpha = 0.05$, which seems very reasonable. But then, the power of this experiment to accept a true H_1 is only about 50%. At $N = 20$ we need 65% of heads to accept H_1 and the power is much better (80%). But when N is further increased to 100 or 1,000 we will encounter a problem. Now we will be compelled to accept the H_1 even when our data clearly support H_0 (the proportion of heads can be as low as 43% to make us reject the H_0 of 40% in favor of H_1 of 70%)! Such rejection is clearly unreasonable as your data are hugely more likely to occur under H_0 than H_1 (see Section 4.11 and Figure 4.12 for a more formal treatment).

In principle a small P value could mean three things (i) H_0 is not true, (ii) sampling error, or (iii) large sample size. So, does the same P value confer the same strength of evidence regardless of the sample size? It can be reasonably argued that a smaller sample size confers more strength of evidence to a P value because its occurrence requires a bigger effect size; but then it is equally reasonable to argue the opposite because a small sample would also increase the chance that the low P value is due to sampling error (Royall, 1997).

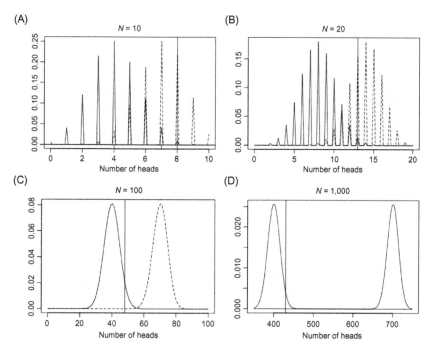

FIGURE 4.10   NPHT of coin tossing experiments at different sample sizes. H_0 is that the frequency of heads is 40% ($H_0 = 0.4$) and H_1 is that the frequency of heads is 70% ($H_1 = 0.7$). (A) When $N = 10$, at least 8 out of 10 tosses have to be heads for the H_0 to be rejected at the usual 0.05 significance level (binomial test). At this sample size, when H_1 is actually true, about half of experimental series would be able to reject the H_0 and accept the H_1. (B) When $N = 20$, at least 13 tosses (65%) have to be heads for the H_0 to be rejected. Now the H_0 and H_1 sampling distributions are better separated and, if H_1 is true, about 80% of experimental series would lead to rejection of H_0 and acceptance of H_1 (power is 0.8). (C) When $N = 100$, we need at least 49 heads to reject H_0 of 40% and accept H_1 of 70% with the power of nearly 100%. (D) When $N = 1,000$, we need only 427 heads (43%) to reject H_0 and accept H_1.

The next line of criticism is that P values sometimes depend on factors extraneous to data or experimental design—like the social hierarchy inside the group of people running the experiment (Sober, 2008). To see how, we will return to the example where a coin was flipped 20 times with the result of 6 heads and 14 tails. The calculated P value was 0.114.

But this is not necessarily so. I failed to mention earlier that there were actually two scientists, a professor and his student, doing the experiment. The $P = 0.114$ was calculated by the student. The professor also did a calculation and got $P = 0.032$, which suggests significantly more evidence against H_0. After checking each other's calculations they agreed that their calculations used exactly the same data and are both statistically correct. The difference in the P values comes from different expectations at the start of experiment. The student, who actually flipped the coin, believed that her task was to do 20 measurements, collect the data and calculate the P value. On the other hand, the professor, who designed the experiment, believed that he had told the student to flip the coin as long as she needs in order to record 6 heads, and then stop—but, as befits a professor, he forgot.

In the students' case the H_0 sampling distribution is a symmetric binary distribution from 0 to 20 and the P value is two-sided. In contrast, the professors' H_0 is non-symmetrical, starting at 6 and going to infinity, and his P value is one-sided. Although the student and the professor had identical substantive hypothesis, RH, and EH, all centering on the fairness of the coin, they asked a different statistical question under the null. The student's question was: what is the probability of 0–6 or 14–20 heads, given that there are 20 tosses of the coin? The professor's question was: what is the probability of 20 or more tosses, given that there are exactly 6 heads?

There is no scientific or statistical reason to prefer one to the other. So it appears that—assuming that the point of a P value is to provide an objective assessment of the strength of evidence—the value of the whole procedure depends on which of the two scientists would win in a shouting match! Assuming that the professor wins, does that mean that what could have happened, but didn't (the professor successfully conveying his message to the student), changes the objective nature of evidence? What makes this especially galling for the supposedly objective frequentist school is that such problems do not arise in the avowedly subjective Bayesian statistics (Dienes, 2011).

Thus it can be argued that P values are not a good measure of strength of evidence. This is because they confuse strength of evidence with the probability of obtaining that evidence (Royall, 1997). Probability is determined by factors independent of strength of evidence, like stopping rules and other sources of multiplicities. In other words, P values are not only determined by what the experimenter did and what he observed, but also by what he would have done, had the observations been different.

A pungent criticism against the rather modest view that a small P value must provide at least *some* evidence against H_0 comes from Cohen (1994). This is how small P values are often implicitly interpreted in the scientific literature:

If H_0 is true, then this result would probably not occur
this result has occurred
therefore, H_0 is probably not true

What follows is a logically equivalent structure:

If A is an American, then A is probably not a member of the US congress
A is a member of congress
therefore, A is probably not an American.

People, who implicitly depend on the former argument to bolster their scientific theories, should be willing to subscribe to the latter.

Our last criticism of P values is pedagogical rather than statistical. It is that scientists typically want (i) to do simple and automatic calculations and (ii) to know about the truth of their theories. Thus, many would like to calculate the frequentist P value—$P(\text{Data} \mid H_0)$—and interpret it as a Bayesian probability $P(H_0 \mid \text{Data})$, or even $P(\text{EH I data})$. Or, in other words, although the statistical method gives the scientist an estimate of the long-run relative frequency of observing certain kinds of data under the null, a popular "reinterpretation" is that P value gives the probability of the truth of a single H_0. Such misinterpretations of the meaning of the P value can even be found in statistical textbooks targeting science students.

To grasp the folly in this, ask yourself: what is the probability that you are dead, given that your head was bitten clean off by a shark—$P(\text{dead} \mid \text{shark})$? Now ask a subtly different question: what is the probability that a shark attack will ultimately kill you—$P(\text{shark} \mid \text{dead})$? The answer to the first question is very close to one and to the second question it is presumably very close to zero. The reason is that the prior probability of the hypothesis (shark attack) is very low.

But then, for someone who swims to school every day through shark-infested waters, the second probability will be much closer to the first. The first question can easily be reframed in the frequentist setting: what is the long-run relative frequency of people being dead after they were attacked by a shark? The second question is Bayesian: what is your personal probability of ever being killed by a shark? To this question there is no frequentist probability answer.

4.10   THE "ERROR BARS"

In the previous chapter it was suggested that a major problem with P values is that they give no information about effect sizes. This is not strictly true. There is an application of the P value, whose aim is to do just that: to estimate the level of uncertainty that is conferred by the sampling error to our estimate of the true value of the population statistic (for instance the population mean). This estimate of sampling error is called the CI.

Many scientists describe the uncertainty surrounding their results with "error bars," the semi-mystical vertical bars on graphs, whose presence is widely thought to suggest that the authors have shown due diligence in repeating the experiment and found consistent results. The error bars have even been described as a kind of demarcation criterion between science and the rest:

Every time a scientific paper presents a bit of data, it's accompanied by an error bar – a quiet but insistent reminder that no knowledge is complete or perfect. It's a calibration of how much we trust what we think we know. If the error bars are small, the accuracy of our empirical knowledge is high […].

Carl Sagan (The Demon-Haunted World)

Here the error bars are seen as a public pledge of openness to criticism and incorporation of further results, which is supposed to set scientific theorizing apart. The scientist who uses them admits that he could be wrong and, furthermore, shows the range of uncertainty in his results, which are consequently open to further clarification. As the number of experiments increases, the error bars grow shorter and the results thus converge on the truth.

Now, isn't it almost magical how a simple estimate of sampling error can almost single-handedly drive science into convergence with truth?

In fact, there is more than one type of "error bar" and also other types of bars on graphs, that are sometimes erroneously called that. The error bar, like the average, is an everyday non-scientific concept—there are specific statistical measures of uncertainty that fall under the umbrella term "error bar." The first, which is sometimes mistaken for a measure of error, is the SD. SD does not estimate error, it shows variation in the data. However, the SD serves as a starting point for calculating our first real error bar—the SEM. SEM equals SD divided by the square root of the sample size (n). This means that as the n grows, the SEM shrinks.

SEM is currently by far the most popular real "error bar" in the molecular biology literature and this is a great pity. Firstly, SEM obviously depends on the accuracy of determining the SD and, as we saw in Section 4.1, at small sample sizes SD calculation systematically underestimates true variation in the data. This means that when n is small, the SEM "error bars" will systematically underestimate true uncertainty about how well the sample mean corresponds to the population mean. The real problem with SEM, however, is that it has no non-mathematical interpretation. The closest thing to an interpretation on offer is that one standard error corresponds to a 58% CI if $n = 3$, and to a 65% CI if $n = 9$. At this point I would like to suggest that if the reader (i) finds this definition incomprehensible or useless and (ii) cannot produce a better alternative, then he or she should never publish a SEM again. I promise that I won't.

The true function of the SEM is to serve as a way station on the road that leads to CIs. In calculations starting with SEM the mathematical inefficiency of SDs at low sample sizes is normalized by the use of Student's t distributions (every n has its own t distribution and a small n leads to a widening of the tails of the distribution and thus to a larger fraction of probabilities residing under the tails). Thanks to this normalization, the CIs have a non-mathematical interpretation. But before we come to that, we will explain the equivalence of the CI and the P value.

Basically, CIs are a range of statistically non-significant P values around the sample mean (around the sample statistic, to be precise). Every value around the sample mean that gives a P value larger than the significance level α, resides in the CI. If α is set at 0.05, then the resultant CIs are called 95% CIs. If α is set at 0.01, then we calculate 99% CIs.

95% CI is the range of values around the sample mean, which have P values greater than 0.05.

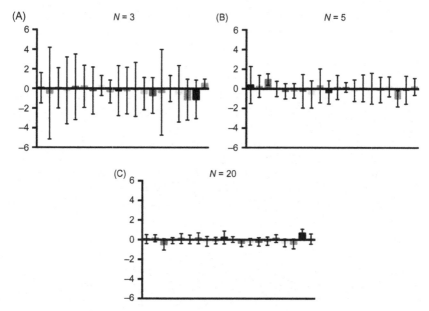

FIGURE 4.11   CIs calculated for a set of 20 simulated experiments. Data come from a standard normal distribution (mean = 0, SD = 1). (A) $N = 3$ (B) $N = 5$ (C) $N = 20$.

If a 95% CI does not contain the point value of the H_0, then the result is statistically significant at the $\alpha = 0.05$ level. If a 95% CI does contain the H_0 value, the result is not significant. It is a mathematical necessity that P values and CIs always agree.

Now we can give a definition of CI:

calculating 95% CIs is a procedure that, if repeated over many samples, captures the population value 95% of the time (Figure 4.11).[11]

Thus, the 95% probability in 95% CIs is a property of the rule that was used to create the interval, not of the interval itself. Of course, a working scientist will usually draw only a single sample from his population of interest and will calculate a single CI. Unfortunately, a single CI has no interpretation. A single 95% CI will not contain 95% of the sample means from further experiments (Krzywinski and Altman, 2013a). Indeed, there is no frequentist answer to the question: What is the probability that the population mean lies inside a particular CI? This answer lies in the Bayesian credibility intervals, which require an additional ingredient, prior probabilities, for their calculation.

---

11. There are several methods for estimating CIs from a sample. For normal data the most efficient CIs are modelled using t distributions, while for non-normal data combined with small N it can be more effective to calculate adjusted percentile CIs using a nonparametric method called bootstrapping (http://www.r-bloggers.com/im-all-about-that-bootstrap-bout-that-bootstrap/).

Because a frequentist CI is in effect a range of P values, a lot of the criticism of P values is also relevant to CIs. For example, because large samples inevitably lead to small P values, they will also lead to narrow CIs. And yet, because increasing your sample size is no protection against bias, the proportion of bias from the total level of uncertainty (systematic plus random error) will increase. And like it or not, there is some bias in every measurement. Therefore, very narrow CIs, which measure the random error component only, are unlikely to contain the true parameter value (Stang and Rothman, 2011).

So, once again, the choice is between the easy frequentist calculation with its somewhat disappointing interpretation, which is all about sampling error, and the much harder Bayesian calculation and its interpretation, which is about our uncertainty about the completeness of our knowledge. It is the Bayesian credibility intervals, not frequentist CIs, which have the potential to satisfy the high expectations that Carl Sagan put on error bars.

The level of agreement between CIs and credibility intervals depends on prior probabilities. If prior knowledge puts the true population statistic close to the calculated sample statistic, then CIs are simply wider than the credibility intervals and giving them the Bayesian interpretation is no big sin. On the other hand, if the prior knowledge suggests a very different result from the one obtained, then the CIs do not even come close to credibility intervals and any Bayesian interpretation of them is likely to be very misleading. So once again, whatever it is that you choose to do with frequentist CIs, please don't put any confidence in them. Confidence is subjective and should be left to those who bet on horses, and to other Bayesians.

There are four recommended uses for the CIs, all assuming Bayesian short run interpretation of the CI. Firstly, CIs tell us about the sensitivity of the experiment. If both the mean of H_0 and some interesting values of H_1 are included in the interval, the experiment was not sensitive enough.

Secondly, CI tells about effect sizes. When you are comparing measurements of the experimental condition with the control condition, then for assessing effect sizes, it is important to calculate the CI of the difference of the two conditions, as opposed to two CIs for both the experimental and control values (Table 4.8). The second method is much less sensitive and may lead to missing important real effects.

Thirdly, a pre-defined CI value can be used as a stopping rule. In lieu of doing a formal power analysis, the researcher simply decides at a minimally interesting effect size and defines a width for CIs that would leave this effect outside the CI of the mean for H_0 (Dienes, 2008). After this, all that is needed is to run the experiment until the pre-set width of the interval is achieved, and then to stop. It is alright to peek at data and to calculate the CIs after each experiment, as long as the stopping rule was pre-planned.

And fourthly, unlike the ordinary P values, CIs allow to test for equivalence between two samples (Motulsky, 2010). To test for equivalence, the scientist must first define a "zone of indifference," that is she must define the biggest

TABLE 4.8 What to Decide When Comparing Two Means (μ_{exp} and μ_{contr}) by Traditional Error Bars

Error Bar	If Two Error Bars Overlap	If Two Error Bars Do Not Overlap
SD	ND	ND
SEM	$P > 0.05$	ND
95% CIs for μ_1 and μ_2	ND	$P \ll 0.05$

ND—not decidable (no conclusion is possible).

TABLE 4.9 Percentage of Eligible Papers That Report a Given Statistic

	Science	Nature	NEJM	Lancet
CI	29	11	84	93
Effect Size with CI	0	3	83	86
Value of prospective power	0	3	61	66

scientifically irrelevant effect size. Then she can collect data and calculate 90% CIs. If the entire range of the 90% CI lies within the zone of indifference, then she can conclude with 95% confidence that the two treatments are equivalent.

A common misuse of CIs is in intragroup comparisons, for instance when comparing different points in the same time series. The problem is that the basic assumption of independence of measurements is violated when a researcher measures the same individuals over and over again. In a situation like this, regression analysis can be used to decide if a trend is really going down or if we are seeing a random effect. It is possible to use CIs to compare different experimental treatments in the same time point, not different time points of the same treatment.

How are CIs used in real papers? Tressoldi et al. looked at all empirical studies with human participants published in 2011 in four high-impact journals, related to behavioral, neuropsychological and medical investigations (Hills, 2013). Their findings are summarized in Table 4.9.

As can be seen, specialized medical journals (*New England Journal of Medicine* and *Lancet*) induce very different behaviors in their authors than the general interest journals *Nature* and *Science*; or in another words: in the real world, when given half a chance, people prefer not to append a meaningful measure of uncertainty to their effect estimates.

4.11 LIKELIHOOD AS AN UNBIASED MEASURE OF EVIDENCE

The frequentist methods of statistical analysis were created as objective procedures for testing the data against a fixed null hypothesis—that is, the probability of observing data as extreme or more extreme under H_0. This fixes the hypothesis and leaves the data to vary. It also uses data which were never observed (the "data more extreme") in making the statistical inference. The goal behind the paradigmatic NPHT was to provide a well-characterized decision-making tool for choosing between hypotheses, which would enable to keep the long-run error frequencies under control. However, most experimental scientists are interested in how well their particular data (and never mind the potential data that were not observed!) support their scientifically relevant hypothesis (which is rarely H_0), rather than in some long-term measure of success over many different assays and hypotheses. So it is natural to demand (i) evidentiary interpretations of data in terms of a specific hypothesis and (ii) the probabilities of hypotheses being true under specific data; neither of which comes naturally from frequentist methods. This has led to stretching of methods that are well suited for quality control and long-run decision making, into domains of scientific inference where their accuracy cannot be vouched for and where better methods are available.

Furthermore, the original reason for developing frequentist concepts in statistics, objectivity of frequentist probability statements, has turned out to be unattainable, except perhaps in quality control settings. As we saw above, any interpretation of evidence that uses P values depends not only on the actual data, but also on the data that could have been obtained, but wasn't and, crucially, on the subjective intentions of the researchers planning the study, collecting and analyzing the data. These are incorporated into the null distribution and since the P value incorporates not only the observed data but all data more extreme under the H_0, subjective aspects such as who is the professor and who is the student can make their way into the P value. P value incorporates the sort information about the research situation (what could have happened, but didn't) that the researchers can only guess about. There is no meaningful way to call it "objective."

To overcome the subjectivity lurking in the tails of the null hypothesis we could interpret the probability of observing data under the hypothesis literally as $P(D \mid H)$, the probability of exactly the data, which were observed, under the specific hypothesis. Here the data are fixed and the hypothesis is left to vary. $P(D \mid H)$ was given a technical name "likelihood" by R. A. Fisher (1921). The maximum likelihood is the parameter value, which is best supported by data at hand—the data provide maximum evidence for this value being the true population value. Likelihoods differ from normal probabilities in that the likelihoods of data under two mutually exclusive and exhaustive hypotheses need not sum to one. It is quite possible that some data (observing some white shoes, say) are equally likely under competing hypotheses H_1: "the moon is made of cheese" and H_2: "the moon is made of something else"—and they can even sum as 2.

When we compare the probability of observing our data under two separate hypotheses, H_1 and H_2, we obtain the likelihood ratio (LR)

$$LR = \frac{P(D \mid H_1)}{P(D \mid H_2)}$$

This ratio constitutes the evidence that the data provide to H_1 over H_2: the larger the LR, the stronger the evidence for H_1 relative to H_2 (Royall, 1997). LR = 1 means that evidence is not relevant and cannot help us to choose between the two hypotheses. Therefore, under the likelihood paradigm we cannot talk of evidence in the context of a single hypothesis. There are two important corollaries to LR, (i) evidence is always relative to the particular hypotheses compared and (ii) evidence supports H_1 over H_2 only when $P(D \mid H_1) > P(D \mid H_2)$. The latter point is called "the law of likelihood." The ontological status of this law differs for different schools of statistics. For a Bayesian it is an elementary and mathematically trivial consequence of basic probability theory, while for the frequentists who see probability as a physical phenomenon, this "law" is more like a proposal. The frequentist school does not adhere to the law of likelihood, but likelihoodists and Bayesians do.

The law of likelihood allows separating the strength of evidence, which resides solely in the LR, from the probability of obtaining the evidence. Furthermore, the law of likelihood implies that only the LR is relevant in interpreting observations in terms of specified competing hypotheses. This implication has been named "the likelihood principle." Under the likelihood principle the unobserved values and the sample space in general plays no role in determining the evidence, which the data provide. The same goes for peeking at data and stopping rules: under the likelihood principle such activities have no influence on the evidentiary weight of the data. Many of the annoying commandments of the frequentist (thou shalt not peek at data, thou must have a stopping rule, etc.) are irrelevant for evidentiary interpretation of data.

The P value and frequentist statistics in general are in conflict with both the law of likelihood and the likelihood principle. For example, when $\alpha = 0.05$ and $\beta = 0.2$ and a particular P value fell just above α, then according to NPHT we must keep the H_0, although the evidence in the LR actually supports the alternative hypothesis by about 2.7-fold (if $\beta = 0.5$, then this ratio rises to 3.9) (Dienes, 2008). On the other hand, when the H_{alt} resides very far from the H_0 ($\beta \ll 0.05$) and we reject H_0 with P value just below α, the evidence will support H_0 over H_{alt} (Figure 4.12). Only when $\alpha = \beta$, is the direction to which the evidence points consistently the same as given by the Neyman-Pearson action rule.

The likelihood paradigm has another distinct advantage over NPHT. While the Neyman-Pearson method keeps the type I and type II error frequencies at a constant level, regardless of the sample size, LRs behave quite differently. According to the law of likelihood, the larger the LR, the greater the strength of

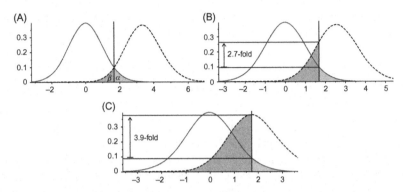

FIGURE 4.12 Ratio of specificity $(1 - \beta)$ to sensitivity $(1 - \alpha)$ of the test determines the maximum discrepancy between the P value and weight of evidence (LR). (A) When specificity and sensitivity are equal, there can be no qualitative discrepancy between the P value and LR. (B) When specificity is 80% and sensitivity is 95%, the evidence in LR of 2.7 in favor of H_1 over H_0 could be misrepresented by P value as favoring H_0. (C) When specificity is 50% and sensitivity is 95%, the evidence in LR of 3.9 in favor of H_1 over H_0 could be misrepresented by P value as favoring H_0.

evidence for H_1 over H_2 (if LR < 1, then the evidence favors H_2 over H_1). But it is also true that the larger the LR, the smaller the probability of encountering misleading evidence (evidence that points in the wrong direction). Specifically, the upper limit on the probability of encountering misleading evidence is 1/LR. Conventionally LR = 8 is considered reasonably strong evidence. At this level of evidence the probability of misleading evidence cannot be greater than 12.5% (and it is below 2.1%, when assuming normal distribution) (Dienes, 2008). At LR = 32, which is often considered very strong evidence, the probability of misleading evidence is always below 3.1% (and <0.4%, assuming normal distribution). As long as we use normally distributed data with known SD and only look at the LRs above 8, then with 98% probability we can collect as large a sample as we wish, without seeing any misleading evidence.

We can (and should) worry about the frequency of weak and misleading evidence at the experimental planning stage, by thinking of two alternative hypotheses and of the sample sizes that would enable collecting strong evidence in favor of one or the other. However, once data collection starts, these worries disappear. We may stop the experiment as soon as we have collected strong evidence for any hypothesis we care about over the null. We can discard the original pair of hypotheses without remorse and look at the evidence *vis-à-vis* any pair of parameter values we wish. This is made possible by the fact that we can calculate a relative likelihood for each parameter value (φ), thus generating a continuous likelihood function. For a normally distributed variable, assuming that the parameter of interest is the mean, the likelihood function is a normal distribution whose mean is equal to the mean of the sample data and whose SD is equal to the SEM of the sample data.

Since nothing except the LR can tell us about evidence, from the likelihood function we can calculate the LR for any pair of values in the parameter space

(Φ). Moreover, the 1/8 and 1/32 likelihood intervals (LIs) show us the range of values, which are consistent with the evidence provided by the data. 1/8 LI denotes the range of parameter values, which are consistent with the evidence in the sense that $1/8 < \text{LR} < 8$. When we require a high level of evidence for deciding that a given value is not consistent with data ($\text{LR} = 32$), we will need wider LIs. For a normal distribution the width of the 1/8 LI corresponds to a 96% CI and 1/32 CI corresponds to 99% CI, but their interpretations are very different.

To illustrate the use of LRs, likelihood functions and LIs, we will continue with the coin-tossing example depicted in Figure 4.9. Still the H_0 states that the frequency of heads is 0.4 and the H_1 that it is 0.7. By using the calculator at http://www. lifesci.sussex.ac.uk/home/Zoltan_Dienes/inference/Likelihood.htm, we draw the likelihood functions for four experimental series ($N = 10$, $N = 20$, $N = 100$, and $N = 1,000$) for the minimum number of heads that by the NPHT procedure would result in the rejection of H_0 and the acceptance of H_1 at this particular sample size (Figure 4.13). For each series we will add the LR of encountering these data under H_1 *vis-à-vis* H_0 and the 1/32 LI for the heads to tails ratio. The results show that as the sample size increases the NHPT procedure and evidence (LR) increasingly point to the opposite directions. As the sample size grows, the likelihood functions (and LIs) will get narrower and we can more confidently say which hypotheses are relatively more consistent with our data.

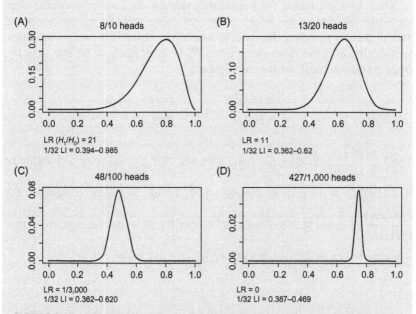

FIGURE 4.13 Likelihood function for binomial data. H_0—the frequency of heads is 0.4. H_1—the frequency of heads is 0.7. LR—likelihood ratio H_1/H_0. LI—likelihood interval. (A) 8 heads from 10 coin tosses; (B) 13 heads from 20 tosses; (C) 48/100 tosses; (D) 427/1,000 tosses.

Unlike the CI, the LI has nothing to do with probabilities of any kind; it only deals with relative evidence and is separated from the probability of finding such evidence. It must be kept in mind that the relativity of likelihoods means that a LI excluding a particular value cannot be used to reject that value (this is how CIs are routinely used). Also, LRs are not adjusted to multiplicities. This has to be done separately and by using other methods. Likelihoods only tell about the strength of evidence; in themselves they say nothing about how to interpret this evidence.

Although a large LR means strong evidence of H_1 over H_2, it says nothing about the truth of either hypothesis. If the truth of H_1 is *ipso facto* much less probable than that of H_2, then, after conducting some experiments, a respectable LR in favor of H_1 can still signify that the truth of H_2 is more probable than that of H_1.

Another, and arguably most serious, limitation of LRs is that they can ordinarily be computed only for mathematically simple statistical hypotheses, wherein the probability distributions for observable random variables are completely specified. Composite hypotheses, whose probability distributions are not fully specified (for example, a normal distribution with unspecified SD), tend not to be conducive to LR analysis. Unfortunately, many scientifically relevant statistical hypotheses are composite. When, for instance, H_1 says that a parameter value >0 and H_2 postulates that it is <0, then we are stuck with two composite hypotheses.[12]

Thus, LRs are meant for presenting evidence *vis-à-vis* two specific simple statistical hypotheses. When researchers want to make claims about the truth of their hypotheses, then LRs should be converted into true probabilities. Specifically, to decide how many times H_1 is more likely to be true than H_2, given the data at hand, we have the formula:

$$\frac{P(H_1|D)}{P(H_2|D)} = \text{LR} \times \frac{P(H_1)}{P(H_2)}$$

where $\dfrac{P(H_1|D)}{P(H_2|D)}$ is called the posterior odds, $P(H_1)$ is the prior probability of H_1 and $P(H_2)$ is the prior probability of H_2. In this, Bayesian, context the LR $\dfrac{P(D|H_1)}{P(D|H_2)}$ is called the "Bayes factor" (LR $\equiv$ B). This introduces a whole new Bayesian bag of tricks, which will be discussed in Chapter 5.

---

12. Composite hypotheses are not a practical or conceptual problem for Bayesian statistics, and traditional frequentist statistics has developed various methods for dealing with them.

4.12   CONCLUSION: IDEOLOGIES BEHIND SOME METHODS OF STATISTICAL INFERENCE

Interpretation of Probability	Statistical Method	Question	Goal	Comments
Frequentist	NPHT	Should I reject H_0?	Control of long-run type frequency of erroneously rejecting H_0	Uses H_0 and H_{alt} in planning, but only H_0 in data analysis
	Fisher's P value testing	What is the strength of evidence against H_0?	Rational opinion on the falsity of H_0 $p = (D \mid H_0)$	Deals with H_0 only
	CIs	What is the uncertainty associated with statistical inference?	Rational opinion on the quality of the test	95% CI predicts that 95% of the future sample means will fall inside the interval
None	LRs	How much the evidence supports H_1 over H_0?	Hypothesis selection by relative strength of evidence	Pairwise comparison of hypotheses
	LIs	Do the data present evidence against a parameter value?	A range of parameter values consistent with the evidence	Uses H_1 and H_2 in planning, but only evidence present in the likelihood function in data analysis
Bayesian	Bayes factor	What are the odds that H_1, rather than H_2, is true?	Pairwise comparison of hypotheses	
	Posterior probability	What is my degree of belief in the hypothesis?	Rational opinion on probability of truth of hypothesis	Can deal with (infinitely) many hypotheses
	Credibility intervals	With what probability does the population value reside inside the interval?	Rational opinion on the true parameter value	95% CI means that the population mean resides inside the interval with 95% probability
	FDR	What fraction of discoveries is false?	Control of multiple testing $q = P(H_0 \mid D)$	Only deals with H_0

REFERENCES

Alexandrov, T., et al., 2012. A review of some modern approaches to the problem of trend extraction. Econom. Rev. 31 (6), 593–624.

Batanero, C., 2000. Controversies around the role of statistical tests in experimental research. Math. Think. Learn. 2 (1&2), 75–97.

Behrens, J.T., 1997. Principles and procedures of exploratory data analysis. Psychol. Methods 2 (2), 131–160.

Benjamini, Y., Hochberg, Y., 1995. Controlling the false discovery rate: a practical and powerful approach to multiple testing. J. R. Stat. Soc. Ser. B. 57 (1), 289–300.

Berry, D.A., 2007. The difficult and ubiquitous problems of multiplicities. Pharm. Stat. 6 (3), 155–160.

Chow, S.L., 1998. Précis of statistical significance: rationale, validity, and utility. Behav. Brain Sci. 21 (2), 169–194. Discussion 194–239.

Cohen, J., 1994. The earth is round. Am. Psychol. 49 (12), 997–1003.

de Finetti, B., 1990. Theory of Probability: A Critical Introductory Treatment. Interscience Publishers, Chichester; New York, NY.

Diaconis, P., Holmes, S., Montgomery, R., 2007. Dynamical bias in the coin toss. SIAM Rev. 49 (2), 211–235.

Dienes, Z., 2008. Understanding Psychology as a Science. Palgrave MacMillan, London.

Dienes, Z., 2011. Bayesian versus orthodox statistics: which side are you on? Perspect. Psychol. Sci. 6 (3), 274–290.

Ellenberg, J., 2014. How Not to Be Wrong: The Power of Mathematical Thinking. Penguin Press, New York.

Filliben, J.J., 2002. NIST/SEMTECH Engineering Statistics Handbook. Gaithersburg: <www.itl. nist.gov/div898/handbook>.

Fischer, K., et al., 2014. Biomarker profiling by nuclear magnetic resonance spectroscopy for the prediction of all-cause mortality: an observational study of 17,345 persons. PLoS Med. 11 (2), e1001606.

Fuchs, C.A., Schack, R., 2013. Quantum-Bayesian coherence. Rev. Mod. Phys. 85 (4), 1693–1715.

Gelman, A., 2013. P values and statistical practice. Epidemiology 24 (1), 69–72.

Gillies, D., 2000. Varieties of propensity. Br. J. Philos. Sci. 51, 807–835.

Goodman, S.N., 1999. Toward evidence-based medical statistics. 1: The P value fallacy. Ann. Intern. Med. 130 (12), 995–1004.

Hacking, I., 1990. The Taming of Chance. Cambridge University Press, Cambridge, UK.

Hacking, I., 2001. An Introduction to Probability and Inductive Logic. Cambridge University Press, Cambridge, UK.

Herbranson, W.T., Schroeder, J., 2010. Are birds smarter than mathematicians? Pigeons (*Columba livia*) perform optimally on a version of the Monty Hall dilemma. J. Comp. Psychol. 124 (1), 1–13.

Herbranson, W.T., Wang, S., 2014. Testing the limits of optimality: the effect of base rates in the Monty Hall dilemma. Learn. Behav. 42 (1), 69–82. Available at: <http://link.springer. com/10.3758/s13420-013-0126-6>.

HillsR.K. (Ed.), 2013. High impact = high statistical standards? Not necessarily so. PLoS One 8 (2), e56180.

Howson, C., Urbach, P., 1989. Scientific Reasoning: The Bayesian Approach. Open Court, Chicago, IL.

Ioannidis, J.P.A., 2013. Research accomplishments that are too good to be true. Intens. Care Med. 40 (1), 99–101.

Jaynes, E.T., 2003. Probability Theory: The Logic of Science. Cambridge University Press, Cambridge, UK.

Keynes, J.M., 2010. A Treatise on Probability, Project Gutenberg. Release Date: February 9, 2014 [EBook #32625] <ftp://ftp.informatik.rwth-aachen.de/pub/mirror/ibiblio/gutenberg/3/2/6/2/32625/32625-pdf.pdf>.

Krzywinski, M., Altman, N., 2013a. Points of significance: error bars. Nat. Methods 10 (10), 921–922.

Krzywinski, M., Altman, N., 2013b. Points of significance: importance of being uncertain. Nat. Methods 10, 809–810.

Krzywinski, M., Altman, N., 2014a. Comparing samples—part II. Nat. Drug Discov. 11 (4), 355–356.

Krzywinski, M., Altman, N., 2014b. Visualizing samples with box plots. Nat. Drug Discov. 11 (2), 119–120.

Levins, R., 1966. The strategy of model building in population biology. Am. Sci. 54 (4), 421–431.

Limpert, E., Stahel, W.A., 2011. Problems with using the normal distribution—and ways to improve quality and efficiency of data analysis. PLoS One 6 (7), e21403.

Marino, M.J., 2014. The use and misuse of statistical methodologies in pharmacology research. Biochem. Pharmacol. 87 (1), 78–92.

Mazur, J.E., Kahlbaugh, P.E., 2012. Choice behavior of pigeons (*Columba livia*), college students, and preschool children (*Homo sapiens*) in the Monty Hall dilemma. J. Comp. Psychol. 126 (4), 407–420.

Motulsky, H., 2010. Intuitive Biostatistics. Oxford University Press, New York.

Pike, N., 2010. Using false discovery rates for multiple comparisons in ecology and evolution. Methods Ecol. Evol. 2 (3), 278–282.

Popper, K., 1959. The propensity interpretation of probability. Br. J. Philos. Sci. 10 (37), 25–42.

Quinn, G.P., Keough, M.J., 2002. Experimental Design and Data Analysis for Biologists. Cambridge University Press, Cambridge, UK.

Ramsey, F., 2011. Truth and probability. In: Braithwaite, R.B., (Ed.), The Foundations of Mathematics and Other Logical Essays. Kegan, Paul, Trench, Trubner & Co., London, Harcourt, Brace and Company, New York, Ch. VII, p. 156–198.

Royall, R., 1997. Statistical Evidence. CRC Press, Boca Raton, FL.

Sober, E., 2008. Evidence and Evolution: The Logic Behind the Science. Cambridge University Press, Cambridge, UK.

Sober, E., 2010. Evolutionary theory and the reality of macro-probabilities. The place of probability in science. LSE, Centre for Philosophy of Natural and Social Science Causality: Metaphysics and Methods Technical Report 21/04.

Sorić, B., 1989. Statistical "discoveries" and effect-size estimation. J. Am. Stat. Assoc. 84 (406), 608–610.

Stang, A., Rothman, K.J., 2011. That confounded P-value revisited. J. Clin. Epidemiol. 64, 1047–1048.

Storey, J., 2003. The positive false discovery rate: a Bayesian interpretation and the q-value. Ann. Stat. 31 (6), 2013–2035.

Streit, M., Gehlenborg, N., 2014. Bar charts and box plots. Nat. Drug Discov. 11 (2), 117.

Thompson, B., 1992. Two and one-half decades of leadership in measurement and evaluation. J. Couns. Dev. 70 (3), 434–438.

Chapter 5

Truth and Belief

In this chapter we will look beyond experimental evidence towards the actual product of science. For most working scientists the goal of their toils is not benefits to the global economy, neither is it some community added value, or abstract increase in knowledge or happiness. These are all incidental benefits, which may accrue from their work, as in a market economy common good is supposed to result from selfish behaviors of many individual actors.

For a scientist, the outcome of a scientific project is often a scientific paper. Publishing papers in the largest possible quantity in the best possible journals will open the coffers of grant agencies, and this in turn opens the doors for tenured academic positions. This is how we are taught and this is the way we operate, should we wish for a scientific career.

The great importance, which publishing has in the lives of individual scientists, leads to the realization that the form and requirements of the scientific paper are instrumental in dictating how science is done and interpreted in the trenches (i.e., at the benches). The journals that matter in experimental biology are not content with merely what the evidence says. Each experimental paper usually has a single main result, which can be found clearly and succinctly articulated in the abstract (always), in the title of the paper (usually), in the beginning of the discussion (often) and at the end of the introduction (sometimes); and as far as the reality of the main result is concerned, the editors are not interested in what the data suggest or indicate: the goal is to publish what is *shown* by the researchers. The main result should be rock-solid and ideally remain in the literature, unchanging, forever.

For the editor the main result of your paper cannot be that you have managed to exclude sampling error as the source of some experimental effect. It has to be a biological fact, which is established (or "shown") by the experimental results. Thus, biological papers tend to be both novel and self-sustaining: they are traditionally expected to stand on their own feet without a need for future replications of experiments or corroboration of the conclusions by other means. This expectation is sustained by the example of physics (or physics-envy?), where, to cite a notable practitioner, "Scientists [as opposed to politicians] can reach agreement quickly because we trust our experimental colleagues to have high standards of intellectual honesty and sharp perception to detect possible sources of error. And this belief is justified because, after all, hundreds of new

Interpreting Biomedical Science. DOI: http://dx.doi.org/10.1016/B978-0-12-418689-7.00005-3
223

experiments are reported every month, but only about once in a decade is an experiment reported that turns out later to be wrong" (Jaynes, 2003, p. 128).

My quick search of 15 research papers randomly chosen from the best journals (*Cell, Molecular Cell,* and *EMBO Journal*) revealed a total of 193 uses of the word "show,"[1] which exceeded the uses of "suggest" and "indicate" combined. The authors are expected to reveal to the reader a slice of the true nature of the world; and as the common formulation is "our results show..." they must present their results to the world as strong enough to stand alone and *show*, rather than *suggest* or *indicate*.

There are two issues with using experimental data as proof. Firstly, as the data are usually presented as averages of repeated experiments with the assorted error bars, our road to certainty seems to lead through a probabilistic jungle of statistical inference. This looks like a hazardous journey.

But then, if the *P* value is small enough or the error bars narrow enough, we might want to claim "practical certainty" for our conclusions ("we show that..."), as a good enough approximation to certainty in most practical situations. This raises the second issue: how to move from the statistical methods dealing exclusively with the null hypothesis, according to which observed data can be explained by sampling error, to some specific biological hypothesis? We shall start by presenting a simple problem.

5.1   FROM LONG-RUN ERROR PROBABILITIES TO DEGREES OF BELIEF

Suppose that a middle-aged woman is invited to participate in screening for breast cancer. The test is 90% sensitive (90 people out of 100, who really have breast cancer, will receive a positive test result). Also, the test is 90% specific (90 people out of 100, who do not have breast cancer will get a negative test result). Now, assuming that she gets a positive test result, how worried should she be?

Since the test is pretty good, the positive result constitutes strong evidence for cancer. There is no doubt about that. Thus most people, including many MDs, would interpret the positive result as making it very probable (perhaps 90% probable) that she has cancer. Because there seems to be a 10% chance of a false-positive result, the physician might recommend a second test to be even more confident in her diagnosis ($0.1 \times 0.1 = 0.01$ or 1% chance of a false-positive the second time around).

However, this reasoning is completely wrong. To see why, let's imagine 1,000 women taking part in the screening (Figure 5.1). The incidence of breast cancer amongst middle-aged women is about 1%, meaning that 990 women of the 1,000 are expected to be cancer-free and the remaining ten have cancer. From 990 healthy people 10% or 99 women will nevertheless receive a positive test result, as will 90% or nine of the ten women with cancer. Nine (the true positives) divided by 108 (all positives) makes 0.083, which means that,

---

1. The uses where "show" refers to a figure, as in "as shown in Fig" and "data not shown," were excluded.

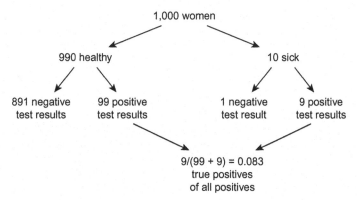

FIGURE 5.1   Diagnosing cancer from general population where the incidence of cancer is 1%. The sensitivity and specificity of the diagnostic test are 0.9.

in the absence of additional information, the probability that a woman, who got a positive test result, has breast cancer is about 8.3%.[2] Doing a repeat test and still getting a positive result raises the probability of cancer to about 45%, which means that it is still more likely than not that she is cancer-free.

The above is, of course, just an example—an intuition pump. So it makes sense to ask about the situation on the ground. Current best evidence indicates that for a 50-year-old American woman screened annually for breast cancer, the probability of getting at least one false-positive test result is somewhere between 49% and 67% (Welch and Passow, 2014). A recent very large study (89,835 women aged 40–59, spanning 25 years) found 500 deaths from breast cancer in the group that received five annual mammography examinations versus 505 deaths in the control group (Miller et al., 2014). Unfortunately, it doesn't stop here. The mammography arm produced 106 excess cancer diagnoses, which means that 22% of invasive breast cancers detected by it were most likely over-diagnosed, representing one overdiagnosis for every 424 women. This means that not only was the mammography useless over the population of women studied, but also its implementation resulted in unnecessary fear and suffering.[3]

Looking back at our cancer test, we can now generalize: (i) Evidence alone does not count: even very strong evidence can be worthless in proving that something is intrinsically unlikely. (ii) Relevant background information, which is not contained in the experimental setup (in our case, the incidence of cancer in the general population, BRCA status, age, ethnicity, etc.), must be taken into account if the goal is plausible reasoning about truth of hypotheses. (iii) The logic behind deriving probabilities from evidence is trivial; there is no room for doubt about its validity. (iv) Sensitivity and specificity are properties of the test, not of the test subject or

---

2. This is the same as to say that the false discovery rate (FDR) is about 0.92.
3. Incidentally, a recent analysis indicates that four out of five diagnostic tests currently in medical use have resulted in no improvement in patient outcomes (Siontis et al., 2014).

her cancer. (v) While incidence of breast cancer in the general population is an objective feature of the world, the relevance of it to us is no such thing. How much weight we will put on it depends on the level of our knowledge about the world.

For example, depending on our additional knowledge of the status of a woman's BRCA genes, the probability of her getting breast cancer during her lifetime could range from below 10% to well over 90%. For example, for a 20-year-old woman with two affected sisters and affected mother a BRCA1 mutation confers a lifetime risk of 92% of getting breast cancer, while a 20-year-old woman with no family history of breast cancer would have a 71% risk (Couzin-Frankel, 2014). For a 40-year-old woman these numbers are 69% and 60%. These numbers are a function of our current knowledge of cancer epidemiology. As our knowledge develops, no doubt so will our risk assessments.

This means that the solid logic behind our calculations will yield us nothing more than subjective guesses about reality, which depend crucially on our background knowledge. In other words, in turning evidence into knowledge, context is everything. Having said that, the kind of logic that we just used does enable us to make rational choices about interpretation of data.

5.2 BAYES THEOREM: WHAT MAKES A RATIONAL BEING?

Ever since Aristotle, scientists have used eclectic mixtures of inductive and deductive methods to draw inferences from data to theories. This may or may not work in specific cases but it surely means that there is no guarantee that different people will reach the same conclusions, even when starting from identical data and background knowledge. In this narrow sense science is not rational and a rational scientist would be akin to a robot, which, when given some information, algorithmically computes its conclusions. Of course, the logic on which the roboscientist's reasoning is based must be both probabilistic and impeccable—it should be the best logic theoretically possible.

To build a rational robot, which would consistently reason from available information to scientific theories, we must base its thinking on general principles, which are beyond reproach in the sense that every reasonable person would naturally agree on them. Our definition of rationality is then that when rational entities are faced with the same problem and have the same information, they must always reach the same conclusions. If our robot only thinks thoughts that are consistent and derived from such general principles, then its conclusions will be rational. The only remaining place for disagreement would be on the quality of input data.

While this rational robot does not yet exist in any sufficiently sophisticated form to be called a roboscientist, there have been serious attempts to base scientific thinking on a firm rational basis, starting in the 1920s with Rudolf Carnap (Gower, 2002). A more recent and promising approach can be found on the 700 pages of E. T. Jaynes' classic book (Jaynes, 2003).

Jaynes' hypothetical roboscientist reasons about propositions based on information that is given to it and its goal is to assign numerical probabilities to propositions, which can be subsequently revised in the light of new evidence.

The probabilities, with which it operates, designate degrees of belief, which are not part of the physical world and thus are, and forever remain, subjective.

This is a very important point, as Jaynes convincingly argues that probabilities as physical manifestations of randomness inevitably lead into logical contradictions. In other words, any causal interpretation of the purely logical relations of the probability theory is doomed to fail (this doesn't mean that one cannot use probability theory to test causal hypotheses; it only means that logic, like mathematics, is not part of the physical structure of the universe). However, probabilistic logic is used to reach a kind of objectivity through rationality.

The whole of Jaynes' robots thinking is based on just three basic principles of thought or desiderata:

1. Degrees of plausibility are represented by real numbers.
2. Rational thinking corresponds qualitatively with common sense. Thus, increasing evidence for a proposition raises its plausibility and vice versa.
3. Reasoning must be consistent. (i) If a conclusion can be reasoned in more than one way, then they all must lead to identical conclusion. (ii) All the available relevant information must be taken into account (principle of total information). (iii) Equivalent states of knowledge are presented by equivalent numbers.

The most telling of the desiderata is (ii)—principle of total information—which implies that a failure to take the totality of available relevant information into account can lead to erroneous interpretation of perfectly good data. This is why a politician can never use statistics correctly, even when given an abundance of good-quality data. In the world of biomedical science, evidence indicates that less than half of lead authors of randomized clinical trials use existing systematic reviews of similar trials in planning of their trials (Cooper et al., 2005). When new trials are conducted, on average over three-fourths of the relevant prior publications go uncited in resulting papers (Robinson and Goodman, 2011). Moreover, it looks like many authors willfully bias their conclusions by preferentially citing sources confirming their claims (Greenberg, 2009; Chalmers et al., 2014).

The Kolmogorov axioms (1933), which are the basis for the whole of probability theory in all its manifestations, are in essence the formalization of Jaynes' desiderata. They are the basis of both frequentist and Bayesian systems, which means that every formally valid argument involving one system translates into a valid argument involving the other. As far as anybody knows, this is the only set of axioms that leads to fully consistent reasoning. The four axioms are:

1. $P(A) \geq 0$ (probabilities cannot be negative),
2. $P(\Omega) = 1$ (probability of a certain event is 1),
3. $P(A \text{ or } B) = P(A) + P(B)$, if A and B are mutually exclusive,
4. $P(A \& B) = P(A) \times P(B \mid A)$.

Axiom 3 states that probability of at least one of the mutually exclusive events equals the sum of their probabilities. Axiom 4 defines the probability of co-occurrence of events by conditional probability (or vice versa).

DEF: $P(B|A)$ is a conditional probability, which should be read as the probability of B, given that A is true.

All probabilistic and statistical thinking (both frequentist and Bayesian) derives from these axioms, which in turn are uniquely consistent with the three desiderata of rational thought mentioned above (Jaynes, 2003). Some of the basic derivations from axioms of probability are

$$P(A) + P(\text{not}A) = 1,$$

if A is equivalent to B, then $P(A) = P(B)$,

if A entails B, then $P(A) \leq P(B)$,

$$P(A) = P(A \,\&\, B) + P(A \,\&\, \text{not}B),$$

and its extension, also known as the "law of total probability"

$$P(A) = \sum_i P(A \,\&\, B_i) \sum_i P(A|B_i) P(B_i),$$

which assumes that B_i is a set of mutually exclusive and exhaustive propositions.

Proof of the Bayes theorem. We will start with illustrating the inevitability of the fourth axiom (some prefer to think of it not as an axiom but as a definition of conditional probability). Let's suppose that we have a box with six objects: two blue balls, one red ball and three red cubes. Now, what is the probability of randomly drawing a red ball from a box $P(\text{red and ball})$? Probability of drawing a ball $P(\text{ball})$ is 1/2, the probability of drawing a red object, given that we draw a ball $P(\text{red}|\text{ball})$ is 1/3 and probability of the red ball is:

$$P(\text{red and ball}) = P(\text{ball}) \times P(\text{red}|\text{ball}) = 1/2 \times 1/3 = 1/6$$

We can also approach this from the other way around: the probability of drawing a red object $P(\text{red})$ is 4/6, the probability of drawing a ball, given that we drew a red object $P(\text{ball}|\text{red})$ is 1/4 and thus probability of drawing a red ball

$$P(\text{red and ball}) = P(\text{red}) \times P(\text{ball}|\text{red}) = 4/6 \times 1/4 = 4/24 = 1/6$$

Thus, to generalize, $P(A \text{ and } B) = P(B) \times P(A|B)$.
From there we can derive the most important result of probability theory:

If $P(A \text{ and } B) = P(A) \times P(B|A)$ and

$P(A \text{ and } B) = P(B) \times P(A|B)$, then

$P(A) \times P(B|A) = P(B) \times P(A|B)$ and

$$P(A|B) = \frac{P(A) \times P(B|A)}{P(B)}$$

This is the Bayes theorem, which is mathematically a result so elementary that some people insist in calling it Bayes' rule to leave the term "theorem" for more seriously headache-inducing stuff. A very entertaining history of the Bayes theorem (which, incidentally, was discovered by Laplace) can be found in McGrayne (2012).

Bayes theorem, which we derived in the text box, provides the only rational way of numerically integrating evidence with background knowledge to arrive at a posterior probability for a hypothesis.

If data is D and the hypothesis of interest is H, then the posterior probability (= updated probability of the of the hypothesis, given the data) is

$$P(H|D) = \frac{P(H) \times P(D|H)}{P(D)}$$

where $P(D|H)$ is the likelihood, convening all the evidence present in data under the hypothesis H, $P(H)$ is the prior probabilities of the hypothesis and $P(D)$ is the prior probability of data. It is important to note that the terms "prior" and "posterior" have nothing to do with time. The whole of probability theory (and logic) is timeless.

All information relevant to the hypothesis, which is not represented in the likelihood, should be integrated into a prior probability. It makes no difference if it was obtained before or after the data, from which the likelihood was calculated. Since there is no penalty for calculating as many likelihoods from as many data as one wishes, it follows that today's posterior is likely to be tomorrow's prior.

Posterior probability is a description of our state of knowledge, not of the state of the world, and it naturally changes with changing knowledge. As such, Bayes theorem is the perfect vehicle for meta-analysis (Jaynes, 2003).

While $P(H)$ denotes prior probability of a single hypothesis given the background knowledge, $P(D)$ is the probability of observing your data under all possible hypotheses. It can be written out for i possible mutually exclusive hypotheses as

$$P(D) = P(H_1) \times P(D|H_1) + P(H_2) \times P(D|H_2) + \cdots + P(H_i) \times P(D|H_i)$$

For a simple situation of two mutually exclusive and exhaustive hypotheses, a useful form of the Bayes theorem is

$$P(H|D) = \frac{P(H) \times P(D|H)}{P(H) \times P(D|H) + P(\text{not } H) \times P(D|\text{not } H)}$$

Logically it does not matter into how many hypotheses we slice the reality, as long as they are mutually exclusive and exhaustive

$$P(H_1) + P(H_2) + \cdots + P(H_i) = 1$$

But, as we shall see, scientifically it matters a great deal.

In the context of a medical test with stable long-run error frequencies we can use the following derivation of the Bayes theorem:

$$P(\text{sick}|+) = \frac{\pi(1 - \beta)}{\pi(1 - \beta) + (1 - \pi)\alpha}$$

where $P(\text{sick} \mid +)$ is the probability that the patient is sick, given a positive test result and π is the prior probability of her having the disease. Plugging in the numbers from our breast cancer example we will get the same result as we did in our combinatorial exercise:

$$\frac{0.01(1 - 0.1)}{0.01(1 - 0.1) + (1 - 0.01)1} = 0.083$$

or the posterior probability of having cancer, given a positive test, is 8.3%. For interpreting the repeat test, our posterior probability value 0.083 will be the updated prior probability of disease.

It must be noted that as α and β are all about sampling error, so will be the posterior probability. Therefore, when this version of the Bayes theorem is used, the literal interpretation of the probability that a patient who has a positive test result is afflicted by the disease depends on there being no bias in the test.

A simple example from genetic counseling (Gelman et al., 2003). Hemophilia is a recessive X-linked disease. Because men have a single X chromosome, we know that if a male child is not affected, he is not a carrier of the gene for hemophilia. In contrast, because women have two copies of the X chromosome, a non-affected woman may be a carrier. Now, we must advise a woman, whose brother suffers from hemophilia, about her chances of being a carrier of the disease-causing gene. From the disease status of her brother, and from the fact that her father is unaffected, we know that her mother must be a carrier.

Our hypothesis space consists of two hypotheses H_1: she is a carrier and H_2: she is not, and our prior probability $P(H_1) = 0.5$ (as because of the brother's condition her mother must have a hemophilia gene). $P(H_2) = 1 - P(H_1)$ is also 0.5. We also know that she has two healthy sons. We therefore need to explicate two likelihoods: $P(\text{son 1 healthy \& son 2 healthy} \mid H_1) = 0.5 \times 0.5 = 0.25$ and $P(\text{son 1 healthy \& son 2 healthy} \mid H_2) = 1 \times 1 = 1$. Note that neither prior probabilities nor likelihoods cover all the information available to us. We know that there is a chance of the woman getting the hemophilia gene not from her mother but by spontaneous mutation. Only, since we expect the probability of this scenario to be very much lower than genetic transmission from the mother, we choose to ignore it.[4]

---

4. *Both priors and likelihoods are statistical models, which are characterized, in addition to the extent of their correspondence to reality, by their ability to make predictions and simplicity/ convenience of use.*

The next step is integrating the information present in priors and likelihoods by the use of Bayes theorem:

$$P(H_1|\text{Data}) = \frac{P(\text{Data}|H_1)P(H_1)}{P(\text{Data}|H_1)P(H_1) + P(\text{Data}|H_2)P(H_2)} = \frac{0.25 \times 0.5}{0.25 \times 0.5 + 1 \times 0.5} = 0.2$$

Therefore, there is a 20% chance that the woman is a carrier of the disease gene, but the accuracy of this estimate depends both on the data and on the model. Should the woman have a third healthy son, we can calculate an upgraded probability by taking the "old" posterior probability, 0.2, as the prior $P(H_1)$, which means that $P(H_2)$ is now 0.8. Running these numbers through the Bayes machine will then give us a new posterior probability 0.11. On the other hand, should the third son be a hemophiliac, the updated posterior will be

$$\frac{0.5 \times 0.2}{0.5 \times 0.2 + 0 \times 0.8} = 1$$

Again, had we decided to take into account the frequency of spontaneous mutation, the posterior probability would be slightly less than certainty. The same would be true had we factored in the probability of misdiagnosis of hemophilia in her sons and the possibility that her hemophiliac brother might not be her brother at all, as babies are sometimes switched by cruel witches.

The question whether our original model was good enough to produce accurate enough posteriors can be addressed at this stage, perhaps even by assessing by frequentist methods the general quality of advice given to many women emanating from Bayesian calculations. Should the model turn out to be unsatisfactory, we can simply start from the beginning by building a new model to better fit the data.

The Bayesian inference is a multi-step process in practice (Gelman et al., 2003):

1. Setting up a probability model for prior information, which is consistent with scientific background information and the data collection procedures.
2. Setting up a probability model for the data—the likelihood.
3. Calculating the posterior probability (this is what the Bayes theorem is for).
4. Checking the fit and the assumptions of the model in the light of the results. For example, how sensitive are the assumptions of the model to natural variation in the data?
5. If needed, go back to step 1 and start a new.

To illustrate the importance of the "correct" model, we will briefly turn to the science of parapsychology. Extrasensory perception (ESP) is a hypothetical phenomenon where people are able to "feel" the future. The quotation marks simply denote the lack of good words for describing such cogitations, as extrasensory by definition means that we have no inkling of the physical or biological basis of it.

It may therefore come as a surprise that over decades scientists have collected strong experimental evidence for ESP (Storm et al., 2010). For example, Soal and Bateman (1954) showed that a Mrs. Stewart could at 37,100 trials correctly guess cards significantly better than predicted by the null hypothesis of a random model (Soal and Bateman, 1954). The null model predicted the frequency of correct guesses at 0.2 and Mrs. Stewarts result was 0.2536. This may not sound like much but the large sample size put her results a whopping 25.8 standard deviations (SDs) away from the null effect, which translates to likelihood ratio (LR) of about 10^{138} in favor of the maximum likelihood value 0.2536 over 0.2 (Jaynes, 2003, p. 120). This evidence is beyond strong: it is mind-bogglingly enormous!

In a more recent work Daryl Bem showed in random subjects statistically significant results in the direction of ESP in four independent experiments (Bem, 2011). A Bayes factor meta-analysis puts a combined LR of about 49 in favor of ESP (Rouder and Morey, 2011) and a later meta-analysis of independent replications of Bem's results in 32 different laboratories presented extremely strong evidence for ESP (LR = 10^9) (Bem et al., 2014). Similar results were obtained in a meta-analysis of older ESP literature (Honorton and Ferrari, 1989).

Even if our prior probability for the existence of ESP is low, say one in a million, the posterior probability for the hypothesis that Mrs. Stewart was a psychic will still come out as extremely high (posterior odds ratio is about 10^{132}) and the Bem data increase the odds for ESP by at least 49-fold. Does this mean that we should seriously entertain the notion of ESP?

Not quite. While the data do convincingly show that the random null model is untenable, they do not indicate that ESP should be the favored alternative. Moreover, as we will soon learn, however great the evidence against the null, it cannot raise the posterior probability of ESP by any significant extent.

This is because although the two-hypothesis version of the Bayes theorem is mathematically correct, it is too simplistic for any meaningful scientific inference. The reason is that every experimental setup comes with a chance of bias.

The probability of bias, be it Mrs. Stewart peeking at cards, the experimenter unconsciously signaling the right answer, or any other of the great many things that can go wrong with an experiment, is relevant information for an experimental study and it should not be ignored in the interpretation of results. As bias is not represented in the likelihood, it must be introduced into the hypothesis space as a specified hypothesis H_{bias}.

Crucially, as long as the prior probability $P(H_{bias}) > P(H_{ESP})$, it is this alternative, not the ESP, that gets the support from the experiments purportedly conducted to study ESP (Jaynes, 2003, p. 123). When we sum the probabilities of all possible types of bias, the posterior probability of ESP becomes

$$P(H_{ESP}|D) = [P(H_{ESP}) \times P(D|H_{ESP})]/[P(H_{ESP})$$
$$\times P(D|H_{ESP}) + P(H_0) \times P(D|H_0)$$
$$+ P(H_{bias}) \times P(D|H_{bias})]$$

This changes the result greatly, as any respectable posterior for H_{ESP} requires that

$$P(H_0) \times P(D|H_0) + P(H_{bias}) \times P(D|H_{bias}) << P(H_{ESP}) \times P(D|H_{ESP})$$

Therefore, as long as the prior probability of bias is higher than that of the scientifically meaningful hypothesis, it makes no sense in conducting the experiment.

A corollary of the above is that even if two scientists receive identical evidence and use Bayesian inference correctly, assuming that for Scientist A the prior probability $P(H_{exp})$ is $<< P(H_{bias})$ and for scientist B the $P(H_{exp}) >> P(H_{bias})$, their posteriors will not converge. This should not be seen as a shortcoming of Bayesian inference but rather as a strong reminder about the importance of keeping systematic errors in check, and of the general futility of testing very unlikely hypotheses.

The need to factor in $P(H_{bias})$ should not be of interest only to parapsychologists, but to anyone planning experiments where the chance of bias is high and/or prior probabilities of individual effects low. Modern omics seems to be in especial danger of unwittingly collecting evidence in favor of H_{bias} (Ioannidis, 2005).

The methods of modern biology have been getting more and more sensitive, which means that they are ideally suited to find the effects, which are truly present in nature. However, high sensitivity of measurement also means that when the null hypothesis of no effect happens to be true, phantom effects are likely to emerge. As the growth of objective scientific knowledge crucially depends on the ratio of plausibility of the scientific hypothesis over the plausibility of bias in explaining the results (Bayes theorem merely quantifies this general truth), there is a metaphysical fly in the ointment of scientific progress. Namely, in order for the scientific method to work in the long run, we need to have reason to believe that many of our proposed theories are close to truth, *before* experimental testing. Should Mother Nature be barren and our hypotheses irrelevant *en masse*, we will still obtain false-positive results indistinguishable from real discoveries. As we learned in the section concerning likelihoodism and confirmed in our Bayesian analysis of the common sense, the facts only speak in relation to clearly stated relevant hypotheses. Should we miss a relevant hypothesis, scientific evidence will mislead.

If a scientist is certain of the truth or falsity of a theory ($P(H) = 1$ or 0), Bayes theorem (or propositional logic) does not allow accumulating new evidence to change his belief by an iota. Where there is certainty, there can be no doubt, no chance of changing one's mind in the light of new evidence, and thus no science. If certainty will not do, all that is left to us is going by trial and error. This makes the fallibility of human reasoning not a limitation of science, but its bedrock, which, when used with integrity and care, may lead to overall growth of knowledge. Thus may a philosophically inclined Bayesian reach the same conclusion as Karl Popper about the trial and error nature of the scientific enterprise and the piecemeal and fallible kind of progress available to us.

5.3 TESTING IN THE INFINITE HYPOTHESIS SPACE: BAYESIAN PARAMETER ESTIMATION

When we test increasing numbers of discrete hypotheses using Bayes theorem, as their number increases and the real-world differences between neighboring hypotheses decrease, we will come to a point where it makes sense to redefine our quest as continuous parameter estimation.

For a frequentist, parameter estimation is what confidence intervals (CIs) were created for. For him the parameter to be estimated has an unknown but fixed value—and because a single value has no frequency, it is not directly approachable by frequentist methods. In contrast, the Bayesian doesn't particularly care if a parameter value can have a frequency or not: it is still unknown to her, and to make a guess she merely estimates the extent of her knowledge about it. While the frequentist "objectively" asks about the state of the world, a Bayesian "subjectively" (but perhaps rationally) asks about her state of knowledge. What is actually estimated by her is not the parameter value itself but rather the best available information about all possible parameter values.

The main outcome of Bayesian parameter estimation is therefore not a point value or an interval, but a posterior probability distribution over all values of the parameter space. From the posterior probability distribution one can easily look up the probability for any particular parameter value—it contains hugely more information than anything the frequentist can come up with.

The Bayesian solution to parameter estimation is, not surprisingly, the Bayes theorem:

$$P(\theta|D) = \frac{P(D|\theta)P(\theta)}{P(D)}$$

where θ is the parameter that is estimated, $P(\theta)$ is the prior probability of the parameter being a particular value, and $P(D|\theta)$ is the likelihood of observing the data, given that the experimental data equal θ. $P(D)$, the expected value (mean) of the likelihood function, standardizes the area under the posterior probability distribution to one. $P(\theta|D)$ is the posterior probability of θ, conditional on the data observed. The mean of the posterior distribution is a weighted average of the prior mean and the sample mean, which makes it our estimate of the true parameter value.

The variance of the normal posterior distribution $(SD_{post})^2 = 1/w_0 + w_1$ where $w_0 = 1/(SD_0)^2$ of the prior distribution and $w_1 = n/SD^2$ of the likelihood function (n is the sample size). We can see that when SD_0 is large (a wide prior distribution indicates uncertainty), SD is small and n is large (signifying lots of good-quality experimental data), the posterior distribution will increasingly look like the likelihood function (vague prior information is essentially ignored). As the sample size increases, the contribution of the prior distribution lessens. Hence the opinions of two rational people, who start even from extremely different normal or binomial prior distributions, will eventually converge, given enough data.

The above formulas are for normal distributions, which, although often used as priors, are by no means the only distributions commonly used. For example, lognormal distributions are often used for informative priors and the beta distribution is a common prior for binomially distributed parameters (Quinn and Keough, 2005).

We now return to our example of coin tossing (see also Figures 4.10, 4.13). Lets suppose that four people pick up four different coins and ask whether their coin is bent. To find out, they will each do a series of 10, 20, 100, or 1,000 tosses and get 8/10, 13/20, 48/100, or 430/1,000 heads. For Bayesian analysis they will next want to construct a prior distribution. A good model for distribution of ratios is the beta distribution and accordingly our gang of four will next use the R package called "LearnBayes"[5] to find a suitable beta prior and to calculate the posterior distribution. They all decide that the most likely frequency of heads is 0.5 and that even if the coin is bent, it would be extremely unlikely to get heads ratios under 0.25 or over 0.75. They now construct a beta distribution that best encapsulates this knowledge, calculate the likelihood function from the actual coin tossing data, and use Bayes theorem to get the posterior distribution (Figure 5.2). Looking at the triplots we will notice that as the sample size increases from 10 to 1,000, the relative importance of the prior distribution diminishes and the posterior distribution will eventually be very close to the likelihood function. Another thing to notice is that the larger the sample size (and the narrower the prior), the narrower will be the posterior distribution. Looking at each posterior distribution we can easily see how much trust we can put on any particular ratio of heads corresponding to the truth of the matter. When the outcome of t-testing of these data (Figure 4.10) was a significant P value and an even more significant headache about the alternative hypothesis; and the outcome of a likelihoodist approach was a continuous function, which enables us to say how well each parameter value is supported by the data (Figure 4.13); here the outcome is a continuous posterior probability distribution,

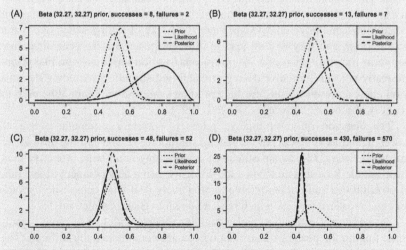

FIGURE 5.2   Bayesian parameter estimation for the same binomial data as in Figures 4.10 and 4.13. The prior is modeled into beta distribution ($a = b = 32.27$).

5. http://cran.r-project.org/web/packages/LearnBayes/index.html.

which accurately integrates our data (as modeled in the likelihood function) and prior knowledge (as modeled in the prior distribution) so that for each possible heads ratio we can say, how well it reflects the actual "bentness" of the coin. For example, after experiments 1 or 2 (Figure 5.2A and B) we can conclude that the ratio 0.6 is well covered by the posterior distribution and can thus be close to correct value, while experiments 3 and 4 (Figure 5.2C and D) result in posteriors, which do not lend credence to the hypothesis that 0.6 is the correct ratio.

What if our prior knowledge is scant? Could we use non-informative priors (like a flat distribution) and still converge on the true value simply on the strength of the data (the likelihood function)? We know that when using a non-informative prior and assuming data from a normally distributed population, the posterior distribution will conveniently be a normal distribution. However, there is a serious problem with using flat priors as a measure of our ignorance (at least when dealing with continuous parameters).

This can be explained by a simple example. Think of a situation where our prior information is limited to knowledge that the side length of a cube is between 1 and 2 units. To express your existing prior knowledge you draw a flat (rectangular) prior from parameter values 1 to 2, signifying that you have no reason to prefer any value in that range. Unfortunately, the same prior information entails that the area of a side of the cube must be between 1 and $4u^2$, which would suggest another rectangular prior. And the volume of the cube, expressed in u^3, would lead to yet another rectangular prior. All these rectangles represent exactly the same prior knowledge, there is no way of saying which one is the "right" prior, and they will all lead to different posteriors. To generalize, any nonlinear transformation of a flat prior leads to a non-uniform prior. Flat priors are really not uninformative. Every probability distribution, including the uniform ones, represents a specific state of knowledge and thus cannot be used to quantify absence of knowledge (Royall, 1997).

The question of uninformative priors is currently a popular object of research, and as yet there isn't a general solution to it. Thus, in the absence of prior knowledge, LRs as measures of evidence may seem more attractive than the available Bayesian methods—the downside being abandonment of any talk of probabilities and not being able to effectively deal with composite hypotheses and nuisance factors (which pose no threat to Bayesian methods).

For parameter estimation under complex models with two or more parameters, finding the posterior distribution becomes computationally difficult. The currently popular solution is to simplify the problem by simulating a great many simpler solutions using Markov chain Monte Carlo (MCMC) methods (Andrieu et al., 2003).

To define the interval where, given our current best knowledge, we may hope to find the true parameter value with a set probability, Bayesian credibility intervals can be calculated. These are also called highest probability regions because any value in the interval has a higher probability of occurring than any value outside. For instance, a 95% credibility interval means that, given the totality of information available to us, it is 95% probable that the true value is inside the interval. A 95% credibility interval corresponds to 95% of the area of the posterior probability distribution. When the prior distribution is non-informative, the Bayesian credibility interval will be numerically the same as the frequentist CI (but the interpretations still differ because frequentists cannot speak of probabilities of fixed parameters without logical contradiction). If an informative prior overlaps with the likelihood function the Bayesian interval will be narrower than frequentist CI, while with an informative prior, which is situated far from the likelihood function, the opposite will be true.

5.4   ALL AGAINST ALL: BAYESIANISM VERSUS FREQUENTISM VERSUS LIKELIHOODISM

In Chapter 4 we learned that while frequentist null hypothesis testing and parameter estimation are difficult concepts to master, their calculations are easy to perform with many user-friendly computer programs (e.g., check out the free online version of GraphPad Prism). You don't need a statistician to calculate the most-used frequentist statistics (but you may need one to properly interpret the results of your calculations). P values, standard errors, and CIs are also the kind of things that reviewers like to see in a manuscript, and reviewers of biology papers are generally no sticklers for statistical detail. Also, most readers of your papers have a familiarity with these methods already from their schooldays.

It is a different matter, whether the ease of calculation and ubiquity of frequentist methods is a good thing for you. The answer depends on how well you understand their applicability and, especially, on the limits of different methods.

There is no single unified frequentist statistics, it is rather a large collection of *ad hoc* methods designed to solve specific questions; and these methods generally work well enough for the types of problems for which they were originally created. For instance, P-value-based significance tests were originally created to assist Roland Fisher in his genetical studies, which he conducted in Rothamsted agricultural station from 1919 to 1933. Fisher's research problems were characterized by very little relevant prior information, relatively large sample sizes and ease of random sampling from many available research plots (Jaynes, 2003, Chapter 16). Thus it can be said that Fisher created the method he needed and it is your business to decide whether the same method satisfies your needs.

There are many situations, which can benefit from frequentist tests, but because each test works well in a limited range of conditions and there really are a lot of different tests, no experimental scientist can be expected to keep track of them, let alone to understand the strengths and weaknesses of the lot. This has led to the dependence of experimental scientists on statisticians, who tend to be educated in the frequentist paradigm by the mathematics departments of our universities, whose insights are mathematical (not grounded in research facts) and who therefore see experimental design as a randomization problem.

This collaboration seems to work reasonably well in large-scale clinical studies, where statisticians are fully integrated into experimental planning alongside medical, biological, and chemical scientists (Berry, 2007, 2012). In small-scale experimental biology, however, the statistician is consulted, if at all, at the data analysis stage when experiments are already over and the researchers have developed questions to which they lack answers. This causes the frequentist statistician to see his job as rescuing what he can from the wreckage of poorly randomized, underpowered studies with unclear and poorly documented experimental designs and uncertain internal and external validity.

The upshot is that the statistician, who gets this thankless job, tries to compensate for the unknown biases and hidden multiplicities (of whom the typical scientist has never been told) and as a result is likely to be overcautious in his approach to the data (Berry, 2012). This can lead to missing out on effects that would be discoverable by just looking at data, or by Bayesian analysis.

When real effects of modest size are missed in the analysis, what are left to discover are the often larger but hallucinatory effects, brought along by sampling effects and various biases. Unlike the Bayesian, the frequentist statistician has no post-experimental analytic tools to address most forms of bias. What he has to offer is randomization and various matching designs; but then, he is consulted long after this has stopped to matter.

What are really important for the long-term health of experimental science are the expected effect sizes. If most effects are modest (for $n = 3$ and normally distributed data, below about 3 SDs), bias could easily constitute most of the published effects.

There are special conditions, under which the frequentist significance tests, LRs and Bayesian estimates closely agree (Jaynes, 2003, p. 550–552):

1. no important prior information concerning the background information to your problem or chance of bias in your experiment;
2. data are normally distributed;
3. you are interested only in the mean value of your dataset, not in any additional structure in the data;
4. data are randomly sampled from a statistical population;
5. sample size is at least 30;
6. there are only a few parameters that you are interested in;
7. there are sufficient statistics (a statistic is sufficient if the sample from which it is calculated gives no additional information as to the probability distribution of the population from which the sample was taken).

If all these assumptions are met, then it does not matter much, which set of methods you use on your data. However, I'm pretty sure that for most biologists this is not the case most of the time.

A critic might here point out that, assumptions notwithstanding, *P* values and CIs are nearly universally used in biology and so far they seem to have worked just fine. So, why worry? To this one could retort with: How do you know? Many misuses of frequentist methods are not expected to lead to unreal voodoo effects (the ones that do have much to do with very low prior probabilities or high prior probability of systematic error or bias). Instead, they will result in missing of real effects.

On a more philosophical plane, let me ask: Do you believe that a scientist must be committed to take all the relevant information relevant to her problem into account? If the answer is yes, then you might consider freeing your inner Bayesian. It doesn't necessarily mean that you need to use formal Bayesian methods in everything that you do (although it may be tempting to calculate the posterior probability of your boyfriend cheating on you, given that you just have found fresh lipstick traces on his shirt).

What is the probability that your boyfriend cheats on you, given that you found lipstick traces on his shirt? When feeling hot, Bayes theorem is a wonderful tonic. So think of your hypothesis space: H_1—he cheats, H_2—his female co-worker had a birthday party, H_3—he tests your trust in him, H_{bias}—it wasn't a lipstick but your child's colors that made the smear. Now its time to weight these hypotheses by converting them into likelihoods: $P(\text{lipstick} \mid H_1) = A$, $P(\text{lipstick} \mid H_2) = B$, $P(\text{lipstick} \mid H_3) = C$, $P(\text{lipstick} \mid H_{bias}) = D$. Next you must try to assign the probabilities to A–D (these probabilities need not sum to one) and to H_1–H_{bias} (these must sum to one). Now you can plug the numbers into the Bayes theorem and do the calculation to find $P(H_1 \mid \text{lipstick})$.

As the probabilities that went into the equation are merely your best guesses, it is highly recommended to test the robustness of your model by playing around with probabilities and the hypothesis space (formulate additional hypotheses to see how much they change your original conclusions). If your conclusions are robust, you have probably formed a strong belief about the status of your relationship. If the conclusions are not robust, then you probably do not have enough information and/or trust in your relationship.

With a solid understanding of the limitations of traditional methods, the role of background information and the basics of integrating different strands of information, you will be able (with a little mental gymnastics) to informally integrate the results of frequentist tests with knowledge emanating from experimental design and prior information about your research question (Phillips and LaPole, 2003).

So, to recoup, what distinguishes Bayesianism from frequentism? The most important aspect is prior probability: a concept, which the frequentist never uses.

The second important difference is the concept of probability itself, which refers to our knowledge about the world, but does not refer directly to the world.

For a Bayesian: information in, information out. In contrast, the frequentist sees his probability as a part of the physical world. For him: reality in, information out. While Bayesian probability updates information with other information, the frequentist probability tries to convert aspects of the physical world into information. This may seem objective and even faintly scientific, but it also precludes probability statements about anything that does not have physical frequencies, including the truth of scientific theories. It also introduces a concept of random sampling from a population, which excludes most scientific experimentation from the domain of frequentist statistics, unless a weird notion of a statistical population of imaginary experiments, from which real experiments somehow randomly materialize, is brought into play. Even then, the concept of frequencies as real properties of randomness of physical systems has the disadvantage of simply being wrong (except perhaps for quantum mechanics, but see below). The interactions in most of physics and biology are fundamentally deterministic; they are not random.

For a Bayesian, frequencies are just that—frequencies. They are not the result of real randomness in nature but rather of our limited knowledge of its workings, and can thus be easily described by Bayesian probability. A Bayesian should have no compunction in using frequentist methods where appropriate, for instance in model checking.

Next comes random sampling from a population: an essential component of any frequentist method, but often irrelevant for a Bayesian, who sees it as a method of losing relevant information from the data and thus weakening the power of one's inferences (Jaynes, 2003). As the whole science of experimental design (there are many textbooks, e.g., Mason et al., 1989; Seltman, 2013) is built on randomization, this can be seen as a serious problem for the Bayesians, who nevertheless continue to make serious efforts to create their own paradigm of experimental design (Lindley, 1972; Chaloner and Verdinelli, 1995; von Toussaint, 2011).

Modern theories of Bayesian experimental design are concerned with active selection of data. In the real world of limited resources it is often important to know where to look next to learn as much as possible from your experiment, or when to stop the experiment. An early modern instance of the practical application of this approach comes from the search by the US navy for a lost hydrogen bomb off the coast of Spain in 1966 (McGrayne, 2012). The idea was to collect the relevant information on the lost bomb, sea currents, etc., divide the search area into smaller cells and, starting the search from the most likely ones to contain the bomb, to requantify the increasing probabilities for the bomb lying in each of the other cells as the search in the original cell turns up nothing. As the posterior probability of the bomb being in cell A decreases, the prior probabilities of it being in any of the other cells must correspondingly increase.

Ideally, taking advantage of this should have allowed the navy to conduct its search with maximum efficiency. As it happened, the navy brass was rather hoping that in case of failure its statisticians would prove that everything possible had been done to find the bomb and, therefore, that the Russians would not find it either. Happily for everybody concerned, the bomb was eventually found when a local fisherman pointed out where it fell.

A Bayesian goal is to design experiments so as to obtain the maximal information gain with minimum cost, with selection of the most useful subsets of data from large datasets, and with the optimal combination of various experiments to facilitate the final synthesis. Bayesian experimental design heavily depends on the concept of expected utility, which can be seen as the integral over all possible data weighted by the probability of the data under the design decision and the utility of the corresponding data. Expected utility can be expressed as likelihood and prior distributions combined with a utility function. However, in many (non-normal) cases it is hard or impossible to calculate it analytically (von Toussaint, 2011).

Another difference between Bayesians and frequentists is how they deal with multiplicities. There is no question that Bayesian methods, if properly implemented, deal with the problem fully. The problem is, however, that the *if* is a big one. Every Bayesian application requires serious thought, both scientific and mathematical, to find the formulation that efficiently encompasses the prior information and, as importantly, to decide what the relevant prior information actually is. The first part should be no bother to a scientist as thinking of the scientific background of his problem should be a part of his job description anyway. The mathematical part, however, is a different matter.

For the frequentist, a major way of dealing with multiplicities is to pre-set the experimental protocols and analytic methods and to avoid peeking at data. If we believe that a scientist's job is to collect unbiased evidence and honestly match it to alternative theories, until strong enough evidence has been gathered for one of the alternatives, then these frequentist requirements make no sense. By adhering to them we will limit our ability to analyze data as they stand and bias our analysis to the type of data, which we expected *before* doing our study. Unthinking adherence to frequentist standards can literally kill, as was (almost) demonstrated on March 13, 2006, when six healthy London volunteers were in 10-min intervals injected with an experimental monoclonal antibody. The reason was to test a novel immune therapy in a first-in-man clinical trial and the result was that the volunteers nearly died of extreme immune reactions (Dowsing and Kendall, 2007). The resulting official enquiry made 22 recommendations, one of which was that "In general, new agents in first-in-man trials should be administered sequentially to human subjects with an appropriate interval between dosing of subjects to limit the number of people that may be affected by a severe adverse reaction" (Duff, 2006). This means that peeking at data is now encouraged, if rather mildly and under special circumstances.

Some Bayesians (for instance E. T. Jaynes, who was a physicist) expound the view that the logical unity of Bayesian methods, so unattainable for a frequentist, means that it is worthwhile for any scientist to learn the theory from ground-up and consequently be able to do his own thinking and derivation (and presumably to say good-bye to the statistician). Others, like mathematically educated Andrew Gelman and Brad Efron (Gelman, 2008), expound the view that the complexity of Bayesian methods makes their gifting to experimental scientists akin to a father giving his 10-year-old son the keys to his F16; not very promising result-wise but telling of the perception of the respective roles of statisticians and scientists. Be it as it may, this biologist must sympathize with Gelman and must therefore be skeptical of many instances of practical uses of Bayesian methods.

There are, however, a few simple Bayesian calculators for simple problems, which are usable to a scientist who has some idea about the basics of Bayesian logic.[6]

Here is a simple criterion by which to judge any proposed Bayes-inspired computer programs that are increasingly being offered to biologists (see also Efron, 2013). If the user cannot understand where prior probabilities enter the process and cannot actually provide input to the program, which he recognizes as prior knowledge, he should reconsider using the program. A computer program can do the calculations but it cannot know what you know about the science. If the experience of using a Bayesian program suggests otherwise, I suggest that you make some room in your hard drive.

The main weakness of Bayesianism seems to be identical with its main strength: comprehensiveness. Bayesian logic accurately integrates knowledge. Because our knowledge about the world tends to be complex, so will be the process that feeds it into the Bayesian machine. The question is then: is the Bayesian bag of tricks, as administered by statisticians, better in a given real-world situation than the experience, common sense, and heuristic methods of scientists, who really know their stuff? Obviously there is no single answer.

It may be difficult to decide if or when a Bayesian calculation trumps common sense (as common sense is not exactly a well-defined alternative), but we know well some of the advantages of Bayes over frequentism (Table 5.1):

1. Bayes uses the likelihood principle, which means that data and nothing but the data are used as evidence, while frequentists use hypotheticals (data that were actually not obtained) as evidence against a null hypothesis.
2. Bayesian solutions are much more informative than frequentist or likelihoodist ones (frequentist solution to hypothesis testing only cares about the central value—usually the mean—and likelihoodists cannot talk in terms of probabilities).
3. Bayesianism uses an intuitive concept of probability as a measure of human knowledge, it allows calculating probabilities of scientific hypotheses.
4. Bayesians make full use of the information present in data; they do not need stopping rules and naturally deal with multiplicities.

---

6. www.lifesci.sussex.ac.uk/home/Zoltan_Dienes/inference/Bayes.htm, http://tonyohagan.co.uk/1b/, http://www.mrc-bsu.cam.ac.uk/bugs/winbugs/contents.shtml.

TABLE 5.1 Comparison of Different Inferential Strategies for Hypothesis Testing

	Frequentist	Likelihoodist	Bayesian	Intuitivist
Uses prior information	No	No	Yes	Yes
Probability as	Frequency	NA	Degree of belief	Degree of belief
Deals with multiplicities	Yes (partially)	No	Yes	No
Abides by likelihood principle and principle of total information	No	Yes	Yes	No
Random sampling necessary	Yes	No	No	No
Deals with bias	Pretest	No	Posttest	Poorly
Outcome	Long-term quality of decision	Evidence that the current data provide for H_1 vis-à-vis H_2	Probability of hypothesis vis-à-vis other specified hypotheses	Plausibility of hypothesis
Optional stopping	Not allowed	Allowed	Allowed	Allowed

5. Bayesian calculation can incorporate bias while the frequentists must assume that there is no bias at data analysis stage for their activities to make sense (frequentists fight bias with randomization, which, alas, leads to a loss of information).
6. There is no null hypothesis in Bayesianism, instead there are scientifically relevant alternative hypotheses.
7. Bayesians are better equipped to deal with non-normally distributed data and small sample sizes.
8. Bayesianism makes it easier to deal with unlikely data.
9. Bayesianism is a coherent system of probability theory as logic. It makes sense!

Here are the advantages of frequentist hypothesis testing over Bayes:

1. Frequentist methods provide simple automated procedures for dealing with the multiple testing problem (although not with other types of multiplicities).
2. Frequentist methods provide long-term quality control in hypothesis testing: control of type I and type II error frequencies. (Bayesianism does not do this; but neither does it use the concept of null hypothesis and the assorted type I and II errors, which, I would argue, is a good thing for most experimentalists.)

3. The frequentist paradigm is naturally suited for checking the robustness and quality of statistical models.
4. The long history of frequentist tests provides the user with a wide choice of specific applications suitable for specific problems, making it a good strategy to quickly analyze data.

The advantage of likelihoodism: there is no need to think of any kind of probabilities. As a method of inference, likelihoodism is as "objective" as they come.

The advantage of common sense: It is quick, simple to implement, and in simple experimental designs can convert one's life experience into scientific results.

The main disadvantages of each main ideology of making sense of data are:
1. Frequentism sees your dataset as a randomly drawn member of an infinite set of all possible datasets. Its conclusions are about this infinite hypothetical set (as incorporated into the H_0 sampling distribution), not about the data that were actually obtained. All frequentist probabilities refer to properties of statistical procedures, not of your data. Frequentist analysis relies on stopping rules and penalties for peeking at data, which translates to a price in missed effects paid by the user.
2. Likelihoodism deals solely with evidence, which is a very narrow focus. Evidence, as a concept, is silent on multiplicities and, more importantly, on the probabilities of hypotheses. Strong evidence does not imply truth, nor does weak evidence imply falsity. Likelihoodism is largely restricted to mathematically simple hypotheses.
3. Bayesianism works in the long run, if it is possible to piecemeal incorporate all relevant information into a model. In real life this may turn out to be impossible to achieve in any way that doesn't use approximations and guesses. Bayesianism has the potential to propagate human error and biases under its coat of rationality. Thus it may teach us more about how to be perfect than about how to be good.
4. Common sense is biased by virtue of our evolutionary heritage. Survival is not predicated on truth but on organisms making conservative (= biased) choices. Although most snakes and spiders and strangers one meets in the woods are not dangerous, it may nevertheless be useful to assume for an individual that they all are (Diamond, 2012). Also, very small or large probabilities are unintuitive by virtue of lying outside the natural limits of human knowledge by evolutionary standards. Science is not rarefied common sense. It continuously pushes back the boundaries of human knowledge; and it has been doing so for a long time (Wolpert, 1993). If we are bounded by intuitive methods, the success of science will also be its undoing.

To conclude, let me offer some general advice with the understanding that others might see these things differently: When having significant prior knowledge, prefer Bayes. When you are in the exploratory phase of enquiry and there is no prior knowledge, LRs are often the best way to present evidence. When dealing with massively parallel testing, use FDR methods (and when the number of comparisons is relatively small, use ANOVA or the Holm-Bonfferroni procedure). When it can be believably claimed that your data are randomly and independently sampled, consider frequentist methods. If still in doubt and intuition seems useless, consult your local statistician. Think of her as of your physician and of her fixes to your problems as if they were medicines with potential side effects. Always ask about the assumptions behind statistical tests and about limits of their use and interpretation. I bet that you don't eat random medicines; neither should you use a randomly picked statistical test. As surely as too much medicine will kill you, too many statistical tests will kill your data. Always interpret statistical results with caution: all statistical methods use statistical models, which never quite equate reality (and are not meant to). Therefore, never trust a statistician to tell you what your scientific result is: it is you who is the specialist in science! And finally, in drawing scientific conclusions from your data, do not ignore intuition and experience. In practice no method is fail-safe in analyzing your data. Think!

5.5   BAYESIANISM AS A PHILOSOPHY

Bayes theorem is a valid piece of mathematics that optimally integrates information. Information in, information out—there is no controversy here, but neither is there new information about reality. Both classical and probabilistic logic are deductive and thus empty of factual content, exactly because they are universally valid (Howson and Urbach, 1989).

Bayesianism, on the other hand, is a philosophy whose claim it is that Bayesian methods are optimal to create knowledge from experimental data.[7] Information in, knowledge out. As we saw above, Bayesian methods attempt to deal not only with the data as evidence in the light of background information, but also to bring in everything we know about errors and biases in our research. The aim is to integrate all objective knowledge into the posterior probability function (or into probabilities that individual discrete hypotheses are true), which is a clear and rational measure of the true state of our knowledge on some specific question—like the true value of a parameter.

This is an ambitious goal, but perhaps not as ambitious as may first appear. Bayesianism as philosophy is still all about integrating information; it only adds to the mix the assumptions that the available information really does reflect the state of the physical world, that human biases are sufficiently quantifiable to matter in a formal calculation, and that the formal Bayes rule of integrating information is at the end of the day more successful in practical terms than any informal alternatives.

---

7. There also exists Bayesianism as a psychology, according to which actual human reasoning is best captured by Bayesian models (see Chapter 6).

Recently, Bayesianism as a philosophy of science has begun to gather scientific kudos in quantum mechanics, offering solutions to longstanding problems, like the inability of "objective" physics to separate the moment of now from past and future, which so perplexed Einstein, and to the problem of non-locality of quantum states exemplified in Bell's theorem (Fuchs et al., 2014; Mermin, 2014; Wiseman, 2014). Quantum Bayesianism (QBism) sits on the foundation of Bayesian interpretation of probability and its precepts can easily be extended to experimental biology. Generally speaking, a scientific theory is seen as a tool that any scientist can use to evaluate, on the basis of her prior knowledge, her personal probabilistic expectations for future experimental results (Fuchs et al., 2014). Behind this view of scientific method lie three fundamental assumptions about the meaning of experiment:

1. An experimental result does not preexist the experiment (note that frequentist null hypothesis testing seems to assume the opposite).
2. Assignment of probability P = 1 to an event expresses an individual's belief that the event is certain to happen, not the existence of an objective causal mechanism for that event.
3. Parameters that do not appear in the theory and correspond to nothing in the experience of any potential observer can play no role in the interpretation of results of an experiment.

A manipulation of the experimental system by the scientist cannot reveal a pre-existing state of the world. It is merely an action whose sole relevant outcome is to provide new experience for the scientist. Therefore, what is accessible of reality must differ between different scientists, who by necessity have different experiences. Of course, this subjective picture of reality is constrained by different people comparing notes about each others' experiences (ideally, using Bayes). Thus a scientific theory is nothing but a tool to organize one's personal experiences. Science does not deal directly with the objective world; it merely deals with the experiences of individual scientists of that world (Fuchs et al., 2014). In philosophical terms, the above is an instrumentalist, rather than realist, theory of science (Brown, 1994).

Bayesianism can be seen as a probabilistic logic, which can be fully deduced from first principles (Jaynes´desiderata and Kolmogorov's axioms), and of which the ordinary propositional logic is demonstrably a special case (Jaynes, 2003). As the Newtonian physics only holds at mesoscopic scales and non-relativistic speeds and masses, so is propositional logic a special case of probabilistic logic, which only holds when probabilities are fixed at 1 and 0.

As probabilistic logic, Bayesianism is in the position to solve some of the problems and paradoxes of inductive logic discussed in Chapter 2. For example, Russell's pigs on the wings parable of scientific method (see Section 2.4) can be restated as:

If A, then B
B
therefore, A is more plausible

This sort of thinking is a clear-cut error in propositional logic but Bayesian logic gives it a useful form:

$$P(A|B \& I) = P(A|I)\frac{P(B|A \& I)}{P(B|I)}$$

where I is prior knowledge (note how prior knowledge puts a necessary restriction to all probabilities).

Russell's statement was that from the statement "if pigs have wings, then some winged animals are good to eat," follows that if some winged animals are indeed tasty, then pigs must have wings (or at least they are more likely to have wings, given the evidence). However, now that we understand the role of prior knowledge (I) in probabilistic inference we can easily see that P(pigs have wings $| I$) is very close to zero and that P(some winged animals are tasty $|$ pigs have wings & I) = P(some winged animals are tasty $| I$). Therefore the fact that some winged animals are good to eat has no evidential bearing on the hypothesis that pigs have wings. Moreover, even if it did, the minuscule prior probability that pigs have wings, given the background knowledge, would surely render it insignificant in terms of posterior probability.

Bayesian logic also solves Popper's problem of probabilistic Modus Tollens. To remind you, MT looks like this:

> If A, then B
> not$B$
> therefore, not$A$

In Chapter 2 we reached a valid version of probabilistic MT that assumes that prior probability of your hypothesis is 50%. Here is a more useful version of probabilistic MT that fully uses the power of prior information.

$$P(A|\text{not}B \& I) = P(A/I)\frac{P(\text{not}B|A \& I)}{P(\text{not}B|I)}$$

We have seen that the probability of a hypothesis depends not only on whether the evidence supports it, but on how much the evidence supports your hypothesis relative to alternative hypotheses, and of their prior probabilities. All meaningful probabilities are conditional. This helps us to solve Hempel's paradox. Hempel's thinking goes as follows:

1. Premise: observing a black raven supports the hypothesis that all ravens are black.
2. This is logically equivalent to a statement that all non-black things are non-ravens.
3. Thus, observation of a non-black non-raven (a white shoe) must support the hypothesis that all ravens are black.

We can show with Irving Good that the basic error lies in thinking that in (i) the evidence can support a single hypothesis (Good, 1967). For an intuitive (or counter-intuitive, as it may be) example let's consider two alternative hypotheses. H_1 states that there are 10^6 birds in the whole world, of which 100 are ravens. H_2 states that there are 2×10^6 birds, of which 200,000 are black ravens and all the rest (1,800,000) are white ravens. We then observe one bird and this happens to be a black raven. Since the probability of observing a black raven, given that you saw a bird under H_1 is $100/10^6 = 0.0001$ and under H_2 is $200,000/2 \times 10^6 = 0.1$, it is clear that the evidence supports H_2 a thousand-fold over H_1 and thus observing a black raven gave credit to a hypothesis that all ravens are not black.

For a second example consider the hypothesis that three ravens, all slightly tipsy, left a party while wearing each other's top-hats. Should we observe raven 1 wearing a hat belonging to raven 2, have we added a degree of confirmation to the hypothesis? If you think so, what about the next observation that raven 2 is wearing the hat of raven 1? This observation actually makes it likely that raven 3 is wearing his own hat and thus disconfirms the hypothesis (Howson and Urbach, 1989). It is again apparent that the whole range of alternative hypotheses must be considered when assessing evidence.

Royall provides another distinction to bring sense into the world filled with black ravens and white shoes (Royall, 1997, pp. 177–179). It turns out that the fact that you observed a white shoe can be evidence for the hypothesis that all ravens are black, or it can be totally irrelevant. Which it is going to be depends on whether the shoes and ravens are sampled from the same population or not. If (as it seems sensible) ravens and shoes come from different sets of observations, we can randomly pick a shoe to observe its whiteness and this provides no evidence for or against any pairs of hypotheses concerning ravens. On the other hand, should we randomly pick a white object and then observe that it is a shoe, it actually is relevant evidence. This is because the object that was picked could have been a white raven, but wasn't. In other words, a randomly selected non-black object that proves to be non-raven is evidence for all ravens being black, while a randomly selected non-raven that proves to be non-black cannot serve as such.

The importance of Hempel's paradox is to point to the limits of human intuition about probability and scientific method. However, frequentist null hypothesis significance testing usually looks at the evidence in the light of a single hypothesis (H_0), and this can lead to silliness, like rejecting true H_0s with confidence (think of ESP) and being afraid of white shoes.

Bayesianism can bring light to the branch of inductive logic called abduction or inference to the best explanation; and in the process discredit scientific realism. According to the principle of inference to the best explanation, we should side with Sherlock Holmes in accepting the theory, which is in the best correspondence with facts, however unlikely the theory happens to be in the first glance. Now, it is generally agreed that the best philosophical argument for scientific realism is that it would require a miracle to explain the success of science if its theories weren't true (or getting ever closer to truth) (Chakravartty, 2007; Brown, 1994). As the above essentially states that the truth of our theories is the best explanation for the fact of the success of science, what we have here is an abductive argument. However, Bayes theorem tells us with the righteousness of logical certainty that the truth of our inferences depends on more than correspondence between facts and theory (the likelihood): it also depends on the prior probabilities of the competing hypotheses and on there actually being a true hypothesis amongst them. This makes abduction simply a case of faulty logic, which can at best be used under specific circumstances as a heuristic. It also undermines the main argument for scientific realism, which in combination with the Duhem-Quine problem (each set of facts can be equally well explained by more than one different theories) and negative meta-induction (so far, scientific theories have been replaced with alarming frequency: why should we expect the future to be any different?), should make realists shudder (Chakravartty, 2007).

Another problem, which Bayesianism helps to clarify, is Occam's razor. Many scientists take it as a general principle that one should prefer the explanation for evidence with the least number of unproven assumptions (or of unknown parameter values). For example, when discovering that fish, rabbit, and human embryos all have gill structures, which is the simplest explanation: descent from a common ancestor, independent evolution, or separate acts of creation? Simply put, the simplest explanation for the data is most likely to be true. This is the opposite of what Karl Popper believed: namely that the simplest explanation should be preferred because it is likely to be the easiest to falsify (and thus the least likely to be true).

However, Bayes theorem shows us that neither view is generally true. According to the razor $P(\text{simple theory} \mid D) > P(\text{complicated theory} \mid D)$. Yet, from the data we do not get directly such probabilities; instead we get likelihoods $P(D \mid \text{theory})$, and to get from the likelihoods to priors we can implement the Bayes factor:

$$\frac{P(\text{simple theory} \mid D)}{P(\text{complicated theory} \mid D)} = \frac{P(D \mid \text{simple theory})}{P(D \mid \text{complicated theory})} \times \frac{P(\text{simple theory})}{P(\text{complicated theory})}$$

We can immediately see that an epistemic principle alone cannot in principle give us advice on the relative probabilities of competing theories, given the data. For this we also need prior probabilities, which are essentially quantifications of our empirical background knowledge (Sober, 2010). If Bayes has taught us anything, it must be that we cannot in principle speak about truth of theories without taking into account all relevant prior knowledge AND all relevant alternative theories.

Where does all this leave Occam's razor? One solution would be to relegate it into an aesthetic principle, like "small is beautiful." This will, of course, beg the question: why should we care about the aesthetics of scientific discovery? Maybe an answer is that since simpler theories are easier to understand, they are also easier to test. This, however, has nothing to do with them being more likely to be true and invites comparison with the drunk who keeps on looking for his house keys under the lantern because that is where all the light is.

Another way is to redefine Occam's razor as the principle that greater parsimony—while not pointing to greater truth-likeness—does point to greater predictive accuracy (Forster and Sober, 1994). The goal behind predictive accuracy is to model today's data so that the resulting model would accurately predict tomorrow's data. The end result is thus not a true theory but a predictively accurate model (an instrumentalist, rather than realist, view of science). This version of the razor is explicated in Akaike's theorem (see Chapter 2), which provides an unbiased method to choose from a list of models, the one where fit of present data to the model is best combined with the least number of variables in the model (basically, the aim is to choose the simplest predictively accurate model).

The Bayesian may claim with some justification that Bayesianism is the logic of science. Everybody is comfortable with the proposition that deductive logic is only as good as are the premises that we use. Logic just provides the fail-safe algorithms, which transform the information present in the premises into conclusions. This is the same in Bayesian logic, which transforms two conditional probabilities (probability of the hypothesis, given the prior information and probability of the data, given the hypothesis) into a single conditional probability (the posterior probability). Although Bayesian logic is often seen as inductive, its method of integrating information is deductively valid. The inductive (and extra-logical) part of Bayesianism is to suppose that the posterior probability can inform us about the validities of general scientific theories.

Here it must be reminded that all Bayesian probabilities depend on particular mathematical models of data and of prior knowledge, and thus on a model of the world. Therefore Bayesianism cannot overcome Humean skepticism concerning causal inferences and the uniformity of nature. The uniformity of nature is already built into the models of which Bayesian probability statements depend.

Another consequence of the relative nature of probability is that we cannot really say which hypothesis from all possible hypotheses concerning a phenomenon under study, is most likely to be true. This is because we restrict our conclusions to the hypotheses that are explicitly tested in the Bayesian analysis (the hypotheses whose prior probabilities and likelihoods are explicated in $P(D)$).

For example, in solving a murder mystery we may test the data (a bloody knife in the butler's cupboard) in the light of two hypotheses H_1: the butler did it and H_2: someone else did the deed, and find that H_1 is much more probable than H_2. At the same time the true hypothesis—H_3 (or H_{bias}): there was no crime but a suicide, and the shiny knife was transported to butler's cupboard by a white crow—is never even considered. It is entirely possible that available information gives the truth (which is H_3) a very low prior probability and it doesn't thus really matter how well we estimate the priors for H_1 and H_2. The result, after application of Bayes theorem, will be entirely rational and entirely wrong. We may even be unaware of the level of uncertainty in our determination of H_1 if the determinants of H_3 fall, for us, into the class of the unknown unknowns. This is not a weakness in Bayesian inference: all it shows is that limited information does not necessarily beget truth.

To generalize, it is possible to slice the cake of all the possible hypotheses in many ways and the posterior probabilities crucially depend on this slicing. There is no reason to say that our winning hypothesis is the most likely to be true (it is only the most probable hypothesis from the lot that we tested, given the statistical modeling). Bayesianism as a philosophy of science was ably described by Niels Bohr in 1929: "In our description of nature the purpose is not to disclose the real essence of the phenomena but only to track down, so far as it is possible, relations between the manifold aspects of our experience" (cited in Fuchs et al., 2014).

Nevertheless, in the Bayesian mode, we can with some confidence exclude the hypotheses whose posterior probabilities are strongly pushed down by accumulating evidence. Bayesianism, seen in this light, is not confirmatory—it cannot confirm the truth of a single scientific theory. Instead it is falsificationist, which means that it should be seen as an optimal way to weed out hypotheses, which are not supported by evidence (Jaynes, 2003). Bayesian analysis of discrete hypotheses is thus a more general case of Popperian falsificationist logic expanded to probabilistic hypotheses and inferences (which is something that Popper himself tried hard and unsuccessfully to do within the frequentist paradigm (Popper, 1992)). A major improvement of Bayes over Popper is that in Bayesian analysis we can not only falsify theories but we always determine which theory of the specified alternatives is the most likely one, after considering both background knowledge and the new data. This allows us to retain successful theories for further use in a more principled manner than it was possible for Popper, whose brand of falsificationism essentially deals with one theory at a time, leaving it vulnerable to paradoxes like Hempel's ravens. Because in the Bayesian scheme we always have more than one theory at play, it makes a lot of sense that all theories with non-negligible prior probabilities should be monitored for signs of life (i.e., increases in posterior probabilities) as new evidence comes in (Chang, 2014). Thus Bayesianism should be seen as a more pluralistic system of thought than classical Popperian falsificationism.

Bayesianism is often taught as having the ability of converging on the true parameter value in continuous parameter estimation. This convergence is, however, far from preordained as it depends on statistical models of the prior probabilities and likelihoods. The major difference between Bayesian hypothesis testing and parameter estimation is the size of the sample space. In hypothesis testing the sample space consists of the discrete hypotheses under consideration, while in parameter estimation it consists of an infinite number of potential parameter values. Be the resulting probability distribution discrete or continuous, the possibility of committing errors in drawing it remains. In case of hypothesis testing the hard part is hypothesis specification (determining the sample space) and with specifying continuous distributions it is choosing from the many different probability functions available.

5.6 BAYESIANISM AND THE PROGRESS OF SCIENCE

The version of Bayes theorem that uses frequentist long-run error rates, which is often used to interpret results of medical testing, can be tempting to use in estimating the success of hypothesis testing in science (Weitkunat et al., 2010).

$$P(\text{true effect}|\text{data}) = \frac{\pi(1 - \beta)}{\pi(1 - \beta) + (1 - \pi)\alpha}$$

$P(\text{true effect} \mid \text{data})$—also called true report probability and positive predictive value in some methodological and clinical contexts, respectively—is the posterior probability of a single measured effect being true, π is the prior probability of said effect and α and β are the long-run error properties of the measurement (namely, type I and II error rates). This approach takes into account the sample size, effect size, variation and specificity of measurements, but it ignores all sources of systematic error or bias. Therefore the strong assumption here is that experimental design is perfect and there is no bias. The logic of Bayesian testing tells us that if we think that the probability of our effect being caused by bias is anywhere near the prior probability of the hypothesis under test, then we must submit to the principle of total information and withdraw our support to the above equation.

If $\alpha = 0.05$, $1 - \beta = 0.8$ (on average 80% of real effects are duly observed in the experiment), and $\pi = 0.5$, then the posterior probability of the effect not being caused by sampling error comes out as 94%. This number corresponds to a perfectly planned and executed (sufficient sample size) but rather mundane experiment. Should the experimental result be described as "surprising" the prior probability goes down, to 0.1 say, and the posterior probability becomes 0.64, meaning that it is still more likely than not that the effect is real, but you probably would not bet your career on it. If, in addition, we consider that many pre-clinical studies have power at the range of 0.25, the posterior becomes 0.36, meaning that it is now more likely than not that the effect is due to sampling error.

This probability is still considerably higher than the 0.11 estimate of true positives for high-profile pre-clinical cancer studies (Begley and Ellis, 2012), but then, explanations for their data concern (at least) as much bias as they do random error (Begley, 2013). To arrive at a more realistic model for the growth of scientific knowledge, we must incorporate bias into the equation:

$$P(H|D) = \frac{P(H)(1 - \beta)}{P(H)(1 - \beta) + P(\text{bias}) + P(\text{sampling error})\alpha}$$

Here $P(H \mid D)$ is the posterior probability of seeing a true effect, given the experimental data, $P(H)$ is the prior probability of true effect of a given size, $P(\text{bias})$ is the prior probability of the effect being due to bias and $P(\text{sampling error})$ is the prior probability of it being due to sampling error. Note that as every observed effect can be fully explained, given the presence of bias, the likelihood of bias $P(\text{data} \mid \text{bias})$ is always 1 and is thus not explicated in the equation. Fixing α at 0.05 and changing the probability of bias (a single measure for the quality of all aspects of experimental planning and execution), power (a measure of sample size and intra-sample variation), and prior probability of the tested scientific hypothesis (a measure of the novelty and potential impact of the research, given a positive finding), we can develop a feeling about the extent that these parameters influence the post-study reliability of any positive findings (Table 5.2).

This type of analysis was famously used by John Ioannidis to make a case that most research findings in biomedicine are false (Ioannidis, 2005).[8] Here it is important to note that while both bias and low power have the potential to do serious harm to "ordinary" biology studies (and, as discussed in Chapter 3, low power begets bias), the really low posteriors appear in omic settings, where the number of tested associations can be in the millions and there is no prior expectation as to the identities of true associations. Here apparently even a full replication cannot help.

Another corollary from Table 5.2 is that a series of low-sample-size, high-bias exploratory studies cannot in the long run increase scientific knowledge. In principle, they can only reduce it. Moreover, even well-conducted explanatory studies are in danger of being irrelevant for the growth of science (when

TABLE 5.2 Posterior Probability That Scientific Studies Reveal True Effects, $P(H \mid D)$, Depends on Prior Probabilities of Competing Hypotheses and on Statistical Power

$P(H)$	P(bias)	P(sampling error)	$1 - \beta$	$P(H \mid D)$	Comments
0.5	0.1	0.4	0.8	0.77	Well-powered and -conducted clinical trial
0.2	0.2	0.6	0.8	0.18	Well-powered and -conducted exploratory study
0.1	0.4	0.5	0.3	0.066	High-impact low-sample-size exploratory study that uses a complicated experimental system
0.5	0.4	0.1	0.3	0.27	Low-impact low-sample-size exploratory study that uses a complicated experimental system
0.001	0.2	0.8	0.9	0.004	A well-planned and conducted omic study that looks at 10,000 genes and expects to find true effects at ten genes
0.004	0.2	0.8	0.9	0.015	A full replication of the above study

Source: Modified from Ioannidis (2005).

8. As Ioannidis used an *ad hoc* calculation in lieu of Bayes theorem, his posteriors are slightly different.

$P(H) \approx P(\text{bias})$)! The big *if* here is that $P(\text{bias})$ is really a catch-all hypothesis lumping together all types and instances of bias. Therefore it is very difficult to assign a single probability to it—and I'm not trying to imply that it is easy with $P(H)$. However, our inability to fully quantify the problem does by no means reduce the seriousness of it. Indeed, it can be argued that a greatly improved understanding of various biases will be a major component in any serious approach to the problem of the growth of biological science.

The conclusions thus far are:

1. High-impact, low $P(H)$ science, that uses complicated experimental systems (high $P(\text{bias})$) to produce unexpected and surprising results, is worse than useless. It may actually reduce the usefulness of the boring everyday science, which uses well-understood methods, by poisoning the well with false leads and thus corrupting the prior probabilities for all scientists—high flying or not.
2. High-throughput omic studies are useless even as exploratory hypotheses-generating mechanisms and will remain so under any realistic estimates of bias. At any rate, this is pretty much what John Ioannidis implicates in his provocative piece (Ioannidis, 2005) and what the empirical results of Begley and others seem to support (Begley and Ellis, 2012). Yet, John Ioannidis hasn't abandoned science as a hopeless endeavor (and indeed published well over 400 papers from 2006 to 2013). One has the impression that a major focus of his and others' methodological work has been on finding optimal ways to use traditional statistical methods to ensure reproducibility of high-throughput studies. A great deal of effort in omic studies thus seems to be directed to generating huge sample sizes, which would help to reach P values below 10^{-5} to 10^{-8}, which are thought sufficient for reproducibility of the results (Broer et al., 2013; Dudbridge, 2013).

There is also some debate about whether we should prefer a single large well-powered study (which is good for excluding sampling error as a cause for observed effect) or many small and thus underpowered studies (in the hope that many different biases introduced by many different groups will cancel themselves out)—a delicate discussion where some participants seem to be holding opposing positions simultaneously (IntHout et al., 2012; Batterham and Hopkins, 2013; Ioannidis et al., 2013).

The problems with using extremely small P values in conjunction with reproducibility of effects as an argument for the validity of real-world omics results are threefold. Firstly, Table 5.2 shows that successful replication of results, which are very unlikely to begin with, is far from being an indicator of truth. Moreover, if the biases inhabiting your experimental system are greater than the prior probability of the results of being true, then it doesn't even make sense to attempt a replication!

Secondly, the extreme sample sizes (often in the tens of thousands) required to achieve such P values make it hard to interpret the evidentiary meaning of the resulting P values. Remember that as the sample size increases the SD of the statistical null model, under which the P value will be calculated, decreases proportionally to the square root of the sample size. Thus a very large sample will result in a null model approaching the point value of the null hypothesis, and the problem with this is that, mathematically, the probability of our experimental value *exactly* equating the null value is itself null. This, of course, will result in a mandatorily small P value.

And thirdly, because bias does not go down with increasing sample size (as the sampling error does), one must suspect that the smaller the P value, the smaller will be the observed effect size, which will then be correspondingly more likely to be an instance of bias. Thus, when the sample is large enough, even a tiny bias can lead to an extremely small P value. As the samples grow, we will be increasingly making highly statistically significant discoveries of dubious biological worth, until at some arbitrarily small effect size we can be virtually certain that all observed genetic associations are instances of bias. This happens for the simple reason that as hard as we might try, we will never be *completely* successful in eradicating bias from our experimental design. Indeed, increasing our sample sizes leads in practice to an exponentially increasing number of statistically significant associations (Flint and Munafo, 2014).

To see how deeply the null hypothesis testing paradigm is entrenched in the GWAS community, suffice it to cite a recent *Nature* expert commentary, whose aim seems to be to convince us non-experts that a very large GWAS of schizophrenia—rather than its smaller predecessors—is the paragon of truth:

> *Not unreasonably, many would ask why we should be any more confident that the consortium's geneticists have now got it right. The short answer is that the tests for associations between each gene in the human genome and disease are now mature. The correct criteria for determining significance in these tests are as familiar to human geneticists as their two-times table....*
>
> (Flint and Munafo, 2014)

The study in question (Ripke et al., 2014) identified 128 statistically significant genetic associations with schizophrenia at 5×10^{-8} significance level and the only serious criticism of it in the accompanying *Nature* commentary was that "Having climbed a mountain of genetic analyses, it is perhaps a shame that the authors do not offer much of a view from the peak—no new biological hypotheses are presented" (Flint and Munafo, 2014). To this we might add that Ripke et al. combined their primary sample from 49 different ancestry-matched, case-control samples (46 of European and three of east Asian ancestry, 34,241 cases and 45,604 controls), to which they added three family-based samples (1,235 parent affected-offspring trios). Now, there must be a great deal of trust involved in this sampling procedure, because at the resulting huge sample sizes and correspondingly small P values even a small bias in sampling can conceivably bias the results.

What would then be the best way out of this mess of low priors leading to low posteriors in massively parallel testing? The answer suggests itself from our breast cancer diagnosis example in the very beginning of this chapter. The devil is in the reference population. In the mammography scenario the reference population, whose incidence of cancer we used as our prior probability, is the general population of women that are invited to participate in the screening—unless of course the woman who is screened has a pre-diagnosed BRCA1 mutation! In this case the reference population is the subgroup of the first containing individuals carrying the same mutation, and the incidence of breast cancer in this population is very much higher. And so is our prior probability.

Thus, adding a little further knowledge to the mix dramatically changes the odds of a positive test result showing the presence of disease. The same logic applies in addressing the validity of omic studies. If we use additional biological knowledge to differentiate between candidate genes, RNAs, or proteins, then the prior probabilities will be vastly different for different units (be they genes, RNAs, or proteins) of the massively parallel analysis and we will find ourselves in the position where evidence of very similar strength can lead to completely different interpretations concerning different units of analysis.

Specific tricks for increasing the prior probabilities in massive parallel testing situations include sorting hypotheses to sets in which they are likely to be true or false together, weighting the hypotheses tested by their differing importance or different prospects for showing effects, and putting the hypotheses into a hierarchical structure where a subfamily of hypotheses at a branch is tested only after the node from which it branches is tested and rejected (this approach assumes that the hypotheses in a branch tend to be true or false together) (Benjamini, 2008).

All proposed fixes boil down to a common exhortation: you should fully use the background knowledge. Massive datasets do not by themselves contain the information needed to analyze them—the key to understanding lies in your knowledge of your scientific problem and of the workings of the experimental system.

5.7 CONCLUSION TO PART II

To summarize the roles played by method and methodology in scientific discovery we need to state the main influences to it and their role in introducing various types of errors (Figure 5.3A). In the classical view of Popper and Hempel, hypothesis testing is a conceptually simple linear affair where we proceed from a general hypothesis to experimental testing of some logical deduction from it, and from there either to falsification (in the logical path) or corroboration (in the statistical path) of the hypothesis. A more modern picture shown in panel B reveals three main contributors to hypothesis testing, one of which—perhaps unsurprisingly—is the world we live in. Experiments are informed by theories under testing, by our background knowledge, and by objective reality. We know that our theories and background knowledge are made of language, but we are more ignorant of the stuff that the world is made of. This is because as scientists we have no direct access to the world, except through the effects that it has on our experimental systems. Because we use language cast into scientific theories

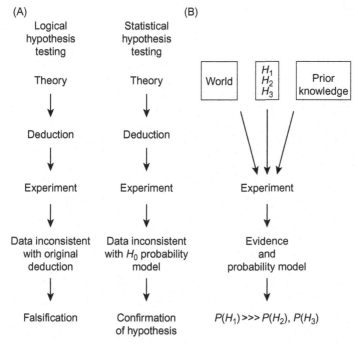

FIGURE 5.3   A comparison of classical explanatory schemes of hypothesis testing (A) with a more modern one (B).

and mathematical models to create and interpret our experiments, we cannot unambiguously infer from their results back to the single true theory of the world, which we haven't explicated and of whose structure we have no idea. This is the fundamental difficulty of science: scientific method enables us to compare theories but gives no hint as to closeness of the structure of our theories to the structure of the world.

Of course, the output of our experiment depends on the quality of its inputs, and with each input come their own errors. From the world comes random error caused by limited accuracy of measurement and by biological variation. We have good statistical methods for dealing with this type of error. From imperfect background knowledge comes systematic error or bias, which misleads us in a directional way. Bias can be quantified and even corrected for, depending on our ability to understand and quantify not only our level of knowledge but also our level of ignorance. And finally, from the mismatch between theories and the world they try to describe comes another type of error, which I hereby christen as "metaphysical error," simply to draw attention to the intractability of it by scientific method. To even begin to address this type of error we need to turn to philosophy as much as to science.

The corollary is that simply by studying methodology of science, we can never hope to explain good science, nor learn how to do it.

REFERENCES

Andrieu, C., et al., 2003. An introduction to MCMC for machine learning. Mach. Learn. 50 (1–2), 5–43.

Batterham, A.M., Hopkins, W.G., 2013. Emergence of large treatment effects from small trials. JAMA 309 (8), 768–769.

Begley, C.G., 2013. Six red flags for suspect work. Nature 497 (7450), 433–434.

Begley, C.G., Ellis, L.M., 2012. Drug development: raise standards for preclinical cancer research. Nature 483 (7391), 531–533.

Bem, D., Tressoldi, P.E., Rabeyron, T., 2014. Feeling the future: a meta-analysis of 90 experiments on the anomalous anticipation of random future events. Available at SSRN: <http://ssrn.com/abstract=2423692> or <http://dx.doi.org/10.2139/ssrn.2423692>.

Bem, D.J., 2011. Feeling the future: experimental evidence for anomalous retroactive influences on cognition and affect. J. Pers. Soc. Psychol. 100 (3), 407–425.

Benjamini, Y., 2008. Microarrays, empirical Bayes and the two-groups model. Stat. Sci. (1), 23–28.

Berry, D., 2012. Multiplicities in cancer research: ubiquitous and necessary evils. J. Natl. Cancer Inst. 104 (15), 1125–1133.

Berry, D.A., 2007. The difficult and ubiquitous problems of multiplicities. Pharm. Stat. 6 (3), 155–160.

Broer, L., et al., 2013. Distinguishing true from false positives in genomic studies: p values. Eur. J. Epidemiol. 28 (2), 131–138.

Brown, J.R., 1994. Smoke and Mirrors: How Science Reflects Reality (Philosophical Issues in Science). Routledge, London.

Chakravartty, A., 2007. A Metaphysics for Scientific Realism: Knowing the Unobservable. Cambridge University Press, Cambridge, UK.

Chalmers, I., et al., 2014. Research: increasing value, reducing waste 1 how to increase value and reduce waste when research priorities are set. Lancet 383, 156–165.

Chaloner, K., Verdinelli, I., 1995. Bayesian experimental design: a review. Stat. Sci. 10 (3), 273–304.

Chang, H., 2014. Is Water H_2O? Springer, Heidelberg.

Cooper, N.J., Jones, D.R., Sutton, A.J., 2005. The use of systematic reviews when designing studies. Clin. Trials 2 (3), 260–264.

Couzin-Frankel, J., 2014. Unknown significance. Science 346 (6214), 1167–1169.

Diamond, J.M., 2012. The World Until Yesterday: What Can We Learn from Traditional Societies? Viking, New York, NY.

Dowsing, T., Kendall, M.J., 2007. The northwick park tragedy–protecting healthy volunteers in future first-in-man trials. J. Clin. Pharm. Ther. 32, 203–207.

Dudbridge, F., 2013. Power and predictive accuracy of polygenic risk scores. PLoS Genet. 9 (3), e1003348.

Duff, G., 2006. Expert Scientific Group on Phase One Clinical Trials Final Report. Stationery Office, Norwich.

Efron, B., 2013. Bayes' theorem in the 21st century. Science 340 (6137), 1177–1178.

Flint, J., Munafo, M., 2014. Schizophrenia: genesis of a complex disease. Nature 511, 412–413.

Forster, M., Sober, E., 1994. How to tell when simpler, more unified, or less *ad hoc* theories will provide more accurate predictions. Br. J. Philos. Sci. 45 (1), 1–35.

Fuchs, C.A., Mermin, N.D., Schack, R., 2014. An introduction to QBism with an application to the locality of quantum mechanics. Am. J. Phys. 82, 749.

Gelman, A., et al., 2003. Bayesian Data Analysis, Second Edition, second ed. Chapman and Hall/CRC, New York, NY.

Gelman, A., 2008. Objections to Bayesian statistics. Bayesian Anal. 3 (3), 445–450.

Good, I.J., 1967. The Estimation of Probabilities. MIT Press, Cambridge, MA.

Gower, B., 2002. Scientific Method: An Historical and Philosophical Introduction. Routledge, London, pp. 1–285.

Greenberg, S.A., 2009. How citation distortions create unfounded authority: analysis of a citation network. BMJ (Clinical research ed.) 339 (3), b2680.

Honorton, C., Ferrari, D.C., 1989. Future telling: a meta-analysis of forced-choice precognition experiments, 1935–1987. J. Parapsychol. 53, 281–308.

Howson, C., Urbach, P., 1989. Scientific Reasoning: The Bayesian Approach. Open Court, Chicago, IL.

IntHout, J., Ioannidis, J.P., Borm, G.F., 2012. Obtaining evidence by a single well-powered trial or several modestly powered trials. Stat. Methods Med. Res. Available from: http://dx.doi. org/0962280212461098.

Ioannidis, J., Pereira, T., Horwitz, R., 2013. Emergence of large treatment effects from small trials—reply. JAMA 309 (8), 768–769.

Ioannidis, J.P.A., 2005. Why most published research findings are false. PLoS Med. 2 (8), e124.

Jaynes, E.T., 2003. Probability Theory: The Logic of Science. Cambridge University Press, Cambridge.

Lindley, D.V., 1972. Bayesian Statistics: A Review. SIAM, Philadelphia, PA.

Mason, R.L., Gunst, R.F., Hess, J.L., 1989. Statistical Design and Analysis of Experiments: With Applications to Engineering and Science, first ed. Wiley, Hoboken, NJ.

McGrayne, S.B., 2012. The Theory That Would Not Die: How Bayes' Rule Cracked the Enigma Code, Hunted Down Russian Submarines, and Emerged Triumphant from Two Centuries of Controversy. Yale University Press, New Haven, CT.

Mermin, N.D., 2014. Physics: QBism puts the scientist back into science. Nature 507 (7493), 421–423.

Miller, A.B., et al., 2014. Twenty five year follow-up for breast cancer incidence and mortality of the Canadian National Breast Screening Study: randomised screening trial. BMJ (Clinical Research ed.) 348 (feb11 9), g366.

Phillips, C.V., LaPole, L.M., 2003. Quantifying errors without random sampling. BMC Med. Res. Methodol. 3, 9.

Popper, K.R., 1992. The Logic of Scientific Discovery. Routledge, London; New York, NY.

Quinn, G.P., Keough, M.J., 2005. Experimental Design and Data Analysis for Biologists. Cambridge University Press, Cambridge, UK, pp. 1–557.

Ripke, S., et al., 2014. Biological insights from 108 schizophrenia-associated genetic loci. Nature 511 (7510), 421–427.

Robinson, K.A., Goodman, S.N., 2011. A systematic examination of the citation of prior research in reports of randomized, controlled trials. Ann. Intern. Med. 154 (1), 50–55.

Rouder, J.N., Morey, R.D., 2011. A Bayes factor meta-analysis of Bem's ESP claim. Psychon. Bull. Rev. 18 (4), 682–689.

Royall, R., 1997. Statistical Evidence. CRC Press, Boca Raton, FL.

Seltman, 2013. Experimental Design and Analysis. Carnegie Mellon University, <www.stat.cmu. edu/~hseltman/309/Book/Book.pdf>, pp. 1–428.

Siontis, K.C., et al., 2014. Diagnostic tests often fail to lead to changes in patient outcomes. J. Clin. Epidemiol. 67 (6), 612–621.

Soal, S.G., Bateman, F., 1954. Modern Experiments in Telepathy, first ed. Yale University Press, New Haven, CT.

Sober, E., 2010. Parsimony arguments in science and philosophy—a test case for naturalism. Proc. Add. APA 83 (2), 117–155.

Storm, L., Tressoldi, P.E., Di Risio, L., 2010. Meta-analysis of free-response studies, 1992–2008: assessing the noise reduction model in parapsychology. Psychol. Bull. 136 (4), 471–485.

von Toussaint, U., 2011. Bayesian inference in physics. Rev. Mod. Phys. 83 (3), 943–999.

Weitkunat, R., et al., 2010. Effectiveness of strategies to increase the validity of findings from association studies: size vs. replication. BMC Med. Res. Methodol. 10, 47.

Welch, H.G., Passow, H.J., 2014. Quantifying the benefits and harms of screening mammography. JAMA Intern. Med. 174 (3), 448–454.

Wiseman, H., 2014. Bell's theorem still reverberates. Nature 510, 467–469.

Wolpert, L., 1993. The Unnatural Nature of Science. Harvard University Press, Cambridge, MA.

Part III

The Big Picture

Chapter 6

Interpretation

In this chapter we will take a step back from the abstract normative accounts of science and ask how real people draw conclusions from scientific evidence. In the actual scientific practice for many scientists the experiments are expensive and time-consuming, which means that the sample sizes tend to be low. The hypotheses that are tested are often not explicated until after seeing the data and trying to rationalize it for publishing. The results of individual experiments can be treated in the framework of propositional logic ("this result shows that ...," "a typical result is shown in figure x") or, alternatively, by probability theory/ statistics (the results are given as average values with "error-bars"), and logical and probabilistic interpretations of experimental results are routinely allowed to mingle. Indeed, it is common to derive nonprobabilistic conclusions from probabilistic statistical inferences.

6.1 HYPOTHESIS TESTING AT SMALL SAMPLES

While the sample sizes in clinical studies can easily be in the thousands, the commonest sample size in molecular biology is three (in experiments involving mice it is typically a little higher—around a dozen by some accounts (Scott et al., 2008)). Interpretation of results from very small samples raises difficulties. Even if in a large sample the average of many experimental results converges to true value, in a small sample, due to random variation, the sample statistic is likely to be substantially different from the population value (Figure 6.1). Accordingly, the size of confidence intervals (CIs) for the mean depends inversely on the sample size. However, confidence intervals also depend on the variance in your statistical population, which in small samples can be as easily under- or overestimated as their means. Thus with small samples individual estimates of the true population value can easily turn out to be both wrong and overconfident. Wherever statistically significance is used to call effects, this property of small samples presents an obvious problem. Another problem with data from small samples is that we cannot be sure of their distribution.

When the sample size is small, the standard deviation (SD) value, as calculated by the traditional method, is likely to underestimate true population SD even when the data come from a normal distribution (Figure 6.2). Therefore, at

Interpreting Biomedical Science. DOI: http://dx.doi.org/10.1016/B978-0-12-418689-7.00006-5

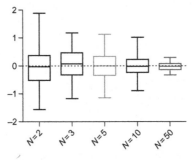

FIGURE 6.1 Variation of the mean at different sample sizes. One hundred experimental series of each sample size were simulated from a standard normal distribution (mean = 0, SD = 1). The distribution of the means are shown as box-plots.

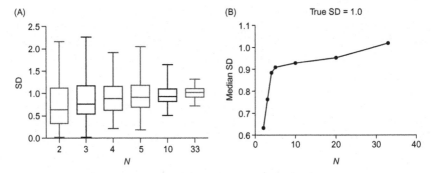

FIGURE 6.2 Small samples lead to large uncertainties and bias in estimating population SD. (A) Medians and interquartile ranges of SDs were calculated from differently sized samples drawn from a standard normal distribution (mean = 0, SD = 1). One hundred and thirty independent samplings were done at each sample size. SD, standard deviation; N, sample size. (B) Calculating SD at very small sample sizes underestimates true variation in statistical population. Median SD calculated at each sample size plotted against sample size. The population SD = 1.

least with small sample sizes, when you only want to look at the variation in the data, there is usually no point in losing hard-won information by modeling the data into a mathematical function of which you have no way of knowing if it really fits. It is better to simply plot the individual data points.

Also, small samples tend to give highly divergent (and thus inaccurate) estimates of the population SD. Each sample will yield its own statistical estimates (like the mean and SD) and with repeated small samples these tend to be further apart from each other than with larger samples. When a small sample happens to overestimate the population SD, the sensitivity of ensuing statistical inferences goes down and interesting biological effects will be missed. When the SD happens to underestimate true population SD, the specificity of the statistical test will go down and biologically spurious but statistically significant effects will

TABLE 6.1 The 95% Confidence Intervals (CIs) for Standard Deviation (SD) = 1 Calculated at Various Sample Sizes	
Sample Size (*N*)	95% CI for SD = 1
2	0.45–31.91
3	0.52–6.28
4	0.57–3.73
5	0.60–2.87
10	0.69–1.83
100	0.88–1.16

become more common. As an example we will look at variation in SDs calculated from many independent samples of varying size, which are drawn from the same normally distributed population (Table 6.1).[1] For example, if our SD = 1 and $N = 3$, then the 95% CI for SD lies between 0.52 and 6.28. When $N = 3$, it is thus quite easy to misrepresent the true SD by several fold.

Sadly, this means that with very small samples, the SD is a pretty useless measure. Because sampling effects are not only relevant to predictive accuracy of descriptive statistics (mean, SD, correlation coefficient, etc.) but also for statistical estimation (CI, regression coefficient, etc.), the problems with small samples are disturbingly universal. Even worse, because the large variation in SDs is a direct consequence of the sampling effect, its harmful effects on scientific inference do not disappear when we ditch statistical analysis altogether and confine ourselves to intuitive data analysis. Intuitively we are likely to grossly underestimate the magnitude of effects of small samples on our scientific conclusions (see below).

This raises the question of whether the methods of analysis, which were created for larger studies, can be successfully implemented in the very small ones, which are the norm in biology. Let's suppose that you are looking at the data from three experimental replications of a mutant versus wild-type experiment, and see a large effect. Do you need to check it by a formal method by calculating *P* values or CIs? This may be a good idea, because what to consider a large effect depends on the level of variation in the statistical populations, which are compared. When variation of results is small from experiment to experiment, the effect size that can present strong evidence for the reality of the effect is correspondingly smaller.

---

1. See http://www.graphpad.com/guides/prism/6/statistics/index.htm?stat_confidence_interval_of_a_stand.htm.

What about the situation where you do not see an obvious effect? Is it reasonable to run some statistical calculations in the hope that perhaps you will find a significant effect anyway? The answer is that unless you suffer from overabundance of pessimism and critical thinking, you should probably trust your intuition about there being no evidence for an effect. (There is evidence that we tend to be too optimistic about evidence to the extent that depressive individuals are more likely to judge evidence objectively than healthy people (Moore and Fresco, 2012).)

A reason for this advice is that it can be difficult to decide which statistical test is appropriate for your data. Parametric tests, like Student's t-test, are very sensitive, but with small samples they assume normal distribution for the data. With small sample sizes it may not be possible to decide if the data are normal enough to justify a parametric test. Data describing biological growth processes or enzyme kinetics are almost certainly log-normal and need to be transformed before running the test (see Section 4.2). The nonparametric tests, like the Mann–Whitney test, do not assume normal or any other distribution for the data, but they are insensitive at low sample sizes ($n < 10$) and should never be used when $n < 5$. However, when data distribution is skewed and $n > 25$, then nonparametric tests can have greatly improved statistical power over the student's t-test (Krzywinski and Altman, 2014). In addition, nonparametric tests apply naturally to data that are already in the form of ranks or degrees of preference. With larger samples (40+) it should not make much difference if a parametric or a nonparametric test is used to calculate P values, as the nonparametric tests are only slightly less sensitive and the parametric ones do not depend on the normality assumption.

To put a little experimental meat on these theoretical bones, let's work through a data analysis problem. The analysis is of an experiment where 14 mice were injected with equal numbers of tumor cells. In seven mice (NC) the tumors were left to grow in peace and then weighed and in the other seven experimental treatment was administered some days before measurement of tumor masses. The results were are shown in Table 6.2.

We will start by drawing a picture of our results with median and each data point shown (Figure 6.3A). Most mice in the experimental group have reduced tumors compared to most mice in the control group. But not all of them. So, let's calculate the P value using unpaired student's t-test: $P = 0.0623$ and the 95% CI for the difference between means of the experimental group and the control group is from -0.70 to 0.02. Thus, whatever our intuition, formally we cannot be sure that there is a difference. But hold on, our data do not look like they came from a normal distribution. Let's next analyze them by a nonparametric Mann–Whitney test. Now $P = 0.053$, which is better, but still not statistically significant.

We know that nonparametric tests are less sensitive than parametric ones, so maybe we should log-transform our data to make it more normal and then do the usual parametric t-test (Figure 6.3B)? Now $P = 0.042$ and 95% CI stretches

TABLE 6.2 Tumor Masses (in Grams) in Controls and Experimentally Treated Mice

Tumor Mass in Grams in the Control Mice (NC)	Tumor Mass in the Treated Mice (Exp)
0.95	0.11
0.82	0.29
1.13	0.39
0.50	0.42
0.65	0.18
0.71	0.16
0.21	1.04

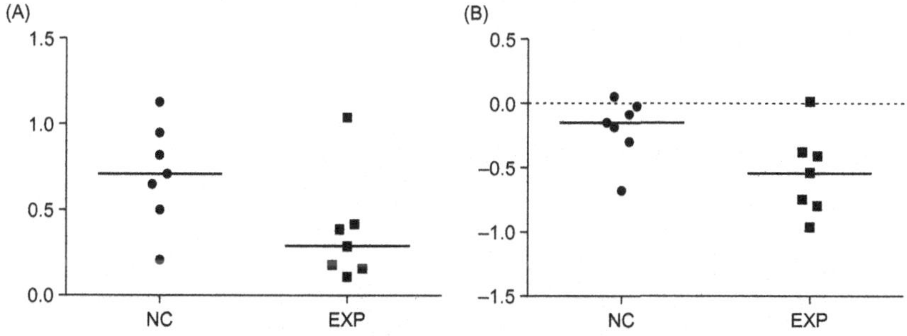

FIGURE 6.3   Plots of (A) linear and (B) log-transformed cancer masses. NC, negative control.

from −0.68 to −0.01, which means that we finally see a statistically significant difference, but the predicted effect size is likely to be scientifically insignificant (as small as a 0.01 gram reduction in tumor mass).

This is not good, so let's take another look at the data. The last value in the control group is much smaller than others, suggesting that mouse no. 7 was poorly injected with tumor cells. If we call this datum an outlier and remove it from analysis, then our t-test for nontransformed data will give $P = 0.0198$ and 95% CI extending from −0.765 to −0.08, a much prettier result. If we in addition remove the outlier (high value of mouse no. 14) from the experimental group, $P = 0.0005$ and 95% CI = −0.77 to −0.30, an excellent result.

So, which P value and CI should we publish? There is a good argument for excluding the outlier from the control condition: we know that injecting tumor cells into mouse veins is not a perfect experiment and 10% error rates are not unusual. If we failed to get the tumor cells into the vein in sufficient quantity in mouse 7, we predictably had not much of a tumor to measure. But hold on—the same argument goes for the experimental group, where a missed vein means a falsely optimistic value (a false cure). Because the experimental treatment was also administered intravenously, the same missed vein caveat applies. So could we just remove the single high value from the treatment group and leave the control group alone? By doing so we will get $P = 0.0059$ and 95% CI from -0.75 to -0.16.

Also, let's not forget the danger of multiplicities: it makes a difference whether we actually conducted one test or six (as we did here). Also note that because the tests that we did are not independent (they all used pretty much the same data) and we had (at least some) rational excuse for each one of them, the ordinary multiple testing corrections do not apply. This adds another layer of intuitive inference into the game.

All this adds up to the realization that if the samples are small and if, based on your data, you intuitively believe in the presence of an effect, then you are probably not going to change your mind if the calculated P value comes as slightly over 0.05. Should you agree with me on this, then neither should you be inclined to change your mind if you intuitively see no effect but are nevertheless able to calculate a smallish P value.

The danger in intuitive approaches is that even perfectly honest researchers tend to be biased in judgment, and that they are so in predictable ways, which is a recipe for overestimation of effects and for generally unwarranted optimism (MacCoun, 1998). For example, we are naturally not very good in thinking about probabilities, especially extreme ones (below 10% or above 90%), and we tend to overestimate the frequency of events, on which we habitually think, like obtaining positive evidence for our pet theories or getting murdered on the street (Kahneman, 2011). In addition, anchoring effects lead people to bias their assessment of unknown parameter values by shifting their estimates toward any numerical value, relevant or not, which is on their mind at the time (Tversky and Kahneman, 1974). When making decisions, people thus often rely on the first piece of information offered to them. An example of anchoring is that when experts first guess at a point value for a parameter, their subsequent estimates of the range of uncertainty become narrower than is the case when the range is estimated before estimating a mean value (Morgan, 2014). A neat piece of research on the power of anchoring presented subjects with two alternative exercises: to estimate, within 5 s, the product of $1 \times 2 \times 3 \times 4 \times 5 \times 6 \times 7 \times 8$, or the product of $8 \times 7 \times 6 \times 5 \times 4 \times 3 \times 2 \times 1$. When the sequence started with the small numbers, the median estimate was 512 and the large numbers produced the median estimate of 2,250 (Montier, 2007). The correct answer is 40,320.

Even if there is no anchoring, people are grossly overconfident in assigning a numerical range to describe their level of ignorance. This phenomenon has been extensively studied by asking subjects to assign 98% credibility ranges to their esti-mates of fact questions, the correct answer of which is unknown to them (a typical example: "please give an upper and lower bound to the length of the Nile so that you are nearly sure that the true value lies inside the interval"). In this sort of exer-cise one would expect that only about 2% of proposed intervals would not contain the true value. However, in most cases 20–60% of the actual estimates exclude the true answer (Morgan et al., 1990). Moreover, there is no natural numerical interpretation of words like "likely," "possible," "unlikely," or "improbable." Even experts of the US Environmental Protection Agency Science Advisory Board, when asked to assign numerical probabilities to words "likely" and "not likely," which were intended for use with EPA cancer guidelines, spread the minimum prob-ability associated with the word "likely" over four orders of magnitude and the maximum probability associated with the word "not likely" over more than five orders of magnitude (US Environmental Protection Agency, 1996). The difficulty of using probability language in an intuitive way seems to be especially pronounced in medicine (Morgan, 2014).

We are also insensitive to prior probabilities of outcomes when, in addition to bare statistical facts, we are provided with an additional description of the situation, regardless of how uninformative (Tversky and Kahneman, 1974). In an interesting study two groups of undergraduates were given a description of a fictitious graduate student Tom W., who was described as intelligent, orderly, not terribly creative, and a loner. One group was then asked to rank a number of specialties by suitability for Tom. The second group was asked to rank the same specialties by probability that Tom is specializing in them. The result was that both groups produced nearly identical rankings (Kahneman and Frederick, 2005). However, because the fields, which were closest to the stereotype described as Tom W. (like library science) were considerably less populated than the more social fields, like social studies—a fact that the undergraduates participating in the study were fully aware of but weren't instructed on—it is far more likely that someone like Tom W. is enrolled in a social studies graduate program than in a library science one. The failure of the subjects to use relevant information to arrive at a more reasonable conclusion shows that people ignore base rates (background knowledge) in favor of new evidence.

Even more famous is the Linda problem: Linda is 31 years old, single, out-spoken and very bright. She majored in philosophy. As a student she was deeply concerned with issues of discrimination and social justice and also participated in antinuclear demonstrations. Surprisingly, more than 80% of undergraduates, when asked point blank whether Linda was more likely to be "a bank teller" or "a bank teller who is active in the feminist movement," chose the latter and thus committed a logical fallacy (Kahneman and Frederick, 2005). In a large sample of University of Arizona students, who were asked to estimate the number of murders in Detroit or Michigan the median estimate of the number of murders was 200 for Detroit and 100 for Michigan (Detroit is situated in the state of Michigan) (Kahneman and Frederick, 2005). In one study even professional forecasters assigned a higher probability to "an earthquake in California causing a flood in which more than 1,000 people will drown" than to "a flood somewhere in the United States in which more than 1,000 people will drown" (Kahneman, 2011).

Most importantly, we are naturally insensitive to sample size, erroneously believing that even small samples closely resemble underlying populations (Tversky and Kahneman, 1974). Subjects assigned the same probability of obtaining an average height for men of more than 6 ft (1.83 m; mean US height is 1.76 m) for samples sized 10, 100, and 1000. When asked to imagine four coin tosses, people tend not to think of all heads or all tails, imagining instead something closer to the 50:50 expectation. The 12.5% probability of four identical coin tosses contrasts with nearly 0% observed in subjects' "fake" tosses. There is also a widespread intuitive expectation that the essential characteristics of a random process will be represented both globally in the entire sequence, and locally in each of its parts. A consequence of this belief in local representativeness is the gambler's fallacy. After observing a long run of red on the roulette wheel, most people believe that black is now due, since the occurrence of black will result in a more representative sequence. Chance is thus seen as a physically real self-correcting process, where a deviation in one direction induces a deviation in the opposite direction to restore the balance. However, in true chance processes deviations are not corrected or correlated, they are merely diluted. The "law-of-small-numbers" underlying the gambler's fallacy entices us to put too much faith in results, which are obtained from small samples and to grossly overestimate the replicability of such studies.

A well-studied type of bias is the confirmation bias, which comes in three basic flavors (Nickerson, 1998). Firstly, ambiguous information is readily interpreted to be consistent with one's prior beliefs. Secondly, people tend to search for information that confirms rather than disconfirms their preferred hypothesis. And thirdly, people more easily remember information that supports their position.[1] A possible physiological explanation for the ubiquity of confirmation bias, which has received some experimental confirmation (pun intended), is that limited capacity of working memory makes it hard to consider more than one hypothesis at the same time.

A related type of bias is preference for early information. When the first replications of an experiment give positive results, which are later diluted with increasing sample size, it is natural to believe that the first results were somehow more meaningful than the later ones. Such bias has been suggested to be behind a great deal of wasted resources in commercial drug development (Berry, 2007).

People also tend to use internal consistency of their knowledge as a major determinant of their confidence in any generalizations based on this knowledge. The problem with this intuition is that highly consistent patterns are often observed when the input variables are highly redundant or correlated. Therefore, we tend to have inflated confidence in predictions that are based on redundant information.

Even if you are a victim of none of the abovementioned common biases, it would be very difficult for you to validate your intuitive decisions about research data without recourse to formal statistical methods. In the absence of such validation any disagreement about interpretation of data would have to be solved based

---

1. Historical analyses of the nineteenth-century physics and studies of undergraduates testing hypotheses have revealed that people actually use a mixed strategy where they initially seek evidence to confirm their new hypothesis, but after some evidence has been gathered, they try to delimit their hypothesis by seeking disconfirmatory evidence (Dunbar and Fugelsang, 2005).

on something else than a rational argument. Unfortunately, most people who are not genuinely competent in a given task, including both university students and their professors, tend to grossly overestimate their competency in as wide a variety of abilities as their factual knowledge, sense of humor, grammatical ability, logical ability, and worth as lecturers (Kruger and Dunning, 2009). It is a lack of insight into their own errors and not mistaken assessments of others that leads to over-optimism among poor performers (Ehrlinger et al., 2008).

There seems to be little choice for us but to return to statistical methods of data analysis. Assuming that the data come from normal distributions with equal variance, what are the minimal effect sizes that we can interpret with confidence as being real (i.e., not resulting from sampling error)? For the Neyman–Pearson procedure, if $\alpha = 0.05$ and $\beta = 0.20$, a convenient formula is:

$$N = \frac{6.185 \times \text{SD}^2}{\text{EZ}^2}$$

where SD is standard deviation and EZ is effect size (H_1 mean $= H_0$ mean $+$ EZ). This equation can be used to set the long-run type II and type I error levels, but it does not imply that a single dataset, which was obtained at the calculated sample size, presents strong evidence for the reality of the observed effect (Royall, 1997). To arrive at a 95% probability of your data providing at least moderate strength of evidence for the reality of the observed effect, the formula is:

$$N = \frac{18.191 \times \text{SD}^2}{\text{EZ}^2}$$

And for strong evidence thereof, it reads:

$$N = \frac{22.557 \times \text{SD}^2}{\text{EZ}^2}$$

Alternatively, conventional power analysis could be done at $\alpha = 0.05$, $\beta = 0.004$ level for a specific rejection of H_0 to convey moderate evidence against H_0; and for $\alpha = 0.05$, $\beta = 0.001$ level to convey strong evidence (Royall, 1997). A more recent "objective Bayesian" analysis reached similar conclusions by recommending the use of $\alpha = 0.005$ significance level as a measure of moderate strength of evidence and $\alpha = 0.001$ for strong evidence (without making

TABLE 6.3 Minimal Effect Sizes Whence Data Convey Strong Evidence with 95% Probability for the Reality of the Effect: H_0 Mean $(\mu_{H_0}) = 1$, $\alpha = 0.05$; $\beta = 0.001$

Standard Deviation (SD)	Sample Size (N)	Effect Size (EZ)[a]
0.1	2	0.4
	3	0.3
	4	0.3
0.2	2	0.7
	3	0.6
	4	0.5
0.3	2	1.1
	3	0.9
	4	0.8
0.4	2	1.4
	3	1.2
	4	1.0
0.5	2	1.8
	3	1.5
	4	1.3

[a]$EZ = \mu_{H_1} - \mu_{H_0}$.

explicit power recommendations; Johnson, 2013). The effect sizes suitable for evidentiary interpretation at small sample sizes are given in Table 6.3.

As can be seen from the table, the effect sizes open for evidentiary interpretation with very small samples ($N = 3$) are about three times the SD. This would be very bad news in medicine, where effect size of a drug of 1 SD is usually considered "large," but it is not necessarily so in experimental biology. There are two reasons for it. Firstly, drugs are tested in patient populations that are typically heterogeneous in respect to susceptibility to the drug, thus having a lot of variation and correspondingly large SDs. Biological experiments are often conducted in much more homogeneous populations, which enable better to recognize the signal from the noise. And secondly, while a 10% reduction in mortality can translate into substantial medical gains and large beneficial drug effects are relatively rare, in experimental biology the experimental systems are often designed with large effects in mind. For instance, when we mutate an enzyme in the hope of uncovering catalytically active amino acids, any mutation that doesn't practically abolish the activity will be discarded from the list of contestants.

People who wish to see smaller effects, but cannot afford to raise their sample sizes to levels required by frequentist methods, might seek solace from Bayesian methods. Because they use information more efficiently, Bayesians require smaller sample sizes to identify an effect of a given size (van de Schoot et al., 2014). Even when using noninformative priors, the required sample size in a Bayesian parameter estimation of normally distributed data can be reduced to only two or three times the number of parameters in the model tested (from four to five times in frequentist Maximum Likelihood Estimation; Lee and Song, 2004). Having said that, when using informative priors, the use of small samples will put extra weight on the quality of prior knowledge that is incorporated into Bayesian analysis. If great care is not placed in choosing the informative prior, it is advisable to use a noninformative prior instead (with the assumption of normality of the data).

When the stakes are high, as in drug discovery, it is customary to use a formal process of elicitation for turning the expert knowledge into workable priors (Morgan, 2014; Kadane and Wolfson, 1998). Bayesian elicitation often involves (i) identifying and recruiting a group of relevant experts, educating them in the basics of probability thinking; (ii) familiarizing the elicitator, who is a statistician trained in the psychology of getting unbiased opinions out of people prone to fallacies of thinking summarized above, on the literature on the background of the scientific subject; (iii) in collaboration with the experts determining and ranking in the order of importance the factors affecting both the parameter value of interest and uncertainty concerning that value; (iv) asking the experts for plausible upper and lower limits for the parameter whose value is assessed, and checking if the experts really cannot imagine a plausible mechanism for even more extreme values; (v) asking for point values; (vi) integrating the opinions of different experts; and (vii) constructing a formal probability model, which then serves as a Bayesian prior. When the choice of experts does not cover the full range of legitimate scientific opinion or the opinions of different experts are widely divergent, then new troubles will ensue. For an "off the shelf package for eliciting probability distributions," which comes with elicitation protocols and software, the reader is referred to the Sheffield Elicitation Framework.[2]

It should be obvious from the above that a formal elicitation protocol is not the road that individual scientists wishing to interpret their experimental results will find enticing. There seems to be a need for a less formal Bayesian solution that doesn't require constructing probability distributions but could still be useful in opening up a scientific argument to scrutiny, would allow to test robustness of conclusions and to find weak links, which bring uncertainty to the argument. We will next try to invoke a less formal kind of Bayesianism.

---

2. http://www.tonyohagan.co.uk/shelf/.

Let us begin by imagining a typical experimental biology paper where the authors compare experimental conditions with control conditions, do this in several experimental replications, quantify the difference and statistical significance ($P < 0.05$)—and on the strength of this evidence plus other experiments claim to have shown the existence of some interesting biological phenomenon. Now, we must start our analysis by listing the mutually exclusive and exhaustive alternative hypotheses relevant to the experimental evidence and to their interesting conclusion. As the first hypothesis we might want to start with H_{no}: "there is no effect," which means that the experimental effect we do in fact see is a pseudo-effect, either due to sampling error or bias. Then the H_1 is "true and interesting effect" and, for the sake of argument, lets suppose that we can think of an alternative interpretation for the data, under which the effect is biologically real, but scientifically less interesting: H_2 is thus "true but boring effect." Of course, in reality there might not be an obvious H_2, or there might be several hypotheses ($H_2 \ldots H_n$), which must then be listed.

In Chapter 5 we talked about a separate hypothesis H_{bias}: "the experiment is biased," which means that our effect is an experimental artifact. This hypothesis claims that we have misunderstood our experimental system and are therefore systematically misinterpreting its results. Maybe we are really not measuring what we think we are? Maybe the difference between control and experimental conditions was not caused by our stated experimental manipulation? Since it is logically possible that the experiment is badly biased and yet points to the right direction, H_{bias} is not strictly speaking mutually exclusive with H_1 and H_2. However, since we believe that a good scientific argument should not only lead us closer to truth, but it should also do this in a reasonable and predictable way, there is good reason to treat H_{bias} as a legitimate alternative hypothesis, whose truth precludes any successful search for scientific truth. It thus makes operational sense to further divide H_{no} into H_0, which deals with sampling error and H_{bias}, which deals with bias. The probability of H_{bias} is then determined by considerations of the experimental system.[1]

1. $P(H_0) + P(H_{bias})$ can exceed $P(H_{no})$ by $P(H_2)$, which is the probability of making spuriously or accidentally correct discoveries. $P(H_{no}) = P(H_{bias}) + P(H_0) + P(H_2)$. In most cases $P(H_2)$ is negligibly low. To satisfy the unit probability of the hypothesis space, $P(H_2)$ could be subtracted in equal portions from all the scientifically meaningful hypotheses in the mix (H_1, H_2, etc.).

Having explicated the hypothesis space $\{H_0, H_{bias}, H_1, H_2\}$, we have already done something useful—we now have a clear overview of the possible alternative explanations for our experimental data. If a colleague thinks that a relevant hypothesis is missing from the list, he can now present arguments for including it. If any of the hypotheses is unduly vague, we will soon find out; and we can always go back and start anew.

It is now time to assign to each hypothesis a prior probability. The prior probability of H_{no} depends on the novelty of your claimed scientific conclusion H_1—if it is a surprising one then $P(H_{no})$ could be fairly high. But obviously, so could be $P(H_2)$. Maybe we decided that the prior probabilities should be ranked as $P(H_{no}) > P(H_2) > P(H_1)$. Does a mix $P(H_{no}) = 0.5$, $P(H_2) = 0.3$, $P(H_1) = 0.2$ sound reasonable? If yes, then we will next partition H_{no} between H_0 and H_{bias}. As long as we keep our type I errors under control at the significance level 0.05, we won't go too far astray by starting by assigning a probability to our experiment being biased: $P(H_{bias})$. This probability depends on the experimental system used, on the level of competence of the people who used it to get the particular data under advisement, and on the experimental controls done. Depending on the situation, $P(H_{bias})$ can vary a great deal. Lets suppose that we are dealing with a fairly new and complicated experimental system (which is arguably a bad thing) used in a competent lab, whose members are not averse to running various controls (obviously a good thing). Then perhaps we can take $P(H_{bias})$ as 0.1. This means that we believe that there might be a 1 in 10 chance that the observed experimental results were seriously biased (and a 9 in 10 chance that they were not) and it leaves $P(H_0)$ as 0.4.[3] Now it is time to tabulate the hypotheses and their prior probabilities (Table 6.2).

At the next step we must assign a likelihood $P(D \mid H)$ to each of our competing hypotheses. This can be seen as weighting each hypothesis by how much the data support this particular hypothesis. For example, If we believe that the P values were calculated correctly (the assumptions like randomness and independence are met), then $P(D \mid H_0)$ should perhaps equal the significance level $\alpha = 0.05$. Because P value is part of the long-run frequentist logic, an individual P value, which is smaller than α, does not in itself mean that we have correspondingly more evidence against the result of a single experimental series being not caused by sampling error. We can take the probability of seeing our data if the system was biased as very high (1.0), as the probability of seeing our data if our favorite hypothesis H_1 holds (1.0), and the probability of seeing the data under its uninspiring alternative H_2 as somewhat lower (0.6). These numbers, of course, depend on the particulars of the case and it is entirely reasonable to expect that different people may want to defend different likelihoods here, so that we could see how they influence the final result.

When we have come up with a set of likelihoods, which we think are defensible, we will next weight our priors with the likelihoods as $P(H) \times P(D \mid H)$ and then calculate the posterior probability of H_1: $P(H_1 \mid D) = 0.2/(0.2 + 0.02 + 0.1 + 0.18) = 0.4$, which means that after all is said and done we think that there is still less than a 50% chance of H_1 being true. Using a similar calculation we can see that the posterior probability of bias is 20% and the posterior probability of H_2 is about the same as for H_1 (Table 6.2). Under the parameters of our

---

3. We are here ignoring $P(H_z)$ as insignificant.

TABLE 6.4 Slicing the Hypothesis Space for an Intuitive Bayesian Analysis

	$P(H)$	$P(D \mid H)$	$P(H) \times P(D \mid H)$	$P(H \mid D)$
H_0	0.4	0.05	0.02	0.04
H_{bias}	0.1	1	0.1	0.2
H_1	0.2	1	0.2	0.4
H_2	0.3	0.6	0.18	0.36

example none of the competing hypotheses was strongly favored, so there is clearly room for use of additional independent experimental systems to further increase the posterior probability of H_1 and lower it for H_2 and H_{bias}. The good news is that we can now use 0.4 as the new prior probability for H_1 and 0.36 for $P(H_2)$, while the $P(H_{\text{bias}})$ for the next experiment does not depend on our previous estimate of it or its posterior probability.

The object of such an exercise can never be to arrive at a single true posterior. It is more modestly to explicate the alternative hypotheses in a tractable form, to think about the extent that experimental results support each hypothesis, to find points of insecurity where we cannot decide on the relevance or support of a hypothesis, and to test the robustness of conclusions in terms of different input probabilities. Therefore, what we have here is a semi-formal thinking aid, not an algorithmic substitute for thinking about one's results (Table 6.4).

6.2 IS INTUITIVE REASONING BAYESIAN?

Do people intuitively reason logically? Experimental psychologists, whose arguably two most famous experimental systems were introduced by Peter Wason in 1960 and 1968, have studied this question for several decades. In the conceptually simpler of the two, Wason's selection task (1968), people are presented with the top sides of four cards, each exhibiting a letter or a number, for example A; K; 2; 7. They are told that if there is a letter on one side of the card, there must be a number on the other side (and vice versa). They are then given a hypothesis to test: that if there is an "A" on one side of a card, then there must be a "2" on the other. Each subject is asked to point out, which cards should be overturned to test the hypothesis. Because here we have a test of logical implication: if "A," then "2," the logically informative cards would be "A" (which by Modus Ponens would require "2" on the other side if the theory is correct) and "7" (which by Modus Tollens would require "anything but A"). If people reasoned according to the rules of propositional logic, they would pick the "A" and "7" cards for testing and leave the "K" and "2" cards as uninformative.

And yet, only 4% of subjects picked the right cards ("A" and "7"), while 42% picked "A" and "2" cards, 33% picked only "A," and 7% picked "A," "2," and "7" (Johnson-Laird and Wason, 1970). It seems that, rather than to seek falsifying evidence in the lines of Modus Tollens and Karl Popper, subjects prefer to confirm the conjunction ("A" and "2"). On the face of it, this looks like an instance of confirmation bias.

The second test system, the 2–4–6 task (1960), is even easier to describe. Here we have a string of numbers: 2–4–6 and the subject's task is to devise both the hypotheses and the tests to find the rule that generated this string. To test her hypotheses the subject may propose as many three-number strings as she likes and the experimenter will tell for each if it corresponds to the rule (which he knows), or not. He will also tell if the individual explanatory hypotheses proposed by the subject are correct or not. A typical test is: "8–10–12" (yes, it corresponds to the rule) and the hypothesis it supports is: "increasing even numbers" or "add 2" (both are incorrect). As it happens, the rule is: "any three numbers in ascending order." Again, the most efficient way to arrive at correct solution would be to combine verification (Modus Ponens) and falsification (Modus Tollens).

In fact, about 80% of subjects never finish the task, mostly because they fall into the rut of confirming an unsuccessful hypothesis over and over again. People consistently devise tests of their hypotheses that would give a "yes" answer if the hypothesis is correct and display no interest in falsifications, which would give "no" for an answer. This basic result has been replicated in more difficult tasks where the subjects can plan and conduct computer experiments to discover the "laws" governing particular computer-generated universes (Mynatt et al., 1978). Intuitively, people just aren't Popperians. Even when explained, how falsificationist hypothesis testing works, they are no more likely to use it in actual testing. When falsification of hypothesis by test occurs, in most cases it does not lead to permanent abandonment of the hypothesis: subjects often return to it later, sometimes to revise it in order to accommodate falsifying data or, quite often, to resume testing without any modification. Whether to abandon a promising (and partially correct) hypothesis upon seeing disconfirmatory evidence or whether to modify it to accommodate said evidence seems to be a judgment call, which different people approach in consistently different ways. When testing confirms the hypothesis, subjects tend to immediately retest it without revision of the hypothesis.

So, the obvious question is: why aren't university students, who make up the samples of experimental psychology studies, intuitively using logic when testing hypotheses? When testing hypotheses, what are they actually doing? Are they illogical? Or are they thinking in terms of probabilities instead of propositional logic?

A problem with using propositional logic in scientific hypothesis testing is its insistence on the truth of premises. If the premises are not true then it doesn't matter if the logical inference is used correctly, the conclusion will be invalid.

In addition, a conclusion, which was arrived at by logic, cannot in principle be overturned by any additional information. This property of logic, where adding premises can never overturn existing conclusions, is called monotonicity.[4] Yet, in the real world almost any conclusion can be overturned by new evidence, which means that, technically speaking, everyday reasoning is nonmonotonous. Unfortunately, nonmonotonic arguments are outside the scope of traditional deductive logic and monotonic inference scarcely applies outside mathematics. An example is this simple problem (Oaksford and Chater, 2009):

> Premise 1: when a dollar is put into a coke machine, a coke will come out
> Premise 2: I just put a dollar into the coke machine in my campus
> Question: will I get a coke?

Logically this looks very much like Modus Ponens, which is an instance of monotonic argument. Yet in the real world there are many ways to block this inference. What is the answer you would give to the question? Would MIT students and students of Greendale Community College, for instance, be justified in giving a similar answer or is it perhaps true that in Greendale the coke machine is always out of order, while the more modern machines at MIT work as clockwork? The answer must therefore be probabilistic and will depend on background knowledge.

If the canonical logical form of the problem is

$$A \rightarrow B$$
$$A$$
therefore, B

where A is "a dollar goes into a coke machine" and B is "a coke comes out," then a probabilistic reformulation might be

$$P(B \mid A\&W)$$

where W is the background information or, to save ink, just

$$P(B \mid A)$$

This theory of human judgment where logical implication $A \rightarrow B$ is substituted by conditional probability $P(B \mid A)$ is, of course, a standing invitation to use the Bayes theorem as an explanatory model of how people actually think. Since Bayesian hypothesis testing is always comparative, rather than dealing with falsification of a single hypothesis, its use in contexts like Wason's selection task will lead to seemingly inexplicable answers by the standards of propositional logic. However, when looking at them through Bayesian glasses the seemingly random answers become intelligible (Oaksford and Chater, 2009).

---

4. There exist nonmonotonous logics, but classical propositional logic is not one of them.

The logical form of Wason's selection task is A→2. In its Bayesian form we will transform Wason's problem into two competing hypotheses:

H_1: $P(2 \mid A) > P(2)$

H_2: $P(2 \mid A) = P(2)$

Because the experimental subjects do not actually turn the cards to see what's on the flip side, they must base their decisions on expected information gain (how much would expected information differentiate between H_1 and H_2), taking all possible outcomes into account. According to H_1 it is more likely that a card that has "A" printed on one side will have "2" on the reverse side. H_2 tells us that in terms of encountering "2" on the backside, it doesn't matter what's on the front.

Now the subjects' choices suddenly make sense. The people who pick "A," are thinking on the lines that if H_1 is true it should be more likely to find "2" on the other side and less likely to find anything else. The reasoning behind picking "2" is that if H_1 is true, it is also more likely to find "A" on the flip side (assuming the probability of observing "2," given that the other side is "notA" is identical under both hypotheses and that H_1 entails observing "2" over this base rate on "A"- cards).

Also, it does not make much sense to pick "7" because the competing hypotheses by themselves cannot well differentiate between our chances of finding either an "A" or "something else" on the flip side (for this we would have to know the base rate of "7"). An additional assumption of rarity is needed to justify the preferential picking of "A" and "7" cards. This assumption states that properties that occur in the antecedents and consequents of hypotheses tend to be rare otherwise, lest the evidence be consistent with most conceivable hypotheses (Oaksford and Chater, 2009). In our example most cards not labeled "A" are not expected to carry "2" on their flip side and most cards that are not labeled "2" are not expected to carry "A"s on their flip side.

The reason behind discussing apparently Bayesian solutions to Wason's selection task, which is not probabilistic and where probabilistic solutions are clearly inappropriate, is to point out that if the theory that everyday thinking is probabilistic and specifically Bayesian is true, then everyday thinking seems to be so rabidly Bayesian that propositional logic is virtually never used. Even when people seemingly act in accordance with the principles of propositional logic, they may really be using a probabilistic approach (which in the case of our coke machine example would make perfect sense).

Nobody thinks that test subjects explicitly use Bayesian calculations to select cards. Rather the human brain might be adapted to run mental routines, whose outputs are similar to the ones we would get by formal Bayesian calculation. Of course, the current popularity of Bayesian theories of human inference and learning among psychologists (Chater et al., 2006; Tenenbaum et al., 2011) doesn't necessarily make them true (Jones and Love, 2011; Bowers and Davis, 2012). Here is a counterexample where failures in a reasoning task are interpreted in a decidedly simpler mode: "A bat and a ball cost $1.10 in total. The bat costs $1 more than the ball. How much does the ball cost?" Almost everyone reports "10 cents" (Kahneman and Frederick, 2005). As one dollar more than 10 cents is obviously $1.10, this results simply shows that people are careless and don't check the total before giving their answer (the correct answer is 5 cents): "The bat and ball problem elicits many errors, although it is not really difficult and certainly not ambiguous. A moral from this example is that people often make quick intuitive judgments to which they are not deeply committed. A related moral is that we should be suspicious of analyses that explain apparent errors by attributing to respondents a bizarre interpretation of the question" (Kahneman and Frederick, 2005).

6.3 THE MOLECULAR BIOLOGY LAB AS RESEARCH SUBJECT

Biological science as done in the laboratory is so much more complicated than Wason's test systems that it's certainly fair to question the relevance of simple card tricks played on unsuspecting undergraduate students on what educated people in white coats do in the lab. Fortunately there exist *in situ* studies where biologists have been followed around by curious sociologists and cognitive scientists in their natural environment, and where real (albeit already solved) biological problems have been presented for virtual experimenting to undergraduates (Dunbar, 1995, 1999; Dunbar and Fugelsang, 2005). When the classic biology problem of lactose operon regulation—first solved by Francois Jacob and Jacques Monod in around 1960—was presented to naïve students who were indoctrinated to expect that genes activate other genes, the students not unexpectedly designed experiments whose outcomes did not support this initial theory (the correct answer is that regulation of this operon occurs via inhibition, not activation). From there on, the students had a choice between two possible strategies: to design further experiments to prove the activation theory or to try to explain the cause of their negative findings. None of the students who chose the first approach got anywhere in discovering how the operon is controlled, while the students who began to generate new hypotheses to explain the data did arrive at the correct solution (Dunbar, 1995). Here the secret of success was designed to be the ability of students to change ones initial goal (to prove the theory of activation) to another (to explore the causes of inconsistent experimental results). The important empirical point was revealed in the second study design, where one gene was designed to be activatory (supporting students' initial theory) and two more were designed to be inhibitory (requiring rethinking of initial assumptions). Now, after confirming their hypothesis in the case of the first gene, students were much more likely to think of new inhibitory theories and to find the correct answer. Apparently, the initial confirmation exhausted their original goal of confirmation and suggested that it is now time to switch goals to novel inhibitory hypotheses. Thus it is our goals, which determine when and how inconsistent evidence is used (Dunbar, 1995).

Kevin Dunbar and his students spent many hours meticulously interviewing real biologists from top laboratories, recording their lab meetings, and reading grant applications and manuscripts of research papers. These are some of the conclusions from the analysis of the resulting data trove (Dunbar, 1995, 1999).

First, modern biology is not for loners: much of experimental planning, analysis of results and many actual discoveries are made collectively in lab meetings.

Second, evidence that contradicts the hypothesis is often used to modify or discard the hypothesis. Generally, hypotheses are modified as little as possible to accommodate the data. If the evidence is inconsistent with not only a particular hypothesis but with all hypotheses of a type, then the first reaction is typically to try to find fault with the experiment, not with the theory that inspired it. It is an important role of lab meetings to change the mind of the postdoc working on a specific problem from discarding their experiments as erroneous to falsifying substantive scientific theories by experiments. The strategy after discarding one's theory is to start thinking of alternative causal models of reality. In order to change a scientist's mind about her interpretation of her findings, she must first believe that her results are not manifestations of random error (Dunbar, 1995). Excluding random error seems to be a very strict requirement for most professional scientists for opening the gates for alternative scientific explanations.

Third, in Dunbar's dataset more experienced scientists were quicker to abandon theories in the face of inconsistent evidence than were their younger colleagues. Thus, while younger scientists are prone to confirmation bias, it seems that older researchers are in danger from the opposite "falsification bias," where the baby (confirmatory data) is discarded with the bath water (disconfirmatory data).

Fourth, a key to success was to have people with different but overlapping backgrounds in the lab.

Fifth, this allows copious use of *local analogies* in finding solutions to experimental problems. A local analogy typically points to an experiment in a close field. The unsuccessful experiment is thus mapped to a similar but successful experiment. This allows to spot any differences between them and to modify the experimental protocol accordingly. Using analogies or metaphors in scientific thinking will force people to take a step back and reconsider their assumptions, which is something that happens more rarely if everybody uses the same specialized language and understands precisely what everybody else is saying (as is bound to happen when all the lab members have received their education from a single source). In molecular biology at least 60% of experiments fail and the use of local analogy seems to be the most efficient way to improve on such failures (Dunbar, 1995). The alternative, to systematically change the parameters of the unsuccessful experimental system until a fix is found, is more time-consuming and seems to have a lower rate of success.

To conclude, in a real lab setting the (no longer) secrets of success include diversity of backgrounds of lab members, frequent scrutiny of one's results, interpretations and models by the whole research group, use of local analogies to fix bad experiments, and finding the right balance between keeping the theory by modifying the recalcitrant experimental system and modifying or rejecting the theory in the light of the experiment. Discoveries are not made at the bench by individual scientists; they are made collectively during weekly lab meetings.

6.4 HOW TO WIN FAME AND INFLUENCE PEOPLE

In any research project the results tend to be contradictory to some degree. The controls rarely behave perfectly, there are these unexplainable minor bands in your gel, some replications behave scarily differently from most of the experiments, etc. The list of ordinary problems in experimental science is endless. Yet, good journals prefer to feast their readers on clear and simple stories whose scientific importance is apparent here and now. The authors are thus encouraged to submit strong, unifying, and plausible claims encapsulated in simple, easily understandable models, which are presentable in the form of a "graphical abstract." Ideally such models provide a specific and unifying biological mechanism of action. The best models are also consistent and interlinking with the current crop of models making their rounds through the top journals.

We will now examine some more controversial examples to give the reader a hint about how it's done and of some of the dangers therein. The author must stress here that, as far as he knows, he is not gifted with a truth-sensing faculty and can therefore only point to what he considers methodological dangers in converting messy results into a clean and publishable story. For rhetorical reasons a part of the following discussion is framed as a manual on how to distort results to fit into the mold of a high-impact journal. Of course, this does not imply that the authors of the papers discussed had anything like that in mind. Especially I do not wish to imply scientific misconduct on anybody's part. On the contrary, in Part III of the book I will try to convince you that scientific misconduct, as currently defined, is both rare and irrational.

Our first example is from the field of antibiotic action. To start with some background knowledge, antibacterial drugs can either kill bacteria (bacteriocidals) or stop their growth (bacteriostatics). Most antibiotics have been long known to attack one of three different cellular target systems, whose components they physically bind and inhibit: protein synthesis, DNA replication, and cell wall synthesis.

Now comes the novel part. In a seminal paper titled "A Common Mechanism of Cellular Death Induced by Bactericidal Antibiotics" a group from Boston University proposed that regardless of their specific target, all bactericidals kill both Gram-negative and Gram-positive bacteria by the same mechanism, which involves stimulation of the production of hydroxyl radicals by the antibiotic. Radical formation is dependent on the central catabolic pathway TCA cycle, destabilization of iron–sulfur clusters, and stimulation of the Fenton reaction, which uses free iron ions to produce hydroxyl radicals (Kohanski et al., 2007). To substantiate this beautifully novel and yet wholly plausible theory, they showed (i) that addition of bactericidal antibiotics to the cells leads to change of fluorescence in a cell-penetrating dye (HPF) to a similar extent as adding hydrogen peroxide (a substrate for the Fenton reaction); (ii) that treatment of bacteria with an iron chelator, which removes free iron and thus inhibits the Fenton reaction, led to increased survival under antibiotics; (iii) that thiourea, a reducing agent, was able to increase survival; (iv) as was the deletion of a gene for iron–sulfur cluster synthesis (iscS); and (v) blocking of the TCA cycle; (vi) and that all three bacteriocidal drugs tested induced DNA and protein damage, which in turn induced a specific stress response (the SOS response) via the RecA protein. When the recA gene was deleted, bacteria became more susceptible to killing by the antibiotics; and (vii) transcriptomic analysis of thousands of mRNAs indicated upregulation of mRNAs involved in electron transport in response to bacteriocidal antibiotics.

The paper by Kohanski et al. provided no less than seven independent lines of evidence in support of its conclusions and was very influential in inspiring research, which confirmed and widened the original findings.

Still, a great deal of the medical problem of antibiotic tolerance (inability of drugs to eradicate bacterial infections) is connected with nongrowing bacteria and, especially, with biofilms where bacteria stick together on a surface and excrete a protective polymeric layer of extracellular DNA, proteins, and polysaccharides. To combat antibiotic tolerance in biofilms is therefore a task of clear and present medical importance. The common wisdom was that efficiency of antibiotic action must be a function of the metabolic activity of the bacterial cell. If the cells are nongrowing, they make less protein, RNA, and DNA and thus the inhibition of their molecular targets by antibiotics will have less effect on the bacteria. In addition, protective biofilms reduce the access of drugs to their targets inside the cells. In a situation like that the best bet for sensitizing bacteria to antibiotics would be to disassemble the biofilm.

How to achieve this in a patient-friendly manner? Enter Richard Losicks group from Harvard, which showed that relatively simple and harmless compounds, namely several D-amino acids and a shorter version of the polyamine spermidine called norspermidine, synergistically disassemble biofilms in both Gram-positive and Gram-negative bacteria (Kolodkin-Gal et al., 2010, 2012). They were able to select artificial, chemically modified versions of norspermidine that exhibited over 20-fold increased biofilm disassembly activity (Böttcher et al., 2013).

So, now we have created, by unveiling biological mechanisms both specific and general, several promising lines of approach to the previously intractable problem of antibiotic tolerance? Or maybe not. Let's start by taking a closer look at the Kohanski et al. (2007) paper through the eyes of researchers trying to replicate their results (Liu and Imlay, 2013; Keren et al., 2013; Ezraty et al., 2013).

1. In Kohanski et al. the dye HPF shifted fluorescence when antibiotics were added. This result was replicated for the antibiotics ofloxacin and ampicillin (but not for kanamycin), but only at concentrations described in the original

Kohanski et al. study (Keren et al., 2013). Strangely enough, for each of the three antibiotics used by Kohanski et al. these concentrations happened to lie very near their respective minimal inhibitory concentrations, and well below the concentrations that are actually found in patients. When the concentrations (and lethality) of the antibiotics were increased, the fluorescence shift of HPF actually decreased. Furthermore, Keren et al. found that whether HPF fluorescence shifts in a particular cell is not correlated with the survival of the cell under antibiotics. Liu and Imlay further showed that HPF oxidation occurs by a different mechanism than was supposed and that this process cannot be used to assess cellular hydroxyl radical levels.

2. In Kohanski et al. treatment with an iron chelator led to increased survival under antibiotics. However, when the levels of free cellular iron were directly measured, no increase in the harmful free iron was seen upon kanamycin or norfloxacin treatment (Liu and Imlay, 2013).

3. In Kohanski et al. thiourea increased bacterial survival, presumably by reducing the cellular environment oxidized due to antibiotic action. This result was successfully replicated, but only at the very low antibiotic concentrations used in the original work (Keren et al., 2013).

4. In Kohanski et al. deletion of a gene for IscS increased survival under drugs. To this, Liu et al. retort that while in cells lacking a hydroxyl radical scavenging enzyme (superoxide dismutase) an iron–sulfur cluster containing protein (6-phosphogluconate dehydrogenase) is inactivated because of cluster damage, bactericidal antibiotics have no such effect. In addition, Ezraty et al. demonstrate that iron–sulfur clusters are required for killing only by aminoglycoside class of antibiotics, and that they act by regulating aminoglycoside entry into cells (Ezraty et al., 2013).

5. In Kohanski et al. blocking of the TCA cycle by gene mutations increased survival under drugs. TCA cycle can only generate free radicals under aerobic conditions. Thus, if the TCA cycle-generated free radicals are instrumental in mediating killing by antibiotics, then growing cells unaerobically would be expected to result in increased survival in the presence of bactericidal antibiotics. However, when cells were grown unaerobically, these antibiotics did not lose their potency (Keren et al., 2013).

6. In Kohanski et al. all three bactericidal drugs tested induced DNA and protein damage, which in turn induced the SOS response via the RecA protein. When the recA gene was deleted, bacteria became more susceptible to killing by the antibiotics. However, according to Liu et al., although the recA deletion strain is extremely sensitive to exogenous H_2O_2, it is no more sensitive to antibiotics kanamycin or ampicillin than the wild-type parent strain.

7. In Kohanski et al. transcriptomic analysis indicated upregulation of mRNAs involved in electron transport in response to bactericidal antibiotics. Unfortunately, in their microarray experiment they used treatment with a bacteriostatic antibiotic as the control condition, against which all bactericidal treatments were compared. This means that each mRNA, which is

interpreted as upregulated by a bactericidal antibiotic, can be equally well interpreted as being downregulated by the bacteriostatic antibiotic (or by a combination of both conditions).

Now it is time to see what went wrong with the work of Kolodkin-Gal and colleagues purporting to show biofilm disassembly by D-amino acids and nor-spermidine. Firstly, members of their lab published further results showing that D-amino acids inhibited bacterial growth and the expression of biofilm matrix genes and that the strain they originally used to study biofilm formation has a mutation in the gene for D-tyrosyl-tRNA deacylase, an enzyme that prevents the misincorporation of D-amino acids into protein in *Bacillus subtilis*. In a *B. subtilis* strain, which has a working copy of this gene, D-amino acids did not inhibit biofilm formation. They conclude "that the susceptibility of *B. subtilis* to the biofilm-inhibitory effects of D-amino acids is largely, if not entirely, due to their toxic effects on protein synthesis" (Leiman et al., 2013). This seems to mean that they no longer see D-amino acids as a specific mechanism to disassemble biofilms in *B. subtilis*, but rather as nonspecific inhibitors of growth in some genetic backgrounds. Since bacterial growth is obviously a prerequisite for biofilm formation, everything that disrupts growth also disrupts biofilm formation. Leiman et al. make no comment on the ability of D-amino acids to inhibit biofilm formation in *Staphylococcus aureus* and *Pseudomonas aeruginosa*, as claimed by Kolodkin-Gal et al. (2010).

The norspermidine work by Kolodkin-Gal et al. (2012) was dissected by a collaboration of two groups from Scotland and Texas (Hobley et al., 2014). They show (i) that contrary to previous claims norspermidine actually enhances biofilm formation in *B. subtilis* at low concentrations and inhibits both bacterial growth and biofilm formation at high concentrations, (ii) that norspermidine is not present in *B. subtilis* biofilms, and (iii) that *B. subtilis* does not even have the genes for norspermidine synthesis (and that the gene that Kolodkin-Gal et al. mutated to show reduced biofilm disassembly has been long known to have another unrelated function). Again, the ability of Kolodkin-Gal et al. to replicate their results in other organisms remains a mystery.

Now that we have tried to deconstruct some pretty stories to reveal the underlying darkness, it might be instructive to take the opposite course and offer kindly advice about how to transform your messy experiments into neat self-contained publishable units. I am sorry but I could only think of seven commandments and hereby must invite the reader to provide candidates for the missing three, which will then be included into the 50-year anniversary edition of this book. That's a promise.

6.4.1   Seven Commandments for the Unscrupulous

1. **Thou shalt use experimental conditions that produce desired results.**
   For example, in some of our examples the precise antibiotic concentrations proved to be crucial for obtaining the sort of results dreams are made of.

This is really the same as saying: harness the power of silent multiplicities! As long as the understanding is that you will not stop tinkering with the system until a good result is obtained, you cannot fail.

2. **Thou shalt give chance a chance.** When previous advice was about how to introduce bias, this one is about how to make friends with the sampling effect. For instance, should your antibiotic concentration be close to the minimal inhibitory concentration (as they often are in published studies of bacterial antibiotic tolerance), there will be more variability in the results from experiment to experiment and you are more likely to obtain interesting results. When you combine this with a small sample size, you are quite likely to get strong effect estimates. Thus, in a double whammy, you will lower the sample size (and thus the cost of experiment) and increase your chances of getting statistically significant results of inflated magnitude.

3. **Thou shalt use the right controls to produce the right effects.** Experimental effect is always relative to a control—thus by deciding on a control, you can also decide on the presence or absence of an effect. The use of one experimental condition versus another experimental condition in the microarray experiment discussed above can be seen as an instance of this. The ability of Kolodkin-Gal et al. (2012) to measure norspermidine in bacteria that lack the means of synthesizing it and at the same time to miss the presence of ordinary endogenous spermidine, which is very similar in chemical composition, is a testament to the use of the right controls. As a control they ran an extremely sensitive and accurate liquid chromatography/mass-spectroscopy analysis on a commercially obtained pure norspermidine preparation; and then, in their actual biological experiment, depended on the retention times obtained from a liquid chromatography column—a quite inaccurate method—to "identify" norspermidine from their samples. Scientifically speaking, this may raise some eyebrows, but it was clearly suited for high-impact publishing.

4. **Thou shalt use error-prone experimental systems.** Some experimental systems are kind to the experimenter by virtue of presenting a high rate of false-positives. For example, we saw how every use of HPF seems to lead to great results. Another fine way of getting publishable results is to depend solely on *in vitro* kinase assays to determine functional cellular phosphorylation pathways (for example see Schumacher et al., 2009 and Germain et al., 2013 for rebuttal, but there are many other similar examples).

5. **Thou shalt reduce complexity of the models, which are tested by experiment.** For example Ferris et al. used functional magnet resonance imaging (fMRI) to measure the oxygen level in blood in the brains of (i) lactating female rats and (ii) cocaine-stimulated virgin female rats to discover that in both groups the same brain region, the dopamine reward system, had increased blood oxygen levels. Their conclusion: Pup suckling is more rewarding than cocaine (Ferris, 2005). This is a clear and compelling

message, but only so far as the underlying model is to be trusted. The model looks extremely simple:

cocaine→brain area X activated→pleasure
suckling→brain area X activated→pleasure
therefore, if suckling lights up X more than cocaine, it must give more pleasure.

Of course, we know that in the real world all sorts of things are associated with X (the dopamine reward system), the activation of which may result in other things than pleasure, and we know that the above model is a huge oversimplification of nature. Yet, for the sake of a good story we might as well ignore the complexity.

6. **Thou shalt increase the complexity of your experimental system** (with the understanding that a simple model in combination with a complicated and expensive experiment will truly awe the socks off your referees). A good example might be the full title of the Ferris et al. paper: Pup Suckling Is More Rewarding than Cocaine: Evidence from Functional Magnetic Resonance Imaging and Three-Dimensional Computational Analysis. Indeed, the luster of fMRI frequently leads to journalistic exploits like "Scan images show that watching online 'adult' sites can alter our gray matter, which may lead to a change in sexual tastes" (The Guardian 26.09.2013) or as David J. Linden, who is a neuroscientist at Johns Hopkins University medical school, explains in his book, "The Compass of Pleasure," physical exercise and cocaine (again) light up the same "pleasure centers," which bolster the theory that "Exercise addicts display all of the hallmarks of substance addicts: tolerance, craving, withdrawal, and the need to exercise 'just to feel normal,' " (cited from Nicholas Kristof, NYT 29.10.2011). To further quote from the Kristof column: "Who knew that orgasms, in men and women alike, light up the pleasure centers much like cocaine? (…) Linden argues that there is such a thing as a genuine biological addiction to sex. The public's failure to recognize this, he says, means that people often don't receive treatment." Would such speculations merit attention without the mystique of peering directly into human brains as they "light up their pleasure centers" via a machine that is so complicated that its workings seem literally miraculous to the public?

7. **Thou shalt misrepresent evidence.** If an effect must be erased from the minds of reviewers, then nothing is simpler than to use a logarithmic scale (Figure 6.4A and B). On the other hand, should there be a need to emphasize an effect, just use a linear scale that starts from some arbitrary (but useful) value, rather than from zero (Figure 6.4C and D). When the issue is growth of microorganisms and the order of the day is to find an effect, then the way to go is to run a competition experiment. Here you will grow the control cells and treatment group cells in the same microcosm and, as the growth is exponential, any small differences in their growth rates, or in starting conditions, will be hugely amplified.

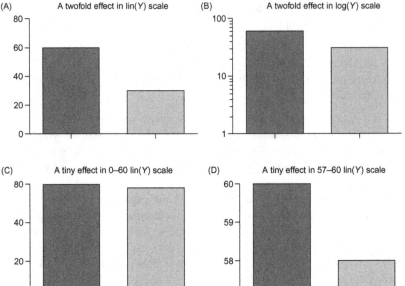

FIGURE 6.4 Graphical misrepresentation of evidence. (A) Twofold effect shown in linear scale— no misrepresentation. (B) The same effect shown in logarithmic scale. (C) A tiny effect shown in linear scale that starts at zero—no misrepresentation. (D) The same effect shown in linear scale that starts at a higher Y-value.

All seven commandments are perfectly legitimate. There is no point in designing experiments that cannot give results, variability is a recognized and necessary part of every biological experiment, right choice of controls is always a matter worthy of scholarly discussion, *in vitro* kinase assays have given a great many useful leads, all scientists are taught that their models should be as simple as possible (Occam's razor), modern experimental systems are complex for a reason (the simple ones have been sucked dry), and using graph designs to turn your results into a presentable story line is a recognized art form (Tufte, 2001). Not to mention competition experiments, which can be indispensable in documenting small but potentially evolutionarily important differences in the fitness of different genotypes. Simply put, there are fifty shades of gray between good and bad science and no black and white methodological advice can change that.

REFERENCES

Berry, D.A., 2007. The difficult and ubiquitous problems of multiplicities. Pharm. Stat. 6 (3), 155–160.
Böttcher, T., et al., 2013. Synthesis and activity of biomimetic biofilm disruptors. J. Am. Chem. Soc. 135 (8), 2927–2930.

Bowers, J.S., Davis, C.J., 2012. Bayesian just-so stories in psychology and neuroscience. Psychol. Bull. 138 (3), 389–414.

Chater, N., Tenenbaum, J.B., Yuille, A., 2006. Probabilistic models of cognition: conceptual foundations. Trends. Cogn. Sci. 10 (7), 287–291.

Dunbar, K., 1995. How scientists really reason: scientific reasoning in real-world laboratories. Nat. Insight 18, 365–395.

Dunbar, K., 1999. How scientists build models *in vivo* science as a window on the scientific mind Model-Based Reasoning in Scientific Discovery. Springer, Heidelberg, pp. 85–99.

Dunbar, K., Fugelsang, J., 2005. Scientific thinking and reasoning. In: Holyoak, K.J., Morrison, R. (Eds.), The Cambridge Handbook of Thinking and Reasoning Cambridge University Press, Cambridge, UK, pp. 705–725.

Ehrlinger, J., et al., 2008. Why the unskilled are unaware: further explorations of (absent) self-insight among the incompetent. Organ. Behav. Hum. Decis. Process. 105 (1), 98–121.

Ezraty, B., et al., 2013. Fe-S cluster biosynthesis controls uptake of aminoglycosides in a ROS-less death pathway. Science 340 (6140), 1583–1587.

Ferris, C.F., 2005. Pup suckling is more rewarding than cocaine: evidence from functional magnetic resonance imaging and three-dimensional computational analysis. J. Neurosci. 25 (1), 149–156.

Germain, E., et al., 2013. Molecular mechanism of bacterial persistence by HipA. Mol. Cell. 52 (2), 248–254.

Hobley, L., et al., 2014. Norspermidine is not a self-produced trigger for biofilm disassembly. Cell 156 (4), 844–854.

Johnson, V.E., 2013. Revised standards for statistical evidence. Proc. Natl. Acad. Sci. USA. 110 (48), 19313–19317.

Johnson-Laird, P., Wason, P., 1970. A theoretical analysis of insight into a reasoning task. Cognit. Psychol. 1 (2), 134–148.

Jones, M., Love, B.C., 2011. Bayesian fundamentalism or enlightenment? on the explanatory status and theoretical contributions of Bayesian models of cognition. Behav. Brain Sci. 34 (04), 169–188.

Kadane, J., Wolfson, L.J., 1998. Experiences in elicitation. J. R. Stat. Soc. 47 (1), 3–19.

Kahneman, D., 2011. Thinking, Fast and Slow. Farrar, Straus and Giroux, New York.

Kahneman, D., Frederick, S., 2005. A model of heuristic judgment. In: Holyoak, K.J., Morrison, R. (Eds.), The Cambridge Handbook of Thinking and Reasoning Cambridge University Press, Cambridge, UK.

Keren, I., et al., 2013. Killing by bactericidal antibiotics does not depend on reactive oxygen species. Science 339 (6124), 1213–1216.

Kohanski, M.A., et al., 2007. A common mechanism of cellular death induced by bactericidal antibiotics. Cell 130 (5), 797–810.

Kolodkin-Gal, I., et al., 2010. D-amino acids trigger biofilm disassembly. Science 328 (5978), 627–629.

Kolodkin-Gal, I., et al., 2012. A self-produced trigger for biofilm disassembly that targets exopolysaccharide. Cell 149 (3), 684–692.

Kruger, J., Dunning, D., 2009. Unskilled and unaware of it: how difficulties in recognizing one's own incompetence lead to inflated self-assessments. Psychology 1, 30–46.

Krzywinski, M., Altman, N., 2014. Points of significance: nonparametric tests. Nat. Methods. 11 (5), 467–468.

Lee, S.Y., Song, X.Y., 2004. Evaluation of the Bayesian and maximum likelihood approaches in analyzing structural equation models with small sample sizes. Multivariate Behav. Res. 39 (4), 653–686.

Leiman, S.A., et al., 2013. D-amino acids indirectly inhibit biofilm formation in *Bacillus subtilis* by interfering with protein synthesis. J. Bacteriol. 195 (23), 5391–5395.

Liu, Y., Imlay, J.A., 2013. Cell death from antibiotics without the involvement of reactive oxygen species. Science 339 (6124), 1210–1213.

MacCoun, R.J., 1998. Biases in the interpretation and use of research results. Annu. Rev. Psychol. 49, 259–287.

Montier, J., 2007. Applied Behavioural Finance. Wiley, Chichester.

Moore, M.T., Fresco, D.M., 2012. Depressive realism: a meta-analytic review. Clin. Psychol. Rev. 32 (6), 496–509.

Morgan, G.M., Henrion, M., Small, M., 1990. Uncertainty: A Guide to Dealing with Uncertainty in Quantitative Risk and Policy Analysis. Cambridge University Press, Cambridge, NY.

Morgan, M.G., 2014. Use (and abuse) of expert elicitation in support of decision making for public policy. Proc. Natl. Acad. Sci. USA. 111 (20), 7176–7184.

Mynatt, C.R., Doherty, M.E., Tweney, R.D., 1978. Consequences of confirmation and disconfirmation in a simulated research environment. Q. J. Exp. Psychol. 30 (3), 395–406.

Nickerson, R., 1998. Confirmation bias: a ubiquitous phenomenon in many guises. Rev. Gen. Psychol. 2, 175–220.

Oaksford, M., Chater, N., 2009. Précis of Bayesian rationality: the probabilistic approach to human reasoning. Behav. Brain Sci. 32 (01), 69–84.

Royall, R., 1997. Statistical Evidence: A Likelihood Paradigm. Chapman and Hall/CRC Press, New York.

Schumacher, M.A., et al., 2009. Molecular mechanisms of HipA-mediated multidrug tolerance and its neutralization by HipB. Science 323 (5912), 396–401.

Scott, S., et al., 2008. Design, power, and interpretation of studies in the standard murine model of ALS. Amyotroph. Lateral Scler. 9 (1), 4–15.

Tenenbaum, J.B., et al., 2011. How to grow a mind: statistics, structure, and abstraction. Science 331 (6022), 1279–1285.

Tufte, E.R., 2001. The Visual Display of Graphical Information. Graphics Press, Chesire, CT.

Tversky, A., Kahneman, D., 1974. Judgment under uncertainty: heuristics and biases. Science 185 (4157), 1124–1131.

US Environmental Protection Agency, 1996. Proposed Guidelines for Cancer Risk Assessment. Office of Research and Development, US EPA, Washington, DC.

van de Schoot, R., et al., 2014. A gentle introduction to Bayesian analysis: applications to developmental research. Child Dev. 85 (3), 842–860.

Chapter 7

Science as a Social Enterprise

The habits of relying on persuasion rather than force, of respect for the opinions of colleagues, of curiosity and eagerness for new data and ideas, are the only virtues which scientists have... [There is no] intellectual virtue called 'rationality' over and above these moral virtues.

(Rorty, 1987)

In previous chapters we argued that neither logical nor statistical hypothesis testing, neither Popper nor Bayes, can explain scientific progress on their own. Popper's theory of the logic of scientific discovery seems to us overly narrow and—because it fails to explain how to choose from the infinitude of prospective theories the (approximately) true ones for testing—ultimately sterile. Bayesians seem to suffer from the same problem. Success of their endeavors depends on the correct slicing of the hypothesis space (the delineation of the mutually exclusive and exhaustive hypotheses) and, as an extension, on how mathematical finesse of their statistical modeling relates to the mundane realities of the material world. Both approaches seem to entail that a prerequisite for the ability of scientific method to discover truth is that a true theory is fed into the methodological machine to begin with—truth goes in, truth comes out. They are in the theory comparison business, where the best of the competing theories is left standing and the rest are discarded by logical or statistical testing. If this is our concept of science, then we have no alternative but to see the function of scientific method as discarding or falsifying untrue theories. Whatever is saved from the axe has no claim to truth, it merely bears witness to the unfinishedness of the task. In the next round of testing, when there is an influx of fresh contestants, there is no guarantee for the survival of the original winner.

Scientific method, thus seen, is not confirmatory. Instead, it is fundamentally critical: critical toward its theories, experimental setups, methods of data analysis and, indeed, toward the scientific method itself. Ideally, there is no dogmatic element in science and as much as necessity demands from us not to criticize everything at once, we are commensurably weakening the power of the scientific method.

Interpreting Biomedical Science. DOI: http://dx.doi.org/10.1016/B978-0-12-418689-7.00007-7

Such a destructive view of science has from the beginning had its critics and several alternative views of scientific progress have been presented. As examples we will now discuss some that originate from the history (Kuhn), philosophy (Feyerabend), and sociology (Merton) of science. What they have in common is that all three authors are authoritative on the history of physics (Kuhn and Merton started out as professional historians of science, Feyerabend did a PhD in theoretical physics) and they all purport to make sense of science as it is actually done by real physicists. It is therefore instructive to contemplate the large differences in their conclusions, which are claimed to emanate from essentially the same source material (plucked from the history of physics from Aristotle to scientific revolution to relativity and quantum mechanics). Incidentally, Popper, another physics major, also used much of the same historical material to illustrate his thesis, only he never claimed history as proof—nor could he, as his aim was to provide a logical basis for the growth of knowledge, or in other words to set the prescriptive laws of scientific methodology, which could be broken by its practitioners only at their own peril.

7.1 THE REVOLUTIONARY ROAD OF THOMAS KUHN

Philosophers long made a mummy of science. When they finally unwrapped the cadaver and saw the remnants of an historical process of becoming and discovering, they created for themselves a crisis of rationality. That happened around 1960.

(Hacking, 1983)

As already mentioned, Thomas Kuhn (1922–1996) started out as an historian of physics (he was trained as a physicist) and used his knowledge of history to propose a general scheme of how science is supposed to actually work, as opposed to how Popper or the logical positivists thought that it should work. Kuhn's little book, "The Structure of Scientific Revolutions" (1962), slowly became an international best-seller, eventually selling over a million copies, and it has influenced the world view of countless scientists. It was the most-cited academic work in humanities and social sciences between 1976 and 1983 and it is very likely to be the philosophy book read by the greatest number of scientists (Kaiser, 2012).

According to Kuhn, scientific process has two very different modes, which alternate in an endless cycle. Each scientific field spends most of its time in a phase called "the normal science," which is characterized by acceptance by every professional scientist of a single "paradigm." A Kuhnian paradigm is a concept with two meanings. Firstly, it is just a fancy name for a situation where everybody more-or-less agrees on the basic principles, questions, and methods of study of their common discipline. And secondly, a paradigm is a set of model problems and exemplary experiments that are likely to be emulated and on which budding scientists practice and sharpen their skills. Scientists learn their trade by working through individual examples, rather than by learning universal rules—hence the need for paradigmatic cases to practice on. Paradigms are born from a specific way of viewing reality and to scientists trained on them they may well become indistinguishable from the reality. "In learning a paradigm the scientist acquires theory, methods, and standards together, usually in an inextricable mixture" (Kuhn, 1962).

The truth of a paradigm is assumed, it is not open to questioning or testing in the confines of normal science. This gives the normal science a level of dogmatism not seen in Popper or in the logical positivists of the first half of twentieth century, but also a measure of stability and cumulative growth of knowledge.

What people do in the confines of normal science is supposed to be akin to solving puzzles. According to Kuhn, puzzle solving is good for two reasons: it is a very effective way of increasing knowledge and without it there would be no scientific revolutions. Solving puzzles is what scientific method, as taught by senior scientists to their younger followers, is good for. If a scientist tries to solve a puzzle and fails, the onus to make it work is on her in the same sense that a failure of a tailor to make a fitting suit is not expected to herald some deep problem in the art of tailoring.

Kuhn even suggested puzzle solving in lieu of Popperian falsification as a criterion for demarcation between science and non-science (Dienes, 2008). For instance, astrology is in principle falsifiable since it makes testable predictions (it is just that the typical astrologist is not keen to falsify his theories). And yet it fails to generate puzzles that people could solve. On the other hand, a theory that is not falsifiable cannot in principle generate interesting puzzles. Therefore, for Kuhn, falsifiability is a prerequisite for the deeper criterion of demarcation—ability of the paradigm to generate interesting puzzles.

In Kuhn's view it is precisely the giving up of criticism in the confines of normal science that transforms a field into a science. After that point critical attitude reappears only at moments of crisis, when it prepares ground for a revolution.

But then, there are anomalies within each paradigm. Anomalies result in puzzles left unsolved despite the best efforts of scientists, and they are then typically set aside for a future date. However, as such anomalies accumulate, a sense of crisis will eventually permeate the field. As a scientist cannot exist without his paradigm, a new paradigm must emerge before it is acceptable to criticize and discard the prevailing paradigm. The crisis thus precipitates a revolution where, for a relatively short time, instead of a single paradigm there are several competing ones. Because real revolutions tend to be bloody and unpleasant affairs it stands to reason that in a scientific revolution the participants want to quickly agree on a new and better paradigm, which leaves fewer unsolvable puzzles on the table and thus restores the general sense of order and normality. Another era of happy puzzle solving ensues until a new crop of unsolvable puzzles precipitates the next crisis. Both periods of normal science and of revolutions are inevitable and, as far as we know, constitute a never-ending cycle.

In Kuhnian theory a paradigm shift cannot occur piecemeal, it is an all-or-nothing affair, leaving no empty theoretical space between revolutionary crises and normal science. The problem here is how to decide which paradigm to accept as the new normal? You might think it not such a big problem since it ought to be possible to superimpose the competing paradigms and see which one covers the most ground (generating the largest number of new puzzles and leaving behind the least number of the really puzzling ones). However, things are made difficult by a very tricky phenomenon called "incommensurability" (postulated independently and a little differently by Kuhn and Feyerabend).

The crux of the matter is that the new paradigm, accepted after the revolution, may not use all the same concepts as the old one, which it replaced. Each new paradigm invents new concepts and discards some of the old ones. Even worse, some of the retained concepts may subtly shift their meaning (this is called "conceptual shift," an example of which is the concept of mass in Newtonian versus relativistic physics). As a result, incommensurable theories are semantically incompatible (Hoyningen-Huene, 2002). Another possibility is methodological incommensurability, whereby we are incapable of deciding on the relative merits of the different research methodologies behind different paradigms, because there exists no overhanging "objective" methodology (Chang, 2014). All this makes it difficult to compare theories by logical analysis *a la* Popper. Different paradigms may even disagree on what entities exist (aether, flogiston, caloric). By using the concepts of one theory, it is literally impossible to formulate the other theory—they are in different and mutually untranslatable languages. How can we then compare the two and say that one is objectively better than the other? If we have no rational way of choosing between incommensurable theories, does this mean that science cannot progress but is aimlessly fluttering from revolution to revolution?

This was certainly not Kuhn's view. For him incommensurability took from the table the possibility of cumulative growth of science, but not the growth of science as such. Firstly, inventing useful paradigms hinges on and is constrained by rationality, which is not itself defined by any single paradigm. Also, not all aspects of the old paradigm are incommensurable with its successor and the ones that aren't will survive the revolution to be carried over to the new normal. To choose between competing theories one must try to take into account their respective accuracies, consistencies, scopes, simplicities, and fruitfulness. Obviously, there is no formal method to weight these rather different considerations, which means that it is quite possible for different scientists to reach different conclusions and for everybody to still be rational. This is not necessarily a bad thing, as the resultant bet hedging by the scientific establishment would confer to science greater robustness and potential to innovate.

Granted that Kuhnian notions have encouraged countless scientists to be brave and try to "change the paradigm," the questions still open include: is there actually such a thing as normal science with a single unquestioned paradigm? Are physics and biology different in that respect? Could it be that while Kuhn is right and some scientists indeed practice uncritical puzzle solving, these individuals were simply badly taught? What about paradigms—do they even exist (Margaret Masterman described 21 distinct ways in which Kuhn used the term in his book (Masterman, 1970))? What exactly does the term *paradigm* mean, how does it differ from *theory*? A search of the science and technology part of the ISI Web of Knowledge (April 15, 2014) revealed that 1,647 papers had mentioned "paradigm shift" in their title (plus, there appear to have been about half a dozen second "Copernican revolutions" in the twenty-first century). Interestingly, only 26 of the 1,647 papers, which claimed to switch the paradigm, have subsequently gathered over 100 citations and about 70% of the paradigm-shattering papers have been cited three times or less.

Incommensurability seems to entail that the old and new paradigms are not mutually exclusive and exhaustive, which means that they cannot be compared via Bayesian analysis (and there is no formal alternative). Also, Kuhn makes it clear that paradigms are examined in the light of evidence singly (if at all), which violates the law of likelihood (as does Popperian falsificationism).

But then, according to Popper, incommensurability is nothing more than "The Myth of the Framework." For Popper, although conversation between different paradigms might be difficult at times, when successful, nothing is more fruitful. In this view, there is no question that Newton's and Einstein's theories of gravitation are rationally comparable.

All this makes the Kuhnian suit of armor a little too restrictive of movement, and a little too far from the formal methods of theory comparison that we actually have, to be worn comfortably.

Some of the most biting criticism of Kuhn comes from Paul Feyerabend:

> [A]re we here presented with methodological prescriptions *which tell the scientist how to proceed; or are we given a* description ... *of those activities which are generally called 'scientific'?* ... *More than one social scientist has pointed out to me that now at last he has learned how to turn his field into a 'science'—by which of course he meant that he had learned how to* improve it. *The recipe according to these people is to restrict criticism, to reduce the number of comprehensive theories to one, and to create a normal science that has this one theory as its paradigm. Students must be prevented from speculating along different lines and the more restless colleagues must be made to conform and 'to do serious work.'* Is this what Kuhn wants to achieve? (6:198).

(Feyerabend, 1970)

Kuhn of course wanted no such thing (Kuhn, 1970). Instead, he aspired to create a theory of science that would be based on the actual science, or at least on the actual history of it. In short, Kuhn believed that he had identified patterns for success in the behavior of scientists. This would be a paradigm shift in itself, as the accepted methodological canons of logicians would be replaced with new ones, which would have been considered aberrant under the old regime.

7.2 THE ANARCHISM OF PAUL FEYERABEND

Ever tried. Ever failed. No matter. Try Again. Fail again. Fail better.

Samuel Beckett

If the historically minded Kuhn can be accused of falling into a narrow prescriptive rut (would Kuhn's theory still be interesting without his rather strictly defined cycle of normal science—crisis—revolution?), nobody can blame Paul Feyerabend (1924–1994) of this sin. His main methodological prescription was: *anything goes!*

Feyerabend's intuition was that general methodological rules do more harm than good and therefore, what a scientist needs is freedom, not restrictions. The same goes for non-scientists and science should not be seen as a privileged means of producing knowledge. Methodological tricks, which could be recommended to young scientists, must be based on research of what scientists actually do, not on rationality or logic. In addition, the fact that a trick worked in the past is no guarantee that it will work in the future. Every methodological rule depends on assumptions about the physical reality; therefore to use that rule to study the world implies the correctness of these untested assumptions (you can see the vicious circle here: you have to assume the truth of what you plan to "scientifically" study).

Considering that before arriving at these conclusions Feyerabend had been a member of the Wehrmacht, who made it from private to Lieutenant, won an Iron Cross in the Eastern front, and was seriously wounded in battle—and all this before becoming a student of Karl Popper—this must be considered a remarkable change of perspective. Feyerabend's first and most famous book is *Against Method* (Feyerabend, 1975). This is how he describes writing it in his autobiography (Killing Time, 1995):

> Against Method *is not a book, it is a collage. It contains descriptions, analyses, arguments that I had published, in almost the same words, ten, fifteen, even twenty years earlier... I arranged them in a suitable order, added transitions, replaced moderate passages with more outrageous ones, and called the result "anarchism". I loved to shock people... One of my motives for writing* Against Method *was to free people from the tyranny of philosophical obfuscators and abstract concepts such as "truth", "reality", or "objectivity", which narrow people's vision and ways of being in the world. Formulating what I thought were my own attitude and convictions, I unfortunately ended up by introducing concepts of similar rigidity, such as "democracy", "tradition", or "relative truth".*

(Feyerabend, 1995)

In *Against Method* Feyerabend argues that the history of science shows that there is no single rational method for doing science. Methods that are used in science have no common structure. In fact, there is no such thing as science as a single entity. Instead there are many sciences, each with their own rules.

A successful scientist cannot and doesn't rely on a single methodology; he relies on different tricks in different times. Big changes, like the Darwinian revolution, affect different areas of science differently and receive different inputs from them. This naturally leads to scientific anarchy. Not only is every methodological rule violated by the scientists, but these violations are actually necessary for the progress of science (Hickey, 2014). Thus science lives in a perpetual state of pluralism where no theory completely fits the facts, all facts contain theory, and incommensurability makes it impossible to rationally pick a winner amongst competing theories. This necessitates the use of *ad hoc* hypotheses and approximations to make the theory fit the evidence.

This procedure was dubbed counter-induction by Feyerabend. The point of counter-induction is that when any methodological rule is considered, however general or seemingly universally valid, it is always possible to show that some scientists have successfully used its opposite. For example, every theory of induction, not to mention popular accounts of the scientific method, claims that facts or experimental results are used to test, verify, or change the plausibilities of theories. If empirical facts are incompatible with the theory, then the theory must go. In contrast, from the principle of counter-induction follows that when facts are inconsistent with the theory, it is as well to keep the theory unchanged and to try to change the facts, so that the contradiction disappears. Because the facts depend not only on the external world but also on our particular observation-language, by which we describe the manifestations of that world, the trick is to invent a new observation-language that would enable to fit the facts into the theory. Thus one can proceed either inductively or counter-inductively, and it is nobody's business but yours, which track you choose.

Another bit of advice is from the start to create and contemplate a host of alternative hypotheses, which are inconsistent either with established theories or the established facts. Because it is often the alternative theories that enable one to unearth evidence relevant to the original hypothesis, Popper's method of falsifying hypotheses singly will not do. From here comes the "principle of proliferation": always invent theories incompatible with the accepted point of view, even if it happens to be universally accepted and highly confirmed. A more recent analysis of history of genetics and physics both fully supports the scientific importance of this dictum and indicates that in reality the scientifically relevant hypothesis space can never be exhausted by an individual scientist (Stanford, 2006).

The consequences of this pluralism include, according to Feyerabend, that (i) the past successes of science do not mean that we can treat our current problems in any standardized way, that (ii) non-scientific ideas cannot be pushed aside by rational argument because there is no clear demarcation between science and non-science, that (iii) success in science can only be assessed after the fact, which means that (iv) there is no good way of weighing scientific promises (like grant applications), which in turn means that (v) inviting the general public to join scientific discussions can do no harm (the only advantage of scientists being that they know more details). The full democratization of science is not in conflict with science; it is in conflict with a metaphysical theory called Rationalism, which purports to offer a single well-defined methodology to science, the learning of which would make you a specialist and preclude the involvement of non-specialists. A good example of Rationalism is Popperian falsificationism, the teaching of which to undergraduates was supposed to help them to more efficiently test scientific-like hypotheses in the cognitive psychology lab, but didn't (Mynatt et al., 1978). More recently, Bayesianism has been seen in a similar light.

This obviously doesn't sound like methodology of science that can be found in the introductions of university textbooks in natural sciences. Science, according to Feyerabend, is sloppy, relativistic, anarchic, ideological, and yet, things couldn't be better! Successful science thus results from a combination of good luck and false beliefs (Hickey, 2014).

Of course, since Feyerabend claims to be ideologically anti-Rationalist and anti-Reason (and not unreasonably so), it begs the question how he should justify his own anarchist ideology without recourse to Rationalist proof. However, Feyerabend's aim is not to establish his theory of science by argument, but rather to make it more plausible by historical examples. His stated ambition is humanitarian, not intellectual. Accordingly, the chapter on Feyerabend in the Stanford Encyclopedia of philosophy written 18 years after his death, concludes: "It is still far too early to say whether, and in what way, his philosophy will be remembered."[1]

7.3 THE COMMUNISM OF ROBERT K. MERTON

Has any thought been given to the number of things that must remain active in men's souls in order that there may still continue to be 'men of science' in real truth? Is it seriously thought that as long as there are dollars there will be science?

(Ortega and Gasset, 1932)

Our next radical idea is that scientists are good. This thought was espoused by the great American sociologist Robert K. Merton (1910–2003), whose lifelong ambition was to turn sociology into a science, and whose major object of sociological study was science itself. His major contributions to the sociology of science started in the late 1930s when the extraordinary successes of science were generally taken as a given. Nevertheless, it was well understood that successful science had arisen in some cultural contexts but not in others, which is not to say that science does not have a logical component or is purely a social construct (Merton, 1974). What it does imply, however, is that studying the logic of

---

1. For example, a more modern less shock-oriented version of scientific pluralism is discussed in Chapter 8, whose main difference from Feyerabend's seems to be that in science not "anything goes," but "many things go" (Chang, 2014).

science is not enough—that study of the social norms and incentives prevalent in scientific circles is indispensable for understanding science.

A major question for Merton was how to explain a profession with practically no cheating behavior in a society which clearly has different norms. How different can the ethos of science be from the ethos of society that harbors it (do we need a democratic society for thriving of science or will totalitarianism of Hitler or Stalin do as well)? What controls the behavior of scientists? What is the relation of truth and social utility in science? How about the roles of politics and ideologies in science? How does the increasing specialization of science and the consequent need for the layman and professional alike to accept its results on faith, affect the scientific enterprise? Is scientific knowledge fundamentally different in democracies versus non-democracies? What is the impact of state control to science? How is the organized skepticism of scientists accepted by other segments of society that abhor skepticism? How is the institutional integrity of science (the purity of science) maintained inside the turbulences of the larger society? What are the roles of societal technological advances in determining the expectations and support the society has for science? How big is the authority of science and what determines it? What are the incentives that lead to social norms of science? And, of course, given that the cultural norms of science and society are different, what are the norms of science?

Merton proposed his answer to some of these questions in a short paper originally published in 1942 (Merton, 1974). He sees the ethos of science as the complex of values and norms, which are binding to the scientist. These can be expressed as prescriptions, proscriptions, preferences, or permissions, and are converted into institutional values of science and into a moral consensus of scientists. These mores are enforced not only because they are seen as pragmatically effective, but also because they are believed to be morally good. Merton saw four basic institutional imperatives in science:

1. *Communism.* This means that scientific goods are in common ownership. They belong to everybody. The scientist extends no claim to the product of his work, except for authorship. But authorship, and by extension scientific priority, is extremely important to him. The sometimes fierce competition between scientists is not oriented toward ownership but toward priority of discovery. This means that scientists are constantly gnawing at each other's throats for the privilege of who can give away the most knowledge! This also means that a scientist who fails to publish his discoveries is thought by his peers to be morally handicapped for withholding his results from the community. The communism of science is underlined by scientists' allocation of credit to their peers (and by their claims of standing on the shoulders of giants), which can be seen as an affirmation of the essentially communal nature of their work. Thus, a measure of humility and collaborative spirit are seen as befitting a scientist. This view of science is incommensurable with technology, seen as "private property" and enshrined in the patent system.

Merton clearly saw communism in science as a good thing. This is how he explains it 46 years later, in 1988:

> ... *Karl Marx adopted the watchword of a fully realized communist society – "from each according to his abilities, to each according to his needs" – this was institutionalized practice in the communication system of science. This is not a matter of human nature, of nature-given altruism. Institutionalized arrangements have evolved to motivate scientists to contribute freely to the common wealth of knowledge according to their trained capacities, just as they can freely take from that common wealth what they need. Moreover, since a fund of knowledge is not diminished through exceedingly intensive use by members of the scientific collectivity – indeed, it is presumably augmented – that virtually free and common good is not subject to ... "the tragedy of the commons": first the erosion and then the destruction of a common resource by the individually rational and collectively irrational exploitation of it. In the commons of science it is structurally the case that the give and the take both work to enlarge the common resource of accessible knowledge.*

(Merton, 1988)

2. *Universalism.* This means that scientific truth is impersonal. As long as methodological rules are followed, it doesn't matter who is making the claim. It doesn't matter if you are a good person, a professor or a student; it is your discovery, not you, that matters. An important expression of universalism is that science is a meritocracy. Free access to the profession of science is a functional requirement for the furtherance of knowledge. Here expedience and morality coincide.

3. *Disinterestedness.* Because the institution of science works for the society (its activity is disinterested), the individual scientists are also pressurized to at least pretend to be disinterested (neutral) in respect to the outcome of their work. This ideally leads to the absence of fraud in science and to an unusual degree of moral integrity. The basis of this demand on scientists is the public and testable character of science, or in other words, the universality of its methodology. The requirement of disinterestedness is enforced by one's peers (other scientists), not by the "clients" of scientists (the larger society). It reinforces verifiability of results and precludes self-aggrandization, thus stabilizing the institution of science. Because scientists communicate with the public through the filter of peer review, they cannot easily exploit credulity and ignorance of laymen for their personal benefit (e.g., by making sensational claims in the press). When this structure of peer control disintegrates, anyone can borrow the authority of science to cloak their mysticism or favorite theory. This is so easy, because only an expert can distinguish a spurious scientific-sounding theory from the real thing. Therefore, in a totalitarian state science can easily be used to justify any racial, eugenic, or economic policy.

4. *Organized skepticism.* This involves public but detached criticism of one's own and others' theories in terms of empirical and methodological criteria.

Organized skepticism is both a methodological and institutional requirement. For a scientist nothing is above criticism, nothing is sacred. Here is an obvious point of friction between science and religion and, science and power structure and, indeed, science and society. This may lead the state or the society to limit the domains in which science is allowed to operate.

To this list *originality, humility, rationality, emotional neutrality*, and *individualism* (against authority) have been later added by Merton and his followers (Vinck, 2010). Note that the norms of originality and humility are at least superficially contradictory, introducing tension to the otherwise coherent Mertonian world view. Originality means that a scientist's job, for which he is rewarded by the establishment, is to expand the frontiers of human knowledge, not to repeat the discoveries of others. Originality leads to the imperative to push boldly into the unknown where nobody else has tread before. To cite the war-time US president Franklin D. Roosevelt, "New frontiers of the mind are before us, and if they are pioneered with the same vision, boldness, and drive with which we have waged this war we can create a fuller and more fruitful employment and a fuller and more fruitful life" (November 17, 1944; http://www.nsf.gov/od/lpa/nsf50/vbush1945.htm). Science is thus an "endless frontier," the pushing back of which is the duty of scientists. Specifically, it is the basic science, which holds the key to solving the practical problems standing in front of society. The analogy here is military and the novelty of research is a fairly strict institutional requirement for continuing financing of a scientist's work, and it punishes scientists who fail to produce original research but "merely" replicate other people's work. Taken together *C*ommunism, *U*niversalism, *D*isinterestedness, *O*riginality, and *S*kepticism make up the value system of science—CUDOS. It is the CUDOS system that gives science its coherence in time, despite the changing paradigms in the content of science, science is still a single thing across its sub-disciplines and across three centuries. Science is given continuity over time not by its methods, which are constantly changing, but by its ethos.

These norms are certainly different from what could be found in American society at large in 1942, or now. Their applicability to science was not decided by Merton after shifting through large amounts of empirical data. CUDOS was rather taken as a given, as something that every scientist (at least those from democracies) would naturally agree on. Seen like this, the question becomes: what makes scientists behave so differently from other people?

Since there is no indication that budding scientists are selected by their moral superiority, it must be assumed that working scientists are morally equivalent to laypeople. Also, it must be that the social norms of science are taught and enforced by other scientists, thus guaranteeing a stable social environment, but the question of the origin of the norms remains. A reasonable theory is that different norms in science versus society were brought about by differing incentives. While normal economy is based on gold, the coin of science is peer recognition. This can take the form of a Nobel Prize, an eponymous law or, more commonly, a citation to one's work. This coin is earned from one's peers in exchange for making the results of one's toils freely available for common use. As gold is freely convertible into fast cars and power over one's inferiors, so is peer recognition converted into research grants and professorships. Because communism is a strict requirement for earning the scientific coin, scientists have no choice but to behave differently as members of their tribe than they do in their free time. However, since the CUDOS-based value system is implicitly instilled into each budding scientist, they absorb it from their peers as if suckling it in by mother's milk—unthinkingly and without doubt.

Another influential theory of how the seemingly selfless cooperation in science arises was proposed in 1962 by Michael Polanyi (1962). Polanyi saw scientists, analogously to the operators in a market economy, as free agents who each set their own research agenda. As is the case in an economy, so is it in science that central planning imposed by the state is doomed to failure. It is thus left to the institutions of science to constrain themselves to supporting individual talent and leave it to the "market forces" of science to sort out which branches of science will blossom and which ones will wither.

And yet, although the individual efforts of scientists are not necessarily coordinated, the outcome of their work is part of a single web of scientific opinion, and it is constrained by it. This vast web covers the entirety of science and it is the sum of expert opinions of a multitude of established scientists, each a narrow specialist but the area of expertise of each overlapping with some of the others'. Thus is the scientific opinion a property of the web; it does not belong to any individual scientist. It is the constraints of the scientific opinion that confer to science a unity of purpose. Because someone can be considered a scientist only if he or she conforms to the mores and prescriptions of the scientific opinion, the efforts of each scientist must be adjusted to the work of the others.

For the in-crowd (people who publish in peer reviewed journals) the scientific merit of a discovery depends on (i) its overall plausibility, (ii) the quality of evidence behind it, and (iii) its originality/practical importance (Polanyi, 1962). Thus the scientific opinion is a tug-of-war between conservative (i and ii) and revolutionary (iii) instincts. To find the right balance is a major question for the peer review process, as should the shifting of this balance over time become a major research focus for the historian of science (Horrobin, 1990). It should be noted here that what we know as methodology of science is overwhelmingly concerned with raising the quality of evidence and thus reducing the error rates in science. It can be argued that the price we have paid for the advantages in methodology has been barring the entry of an increasing number of truly important discoveries into the realm of scientific acceptability (Horrobin, 1990). To exclude half of mundane everyday results on the grounds of insufficient methodological quality can on the whole be good for scientific progress. On the other hand, to discard on the same grounds half of life-changing discoveries in medicine, before they can enter into the consciousness of scientific opinion, can be seen as a tragedy.

A major assumption behind Michael Polanyi's free-market theory of science is that for the self-regulation of science to work, scientific standards of each branch of science must be essentially the same (Polanyi, 1962). If, for example, it would be much easier to publish high-visibility low-quality science in social psychology than in physics, then the psychologists would eventually outcompete the physicists in funding and prestige and the end result would be that the scientific opinion as a whole would no longer reflect useful knowledge. Indeed, recent studies support the existence of an hierarchy of sciences where it is seemingly easier to get and to publish interesting positive results (or harder to publish negative results) in the "soft" sciences, like psychology, than in the "hard" physical sciences (Fanelli and Glänzel, 2013; Fanelli, 2010). The evidence indicates that the more a science is concerned with studying human beings, the "softer" it is (Fanelli, 2010).

While Polanyi's market economy of science led to the optimal use of resources and growth of knowledge, a rethinking by Pierre Bourdieu (1975) presents a darker view of cut-throat capitalism. According to Bourdieu, scientists compete with each other for the finite resource of scientific credit (whose fountain is one's peers), thus building symbolic capital, which can be hoarded and then spent to purchase funding, jobs, or machines. Scientific credit is exchanged for published discoveries, which, outside their exchange-value, do not have any intrinsic value for the scientist. Truth or Mertonian norms are irrelevant, it is the exchange rate that matters in a market economy, and in science demand largely depends on the interest that other scientists have for your discoveries (Vinck, 2010). There is no central authority to mitigate the effects of free market. It is the publication in a top journal that is important in the market of ideas, not the truth of the discovery thus published. For an individual scientist the object of the game is to dominate his field. The vast majority of scientists are of course in the category of being dominated, and their reaction is to strive for control over each other, thus, as a byproduct of mutual criticism, perhaps leading to production of true knowledge.

Although market forces are individualistic, there are no objections against some players congregating into collaborative networks to better fight the others. Also, each scientist is pushed by market forces to choose his topic of interest by its perceived importance to his colleagues, which leads to "herding" (see below).

A lot of water has flown under the bridge since 1942. Whole industries that depend on science have arisen and grown. The frontline of science is more in biological and less in physical sciences than it was before the War. Bombs have been made and exploded. Has all this changed the Mertonian ethos of science? What are the factors that work against the CUDOS system of command and control of scientists?

Perhaps the most clear-cut test of the CUDOS hypothesis would be to study the Mertonian coin of science, peer recognition, and specifically to ask whether citations really work as a form of recognition of other workers in the field for their contributions to the problem at hand. This question has been studied by many and the authoritative systematic review can be found in Bornmann and Daniel (2008). The authors find a wide specter of problems with the quality of individual studies of citing behavior and are unable to draw firm conclusions. Still, the general thrust of the field seems to indicate that erroneous and spurious citing is rife, especially in biomedicine where author surveys have revealed that most citations are of limited or peripheral importance to the citing study (i.e., they are essentially random).

Next we will discuss some factors which are at odds with Mertonian norms.

7.3.1 The Matthew Effect

Our first factor was introduced by Merton himself in an influential *Science* paper and it was proposed to work against the principle of disinterestedness (Merton, 1968). This is the so-called Matthew effect, according to which "For unto every one that hath shall be given, and he shall have abundance: but from him that hath not shall be taken even that which he hath" (Matthew 25:29). In the context of science this means that scientists who have more recognition today (be it in the form of fame, citations, etc.) will in the future receive relatively more recognition for the same quality of work than their less-illustrious colleagues. This leads to a ratchet effect where, once a scientist has achieved a degree of eminence, it becomes very hard for him to fall in prestige. This can be construed to mean that as your fame increases, more and more is expected of you and you will accordingly feel the need for ever greater achievements.

An interesting consequence of the Matthew effect is that the younger co-authors of the Famous Men, while their papers get more attention than they perhaps deserve, are left without the personal credit that they might deserve. Everybody assumes that intellectually the paper belongs to the famous author, and very little will be left for the people who often really did most of the intellectual work. A number of Nobel laureates interviewed in the early 1960s by Harriet Zuckerman were cognizant of this problem and to counteract it, had started to put their names down as the last and by implication the least important author of their papers. With hindsight we all know where this has brought us.

Another carry-over of the Matthew effect is that students of the Famous Men and of Famous Universities will take a disproportionate portion of the academic pie. For example, of the 55 American Nobel Prize winners in Harriet Zuckerman's sample, 34 had worked as young men under a total of 46 Nobel laureates. Moreover, at the time they did the research that landed them the prize, 49% of the future laureates who worked at a university, worked at just the top five of them, which in turn accounted for 3% of all faculty members of American universities (Merton, 1988). Indeed these universities retained 70% of the future Nobel winners they educated (and 28% of other PhDs they had trained).

This leads to the generalization that the people, who will be winners in the ratcheting game of fame, must start their run of success while young. Indeed, quickly obtaining one's PhD and fast early promotion into independent research are the best predictors of later success (Youtie et al., 2013). There is a premium on early manifestation of talent and, consequently, late bloomers are at a disadvantage in science (Merton, 1988). Students with excellent grades get a chance to do their PhD at the best labs, which translates into better postdoc opportunities and better chances of eventually landing an independent academic position. How you divided your time between studies and having a life at 19 may well be more important for your fate when you are 35 than your actual scientific ability at 35! Since we really do not know to what extent the early ability (which, by the way, is measured relative to one's age group) is a predictor of later ability (which really matters in science), this aspect of the Matthew effect seems to be wasteful of scientific talent.

A related contentious issue is whether the (objectively) best people on average really get the best jobs in science. If the answer is no, then the principle of Universality is in danger. Determining which candidate is objectively the most qualified for a science job is obviously a tough nut to crack, so a lot of discussion has turned around the easier question of whether it is the previous publishing activity or the identities of candidates mentors that decide, who gets the jobs in the best universities. The usual answer is that in getting your first real academic job it is your mentor that counts way more than your actual publication record (Long et al., 1979; Allison and Long, 1990) (but see Cable and Murray, 1999). And, people who land at prestigious institutions have a head start regardless of their actual ability, due to better infrastructure, better chances for good collaborations, greater intellectual stimulation and a greater press from the university to ensure their continued motivation. This leads to subsequent upward movement of scientists who start at highly prestigious universities and downward movement (on average, of course) of the ones who start their careers at lower-placed institutions (Allison and Long, 1990; Su, 2009).

7.3.2 Artificial Abundance

The Matthew effect is further amplified by the increasing abundance of scientific papers. Since even the most conscientious scientist can stay abreast with ever lesser portions of the relevant literature, she will increasingly need a heuristic to select what she will actually read. And the simplest heuristic is availability. She will simply search the web with the names that first pop to mind—and they tend to belong to famous people. Since people ratchet up their fame by publishing at increasing volumes, we have here a nice instance of a positive feedback loop. Thus your very need to stick out from the vast crowd of spurious papers encourages you to publish spurious papers of your own, to increase the chances that your really important papers get noticed. This dynamic is further supported by institutions, which use science metrics to cheaply and quickly compare competing scientists. A possible outcome of this trend is the observed increase in collaboration and in the number of coauthors *per* paper of young scientists (O'Brien, 2012).

While it is important for the careers of younger scientists to publish—no matter what—, the older scientists, who already are somewhere, may switch tracks and start suppressing their lesser work in order to focus attention on important messages. To illustrate this, Merton (1968) cites Seymour Benzer (1921–2007)—already a prominent molecular biologist in the early 1960s—who claimed that Max Delbrück (one of the fathers of molecular biology) had saved him from going "down the biochemical drain" by writing to his wife to ask her to tell his husband to stop writing so many papers. This was an advice Benzer took—and in 1967 he ended up switching from phage genetics, which was the classic field of molecular biology in the 1950s and '60s, to the then-new science of behavioral genetics.

7.3.3 Artificial Scarcity

Robert Merton believed that citations to the work of others function as the basic currency of science. Since this coin is given to you by your colleagues in return for your gift to Humanity of your results, its (unintended?) consequence is the CUDOS-based ethos of science.

The coin that you earned for being good, you will spend at your university, which uses it to assess your suitability for your job. The reason for its interest in the citations to your papers is that it wouldn't actually have to decide how good you are as a scientist. Instead it asks two simpler questions: "how many citations?" and "in which journals?" Because the quality of a journal is, if anything, an even more difficult research subject than the quality of a researcher, your university will repeat the trick and define a good journal by the average number of citations to its articles.

Unfortunately for you, in a climate where there are many more journals (and papers) than anybody could possibly read, people have to make choices. Firstly they read the papers from the famous journals, and secondly they read papers of famous authors. However, because both journals and authors are ranked by citation counts, it is in the common interest of famous people and famous journals for the first to publish in the second. In each large division of life sciences as defined by ISI Essential Indicators (each with hundreds of journals), six journals account for 68–94% of the one hundred most-cited articles (Ioannidis, 2006).

Because the journals that count need to keep their averages high, they must severely restrict access to their hallowed pages (Young et al., 2008). To be more concrete, circulation rates among journals that publish clinical research span over three orders of magnitude and their acceptance rates fall on average by 5.3% with each doubling of circulation (Goodman et al., 1998).

The importance of both artificial abundance and artificial scarcity for individual scientists is shown in the work of van Dijk and colleagues, who showed that the best predictors of someone making it into a principal investigator (PI) in biomedical science (excluding clinical science) are publishing in the high-impact-factor journals or, then, just publishing a lot of papers (van Dijk et al., 2014).

(A PI is here defined as someone who has published at least three last-author papers.) Sheer publishing volumes can compensate for not being able to publish in the top journals and the people who distinguish themselves very early in their careers have the best chance of making it to a PI. Importantly, your h index,[2] impact of your papers (as opposed to journals publishing them), and getting ones education in a top university have all considerably less predictive power of your future career (van Dijk et al., 2014). All-in-all, from the PhD holders who have published at least three papers in the biomedical field about 6% will become PIs over the course of their careers; and from the people who have made it into the top 10% of journals (by journal impact factor), about 20% will get to be PIs.

Thus, in the Sea of Artificial Abundance lie the Islands of Artificial Scarcity, and to get noticed and cited you should get your feet down on one of these as early in your career as possible. But to get there, you will need to compete with lots of others who all need a foothold as desperately. To not drown in the process, you have three strategies open to you. Firstly, you might want a Famous Man (or Woman) in your authors list, secondly, you could send out some really spectacular results to woo the editors with, and thirdly, you could put your trust in numbers and send out as many papers as you possibly can with the hope that sooner or later one of them will slip past the reviewers into a top journal.

Both the number of papers a scientist publishes and their impact correlate with the number of collaborations he engages in (Adams, 2012), and, accordingly, in the US collaborative publications have increased by about twofold from 1990 to 2012 (National Science Board, 2014). In the extreme end, the papers with more than 100 authors become a noticeable trend only in the 1990s, the first paper with more than 1,000 authors was published in 2004; and by 2011, 120 physics papers had more than 1,000 authors (44 had more than 3,000) (Adams, 2012). Importantly, collaborations get more citations on average than non-collaborative work by the same scientists (Adams, J. in Roberts, 2006; available at http://go.nature.com/zzwn8z.). This could be caused by higher scientific quality of pooled talent, stronger scientific claims supported by concerted action of more diverse experimental systems, or simply by increased visibility of longer author lists in readers' web searches.

7.3.4   The Winners Curse

Let's imagine a group of children playing a game. The object of the game is to guess the height of their teacher. After making their predictions they will ask the teacher for her height and proclaim whoever made the closest estimate the winner. One kid wins, everybody else loses. Assuming that it is equally likely that individual children over- or underestimate, the winning estimate tends to be close to the average of the kids' estimates.

---

2. H index of h means that you have published h papers, which have been cited h or more times (and possibly some more papers that have been cited less than h times).

Now, consider a Game of Science that is played by professional scientists. Every week scores of scientists are making all sorts of effect estimates. The winning estimates are published in a very small number of top journals and the winners of the game get to define the "real discoveries." The vast majority of scientists never publish in these journals. They are the losers who will be sooner or later pushed away from the frontline of research. Year after year the game goes on.

As you might have already noticed, there are important differences between the two games. In the first game nothing much is at stake and the winner is whoever made the most accurate (and typically average) estimate. In the Game of Science a lot more is at stake and the winner is whoever made the largest estimate. In science nobody knows what the true effect size is, but there is usually no reason to think that it is the largest estimate. This game is very similar to what happens in auctions, where this type of problem is called "the winner's curse." In a competition to outbid rivals, whoever places the largest bid is likely to be overpaying. The steeper the competition (the more bidders for every unique vase or oil field) the more outlandish is the winning bid likely to be. Winner's curse in science is a direct consequence of artificial scarcity, which in turn is a direct consequence of the business model of the journals that profit from it. The result is that you really should be careful about what you read in top journals.

The winner's curse in science can go far beyond the top journals: in some fields, almost no negative results are published at all (Kavvoura et al., 2007; Kyzas et al., 2007; Fanelli, 2010).

7.3.5 Herding

Another consequence of high levels of competition in science is the behavior of herding. The logic behind herding is as follows:

1. The livelihood of both journals and of their authors depends on high citation rates to their papers.
2. The fields that have more members publish more and each publication contains citations to other papers in the same field.
3. Therefore it pays for the individual scientist to do their research in a field where many of his colleagues are already working and publishing in good journals (Wallace et al., 2009). These fields are known as "sexy."

Herding works against the norm of originality because many people studying the same (or very similar) problems will inevitably result in some duplication of effort. At the same time, the norm of originality of scientific contribution is enforced strongly enough by the journals, so that in most cases of duplicated research only the authors that successfully claim priority manage to publish their findings in good journals.[3]

---

3. This is more true in biology than in medicine.

The major consequence of herding is likely to be globally reduced ability to produce truly novel science as it is a bad gamble to work in a less populated field, which has less people who might read and cite your work. This keeps researchers, who are not yet Nobel laureates, from straying from well-trodden paths, although the potential benefits of being a first to arrive in a new field are huge. In herding, the actions of a few Famous Men drive scientists' notions of what the hot fields of study are. Herding occurs when individuals regard others' earlier actions as more informative than their own private information (Young et al., 2008).

How does a field become sexy? It could be that a Famous Man starts to work on something new and publishes his findings in the top journals or simply that a seminal work is published, like the role of Runx3 in cancer of the gut (Li et al., 2002), and many people suddenly want to know if the protein has a role in the genesis of their favorite type of cancer. The 2002 paper of Li et al. has been cited 739 times (by January 14, 2015) and according to ISI Web of Knowledge additional 569 papers, which list both "Runx3" and "cancer" as their topics, have been published since. As a comparison, about 81,000 papers list "p53" and "cancer" as topics.

How well did the subsequent entrants to the "Runx3 in cancer" field do? When we exclude from analysis the papers that have been out less than two years and thus haven't had enough time to collect citations, about 19% of the papers managed to attract 50+ citations (which can be considered a hefty capital), and another 28% gathered between 20 and 49 citations. This may not sound like much, but consider that according to ISI Web of Science (WoS) essential science indicators a paper in biology and chemistry needs only 40 citations to be in the group of top 10% most highly cited papers and 141 citations to be in the top 1%. For someone working with Runx3 in cancer the chances of publishing a paper in the top 10% of biology and chemistry are 25%.

Is herding with its concomitant conformism a good or a bad thing for the fields where people crowd into? It could be that it is exactly the herding behavior that allows the measure of independent corroboration of scientific results that we have, and thus leads to the situation where, in the long run, truth rises to the top and spurious claims are quietly forgotten. But it is also possible that in a crowded field, where several labs are testing the same hypotheses and the positive results are published preferentially, the main local effect of crowding is increased multiple testing problems and saturation of literature with false claims. In addition, it could well be that as herding increases competition in the crowded fields, it will consequently increase the incidence of questionable research practices in them (Pfeiffer and Hoffmann, 2009).

A group from Columbia and Yale Universities tried to address these questions with absolutely shocking results (Rzhetsky et al., 2006). They made use of a database of several million statements about protein–protein interactions, which had been data-mined from scores of research journals and built a rather complicated mathematical model to analyze these data in a time-dependent fashion. In their model the first paper to comment on an interaction is supposed to contain original research on it and every subsequent paper either does some more experiments and interprets them taking into account the previous knowledge (which it weights from 0 to 1), or it simply propagates previous knowledge without its own experimental input. Thus, in due time long chains of reasoning are created where in individual nodes the specific protein–protein interactions are stated to either exist, or not. At the end of each chain lies the current consensus about the presence of an interaction. In addition, the model allows experiments to produce false-positive and false-negative errors, exceptions to the rule, and it allows for a publication bias in favor of positive results (claimed interactions). Running simulations on the model allowed defining a combination of parameters, which lead to the maximal posterior probability that the general consensus at the end of the reasoning chains to actually reflects truth. Sadly, when the real data were plugged into the model, the results indicated that scientists put far too much weight on published statements when interpreting the results of their own experiments. In other words, they were far more conformist than is good for science. The correlation between statements within individual chains of reasoning was near-perfect ($r = 0.99$), showing a nearly complete lack of disagreement concerning individual protein–protein interactions.

Unless we assume that the vast majority of protein–protein interactions ever tested are true *and* that both the actual false-positive and false-negative rates in measuring these interactions are well below 0.05, the numerical results are consistent with a picture of the universe where (i) most tested protein–protein interactions are not really there (which is not surprising, as nobody thinks that most proteins interact with most other proteins), (ii) only a very small fraction (considerably less than 10%) of reasoning chains end with the correct statement about the presence or absence of an interaction, (iii) both false-positive and false-negative error rates are well over 90%, and (iv) any randomly chosen positive statement about an interaction is very likely to be false. In other words, it seems that here we have an instance of a field where the negative results are buried and positive results are uncritically perpetuated to eternity, plus the experimental protocols are extremely error-prone.

In a complementary approach Pfeiffer et al. compared expert-curated databases of small-scale experiments of yeast protein interactions with the much larger databases of yeast protein interactions that were data-mined from omics studies (Pfeiffer and Hoffmann, 2009). The idea here is that the results of small-scale experiments are more prone to researcher biases (manipulation of results and selective publication of effects) than the hypothesis-free omics studies. They discovered that interactions that happen once or twice in the small-scale experiment database are overwhelmingly not replicated by the four different omics methods analyzed (only 8% concordance with at least one omics study), while even in the case of interactions seen in over 50 small-scale studies the concordance didn't exceed 40%. Strikingly, when the small-scale results were ranked by popularity of the interaction partners (assessed by the frequency of their occurrence in the literature), it transpired that the concordance between small-scale and omics experiments of highly popular proteins was about two-fold lower than that of less popular proteins. The authors present evidence that this effect is mostly caused by the multiple testing problem, although the signal of bias was also present. Taken together, this seems to constitute strong evidence that crowding is bad for the health of crowded fields.

7.4   SCIENCE AS AN OLIGOGRACY

There are alternative views to Merton's on how scientists behave and on what motivates them. Let us start with a citation from Lewis Feuer (1963), another American sociologist of science:

> *[The scientists are] becoming just one more of society's interest groups, lobbying for their greater share of the national income, and for the perquisites of power and prestige.*

(Shapin, 2008)

This view of scientists is the exact opposite of the Mertonian ethos, whereby scientists are not supposed to be in direct contact with the public (the paying customer) but get their funding funneled to them through their peers as a payment for their honesty and communism, but also as a check on their integrity. This roundabout way to society's largesse is meant to keep the scientists on the straight and narrow, and away from the temptation of using their expertise for fooling the public for personal gain. Feuer is essentially saying that in the real world things work differently and scientific establishment is an oligogracy. An oligogracy is an oligarchy of demagogues, in which a small group rules by popular prejudices, false claims, and promises.

So, which is it? How should we take Merton's theory of the ethos of science? Is it a factual statement about science *circa* 1940? Ditto 2015? Was it true then and false now, or vice versa? Or maybe the answer doesn't matter and the CUDOS structure should be seen as prescriptive rather than descriptive? If the *role models* (another concept originating with Merton) of science teach CUDOS to their students long enough, then maybe the concept of *self-fulfilling prophesy* (Merton's creation, again) kicks in and what was once phantasy becomes true? Would that be a good thing, or should we be afraid of the *unintended consequences* (guess, whose concept)?

To be able to decide between Mertonian and capitalistic ethos, there are a few aspects of the economics of science that need to be discussed: knowledge transfer, using bibliometrics to rank individual scientists, and the career structure of scientists.

7.4.1   Knowledge Transfer

On December 12, 1980 the US Senate passed a seemingly obscure piece of legislation known as the Bayh-Dole Act that allowed universities to claim title to inventions that had been made with federal funding and to exclusively license their patents to private industry. The initiative for this act originated largely in the research universities (Rasmussen, 2014; Markel, 2013).

This is how Senator Birch Bayh introduced the Bill to the Senate on September 13, 1978:

> *It is not government's responsibility—or indeed, the right of government—to assume the commercialization function. Unless private industry has the protection of some exclusive use under patent or license agreements, they cannot afford the risk of commercialization expenditures.*

(Stevens, 2004)

The counterargument came from no lesser a luminary than Admiral Hyman B. Rickover, who was instrumental in creating the US nuclear navy without resorting to any exclusive licensing of technology to private companies:

> *In my opinion, government contractors—including small businesses and universities— should not be given title to inventions developed at government expense.... These inventions are paid for by the public and therefore should be available for any citizen to use or not as he sees fit.*

(Stevens, 2004)

The passing of the bill was vigorously opposed by the Carter administration, which wanted to give the right to exclusively license discoveries made with public funds not to the universities, but to the taxpayer (in the form of the US government) and prepared a competing bill (Stevens, 2004). Because of a quirk of fate both the senate (where Senator Bayh lost his seat) and the president changed at the time of passing of both of the competing bills, which meant that the choice ultimately lay with the incumbent president Ronald Reagan, who was apparently not thrilled at the prospect of starting as president by creating another government bureaucracy and thus backed the Bayh-Dole Act.

The Bayh-Dole Act was meant to unlock the inventions and discoveries made with the help of taxpayers' money for commercial exploitation and it has been claimed in the process to have changed the face of both American and global science and to have played a critical role in rejuvenating the entire US economic system (Loise and Stevens, 2010).

Before the advent of Bayh-Dole, the US government owned the patents stemming from federally funded research, and it granted non-exclusive licenses to the private sector to develop these into commercial products. This precluded monopolization of research that had been publicly funded and isolated academic scientists from commercial pressures, but it was assumed that the non-exclusivity of the licenses also reduced the private sectors' interest in commercializing university-generated knowledge. By 1978 the US government had licensed less than 4% of the 28,000 patents it had acquired through its financing of academic research (Loise and Stevens, 2010). The Bayh-Dole Act removed a wall between the university and academia and transferred title from the government to universities, which could then grant exclusive licenses to private sector. Even better, it allowed the academic researchers to directly and substantially profit from their discoveries.

This arrangement has clearly been to the advantage of industry: in 2009 sales of drugs originally discovered by public sector researchers were estimated at $103 billion (Stevens, 2010). University-licensed products were claimed to have created over 279,000 jobs from 1996 to 2007 (Roessner et al., 2013) and in industry's estimate between 1996 and 2007 university-based research-licensing agreements contributed from $47 to $187 billion to the US gross domestic product (Schacht, 2012). Indeed, the entire industry of biotechnology has been claimed to have sprouted from Bayh-Dole (although a careful historical analysis fails to bear this out (Rasmussen, 2014)). Half of the papers published in *Nature Biotechnology* from 1997 to 1999, for example, were linked to a patent (Nelson, 2014). Even now at least 50% of biotech firms start from a university license. In 2011, 671 university research spin-off companies were formed in the United States, comprising about 0.1% of all the companies formed in that year (Anon, 2013).

The question of the impact of Bayh-Dole to universities and to individual researchers has spawned a number of studies. It is clear that while in the twenty-first century every decent university has to have a technology transfer department, the vast majority of them never bring in enough money to the university to cover the cost of their upkeep (Abrams et al., 2009).[4] Indeed, 95% of the university patents remain unlicensed regardless of Bayh-Dole and it seems that even US top universities (like Caltech) have taken the road of trying to profit from the discoveries of their researchers by licensing their patents wholesale to patent trolls—entities that exist not to develop the inventions they own but rather to extract money from the real innovators, who inadvertently infringe on their large patent portfolios (Ledford, 2013). To further cite from a *New York Times* story titled Legislation to Protect Against 'Patent Trolls' Is Shelved (May 22, 2014): "A yearlong effort to pass legislation protecting companies against so-called patent trolls was declared all but dead on Wednesday, when the bill was removed from the Senate Judiciary Committee's agenda. Supporters of the bill said heavy lobbying by pharmaceutical and biotechnology companies, universities and trial lawyers prevented the bill from advancing." This leaves one with the distinct impression that thanks to Bayh-Dole, many a twenty-first

---

4. Some university people have countered with an argument that a major function of university IP efforts is not so much gross profit to the university, which is an unreachable goal, but educating science students in the ways of commercial modes of thinking and, specifically, on how to apply for patents.

century university sees its mission as making a profit by actively hindering the commercial development of scientific discoveries. Meanwhile the percentage of patent litigation by trolls has increased from about 25% in 2007 to *ca.* 60% of all patent litigation in 2012—and even this number ignores the fact that the vast majority of patent trolling will be settled outside of court (Feldman and Price, 2014). The cost to companies of such activities has been estimated at $29 billion (*Nature* editorial June 5, 2014).

There is no indication that any of this has dampened anybody's enthusiasm for technology transfer. We also know that only a minority of researchers (maybe a third) ever participate in the process of knowledge transfer but those that do so frequently (maybe 6% of the US faculty) are much better off in their academic careers (Thursby and Thursby, 2004). What should concern the majority of researchers in biomedicine, however, is that they can no longer automatically expect to freely use the intellectual property which is licensed to private entities. The *Madey v. Duke University* ruling argued that any research conducted in a university may be viewed as advancing the business interests of the university and is thus fundamentally commercial, rather than philosophical. This means that, unless the licensing university has taken proper precautions, exclusively licensed product of research may be out of bounds for further use in fundamental research—even in the lab that made it (Boettiger and Bennett, 2006).

The tentative conclusion from the history of the Bayh-Dole Act may be that while to the industry it has been a boost, to most universities it represents a financial loss. It directly benefits a small fraction of researchers, probably to the detriment of the others, who will be at a disadvantage in hiring and in their access to public funds. To the consumer it may serve novel technologies with monopolistic prizes and its impact to the general structure and long-term development of science is unclear. Nearly 35 years after its inception, most university scientists still do basic science, but now many of them have a social norm telling them that it is not only okay, but desirable, to directly benefit from their research as individuals.[5]

For the company that buys the license the scientist has become a business partner, albeit one who does not bear financial responsibility for the correctness of her scientific claims. For better or worse, if communism and disinterestedness ever existed as social norms in science, senators Bayh and Dole helped to drive them to extinction. It may therefore strike you as ironic that some recent communications by members of the pharmaceutical industry (Chapter 1) leave the distinct impression that the industry would prefer to see their partners in academia to be the followers of Robert Merton rather than of Milton Friedman. In the pragmatic view of Bayh-Dole, a scientist, like any other American professional, takes his ethos from business. He makes a product, markets it to other

---

5. Although it must be mentioned that, unlike in the United States, in Sweden the teaching scientists (but not non-teaching researchers) have since 1949 held exclusive patenting rights to their inventions, and the sky has yet to fall on Sweden.

scientists and sells it to the highest bidder. The goal is maximizing profit, and if to do so the scientist needs to create new knowledge, so be it. Whoever is better at all three (production, marketing, and sales), will outcompete the others. Because universities tend to be chronically short of funds, it is in their financial interest to recruit scientists who come with their own funding, from which the university can then collect a tax, called overhead. This will help the university to build more labs and to attract more paying customers (scientists). Thus the university is turned into a hotel for scientists, and scientists are transformed from its members to its clients.

Here we have a picture of unbridled competition where the winner takes all, and indeed stands on the shoulders of the losers to stomp them deeper into the ground. But then, couldn't it be that like the workings of a sausage factory, although shocking to a casual viewer, this way of doing science is good for the science as a whole? If cut-throat competition is supposed to be good for the growth of the economy, then why not for the growth of knowledge? This is how John Ioannides presents his anti-utopia of science that shares its reward structure with the commercial sector:

> *Planet F345 ... is inhabited by a highly intelligent humanoid species very similar to Homo sapiens ... the lion's share of the research enterprise is conducted in a relatively limited number of very popular fields... where empirically it has been shown that there are very few or even no genuine nonnull effects to be discovered ... Young investigators are taught early on that the only thing that matters is making new discoveries ... Universities are practically run by financial officers that know nothing about science (and couldn't care less about it), but are strong at maximizing financial gains ... The members of ... national academies of science are those who are most successful and prolific in the process of producing wrong results.... The main motive is again to get extravagant results, so as to license new medical treatments, tests, and other technology and make more money, even though these treatments don't really work.... Simple citizens are bombarded from the mass media on a daily basis with announcements about new discoveries, although no serious discovery has been made in F345 for many years now.*
>
> (Ioannidis, 2012)

The new kind of translational (from university lab to industry) science seems in many ways to be the antithesis of Mertonian science (Table 7.1). Here scientists are expelled from their ivory towers, curiosity-driven research is substituted by use-driven research and the leading partner (and funder) is now the private industry. For an individual scientist the enforcers of rules of conduct are no longer his peers and the coin of science is now patents and personal wealth. Finally, it must be stressed that whether all this is a good or bad thing for the society is in the final analysis an empirical, rather than ethical, question (although the people who proudly seek to commercialize their discoveries should perhaps ask themselves why the journals they publish in require them to disclose this fact as a conflict of interest).

TABLE 7.1 A Comparison of Principles of Mertonian and Modern Translational Science

Mertonian Norms	Modern Norms
Communism	Privatization (patenting)
Driven by curiosity, peer recognition	Driven by needs of society, individual profit
Disinterestedness	Direct monetary interest
Individualism	Teamwork
Freedom of study	Science is a servant of industry
Projects initiated by scientists	Projects initiated by the industry
Loyalty to scientific community	Loyalty to society

7.4.2 Bibliometrics: Can We Quantify Science?

It should not come as a surprise to anyone who has spent a week in a lab that there are more professionally trained scientists out there than there are jobs in academia. And people who do have a job tend to have more scientific ideas than there is available funding. Things have not always been like this, but since the end of the Cold War they have increasingly desperately been so.

The aspect of the problem, which we discuss here, is the obvious need for someone to somehow decide, who and what gets funded. The someone here is overwhelmingly other scientists and scientists being scientists they like to base their decisions on objective criteria.

To dole out the grants of the US National Institute of Health (NIH), an army of about 24,000 experts annually reviews *ca.* 75,000 applications. The cost of this operation is $110 million (although the experts work for free) and the outcome is that neither the rank given to a grant application by the experts nor the level of its eventual financing predict future publication by the applicant (Mervis, 2014; Berg, 2012). To put it more bluntly, there is currently no reason to think that the grant allocation process does anything but perpetuate mediocrity and waste resources.

But this much it does with zest. According to a recent study "three out of five authors of … influential papers [published since 2001 and having over 1000 citations to them] do not currently have NIH funding as principal investigators. Conversely, … a large majority of the current members of NIH study sections— the people who recommend which grants to fund—do have NIH funding for their work irrespective of their citation impact, which is typically modest" (Nicholson and Ioannidis, 2012).

The general success rate of applicants has fallen by about twofold in 15 years to 17% and the mean age of getting ones first NIH grant has risen from 36 years in 1980 to about 42 years in 2011(http://report.nih.gov/investi-gators_and_trainees/). In fact, more researchers over 65 are funded than those under age 35 and only about 3% of NIH awardees are now 36 or below (down from 18% in 1983). Even after a young scientist has secured a research position in academia, she must wait an average of 4–5 years to receive federal funding, compared with 1 year in 1980 (Alberts et al., 2014).

These dynamics have been accompanied by rising mean age of scientists when they make their Great Discovery, which is now slightly under 40 for those few that do (Jones, 2010). This delay may be caused by preferential picking of low-hanging fruit and the consequently increasing complexity of the scien-tific questions that are still open to scientific enquiry and—incidentally—may explain "why growth of economic productivity did not accelerate through the twentieth century despite an enormous expansion in collective research effort" (Jones, 2009).

How are the winners in the Game of Science picked? The parameters con-sidered are: (i) the bureaucratic quality of the application, (ii) bibliometrics, (iii) the scientific quality and societal potential of the application, and (iv) the quality (past achievements) of the applicant. The current trend seems to be to de-emphasize (ii) and (iii), and to emphasize (iv) (Collins and Tabak, 2014).

Because of objectivity concerns and cost considerations a lot of effort has been placed in devising numerical proxies to assess the scientific success of indi-viduals. These tend to use the volume of publishing, the quality of the journals (journal citation indexes), and/or the number of citations to one's publications. Currently the most popular of these is the h index, which rewards consistency of publishing over a long time (older scientists who have control over large teams tend to have larger h indexes) and punishes people who publish few papers, even if those few constitute scientific breakthroughs. For those of us who are still young or who believe that it should be more than sufficient to make one or two great discoveries during a scientific lifetime, the popularity of the h index cannot be good news.[6]

Bibliometrics as a serious science was born in the 1950s from efforts to objectively measure the long-term trends in scientific output. This proved to be an excellent approach to study the global dynamics of science, as well as to compare the scientific output and quality of different regions of the world, countries, or (arguably) even universities (de Solla Price, 1975). The usefulness of bibliometrics for small-scale comparisons of individual scientists is a differ-ent matter. To gauge the sensibility of using a scientist's publication record for assessing her professional acumen, we must start by making explicit some of the assumptions behind this approach.

---

6. There is also the age-adjusted h index where the normal h index is divided by the length of one's publishing career, but the different publishing opportunities of a postdoc, who runs her own experiments, and the professor, who likes to stick his name into every manuscript that comes across his inbox, do not allow for a clean comparison.

1. Good scientists tend to publish better than their lesser contemporaries. They also tend to gather more citations to their work. In brief: P(publishing | good scientist) is very high.
2. Scientists who publish well are likely to be better than those who publish less well. In brief: P(good scientist | publishing) is very high.

For the sake of argument (and we will return to this later) lets assume that statement 1 is true. However, on a little reflection we will see that the status of statement 2 is somewhat uncertain. From Chapter 5 we understand that the road from 1 to 2 is Bayesian and next we offer a semi-formal Bayesian working out of it.

First we need to explicate the alternative hypotheses under which we might expect some scientists to publish more than others. We must also take into account the stringency of selection of scientists for jobs or funding, to be able to come up with prior probabilities for our hypotheses. For our purposes we might realistically assume a fairly strict level of selection: maybe that one in five candidates will get the financing.

H_1: honest scientists, who are very good at what they do, publish well and are expected to win the competition—likelihood P(publishing | H_1) = 0.9, $P(H_1) = 0.2$

H_2: dishonest scientists, who simply make up their results, publish well— P(publishing | H_2) = 0.9, $P(H_2) = 0.04$

H_3: biased or unscrupulous scientists, who use the formally legitimate "researcher degrees of freedom" to statistically cook their results (Simmons et al., 2011), publish well—P(publishing | H_3) = 0.7, $P(H_3) = 0.3$

H_4: methodologically competent honest but mediocre scientists, who publish well due to Matthew effect, working in good company (collaborating with better scientists), or publishing every dreg of evidence they get— P(publishing | H_4) = 0.2, $P(H_4) = 0.46$

In assigning numbers to mutually exclusive and exhaustive alternatives I tried to be conservative and considered about two-thirds of scientists to be methodologically competent and unbiased and honest, 30% to be influenced by bias or to participate in the sport of statistical data dredging, and 4% of scientists to be involved in scientific misconduct as currently defined (see Section 7.5). I also put a rather low likelihood on the proposition H_4 that a mediocre scientist would punch above his weight and "publish like a pro." The reader is invited to come up with her own numbers, which she finds defensible, and see how much her result will differ from mine:

$$P(H_1|D) = \frac{P(D|H_1)P(H_1)}{P(D|H_1)P(H_1) + P(D|H_2)P(H_2) + P(D|H_3)P(H_3) + P(D|H_4)P(H_4)}$$

$$= \frac{0.9 \times 0.2}{0.9 \times 0.2 + 0.9 \times 0.04 + 0.7 \times 0.3 + 0.2 \times 0.46} = 0.35$$

As a result of this exercise I should now believe that the probability of a candidate, who would win the competition on the strength of her publication record alone, to actually be a worthy winner, is about 35%. The main reason why the posterior probability of H_1 comes out as 35%, as opposed to 80%, for instance, lies in H_3, which lumps together scientists who fall victim to one of numerous biases with those who spend their careers gaming the system. This illustrates a general sociological law, which I hereby christen "tragedy of the proxy."

7.5 TRAGEDY OF THE PROXY

Civilization advances by extending the number of important operations which we can perform without thinking about them. Operations of thought are like cavalry charges in a battle – they are strictly limited in number, they require fresh horses, and must only be made at decisive moments.

A. N. Whitehead

My probably not-so-original law states that whenever a well-defined proxy measure is used to assess some real but poorly defined ability of people; the objects of measurement will soon learn to manipulate the proxy and will pervert the original reason for its measurement. The fact that the measured are on average as smart as the measurers inevitably leads to substitution of the proxy for the real thing. The proxy becoming the real interest of all concerned parties is, in a nutshell, the tragedy of the proxy.

For instance, when a professor conducts a machine-readable multiple-choice test for an exam to save his time and nerves, he can be sure that his students will very soon learn how not to learn the conceptual basis of his subject and will instead memorize the short factual answers to the test questions of yesteryear. Year after year, as the questions repeat themselves, the students' grades will rise as their understanding of the test increases, and the understanding of their academic subject gradually diminishes. Similarly, when scientific ability is assessed by the proxy of h indexes, then pretty soon the proxy turns into the "real thing" and it will be hard to even talk of scientific ability in any other sense than the ability of publishing consistently. Both the scientists and their university bosses will work hard to maximize the value of the index on which they mutually depend, and the people who are more successful at this will be successful in every sense.

To illustrate the tragedy of the proxy in science we shall return to the end of the Cold War in about 1990. The end of the Cold-War effort signified a great crunch in science funding in both the losing and the winning side, which meant intensifying competition for funding and jobs (Goodstein, 1994). Naturally this led to increased emphasis on bibliometric proxies for ability. This in turn led to changes in publication and citation practices. In 1990 only a minority of published scientific papers received any level of peer interest: only 45% of published papers were cited within 5 years. Thus in

1990 an *h* index could be seen as a reasonably informative proxy of the owners' scientific importance. However, this situation wouldn't last—as much as 88% of medical papers published in 2002 were cited within 5 years (Ioannidis et al., 2010). This increase in citing was in part achieved through increasing self-citation rates, with the understanding that self-citations also increase the frequency of "real" citations.

At the end of the day, this is how the bibliometric proxy was tamed by its intended victims: "coauthor more papers (salami slicing, elimination of quality checks, undeserved coauthorship…); ignore unfavourable reviewer comments; keep submitting until you find a hole in the peer review system; self-cite; and expect random citations" (Ioannidis et al., 2010).

Now that the rules have been changed and the people who ride the proxy will triumph over more old-fashioned candidates, it will be them that you will increasingly encounter at academic positions, as peer reviewers, editors, and committee members who decide who gets funded—and in the process sanctify their own standards.

Another instance of the tragedy of the proxy in science is the fight against scientific misconduct. The original ill-defined, albeit important, question is: how should scientists operate to guarantee the best possible outcome, that is, steady scientific progress? This is essentially a question of methodology: when we know which method will most likely lead to truth, then we can literally pass a law to make its use compulsory. This would make science function like any other normal profession where practitioners are constrained by unequivocally stated laws and norms, but under the accepted rules have freedom to operate. However, in science we have neither The Method nor its codification in law or the consequent professional stability. Instead there are increasing doubts about the health of the profession, combined with the need to paint to society a picture of science as the endless frontier (complete with spaceships, the pioneer spirit and huge pay-offs that wait just around the corner) and thus to guarantee its continuing financing.

Of course, at every frontier amidst the sturdy settlers and war-like but honorable Indians you will expect to meet a few gamblers and crooks. And so it is in science. If we can pin our troubles on a few bad apples, which are then removed from the body scientific, then what is left is clean and wholesome, and, given enough support from the society, will lead us to our happy destinies. In science the deviants are involved in something which is called "scientific misconduct" and they are often thought to make up only a few percent of scientists (Fanelli, 2009). Scientific misconduct is an umbrella term for a group of precisely defined behaviors. The US Office of Research Integrity (ORI) defines scientific misconduct as an "intentional attempt by an investigator or scientist to manipulate data or manipulate results." This official definition includes and is strictly limited to: (i) *Fabrication*—the making up of research data and recording or reporting them; (ii) *Falsification*—manipulation of research materials, equipment, or processes, or changing or omitting data or results such that the research is not accurately presented in the research record; and (iii) *Plagiarism*—the appropriation of another person's ideas, processes, results, or words without giving appropriate credit.

In practice these are all defined narrowly so as to include direct, purposeful, and knowing making up of data or plagiarism by researchers and to exclude the more common but also the more ambiguous instances of wrongdoing, such as significance chasing. A typical case of scientific misconduct is somebody not doing the experiment and publishing a picture documenting some other experiment. In 2008 ORI proved 13 instances of scientific misconduct (out of 17 completed investigations). On average they spend 7 months in investigating each case (their record is 8 years) and this tempo has been stable for some time. Based on the number of ORI-confirmed cases, in the US fraud is documented in about 1 in 10,000 to 1 in 100,000 scientists, while data on self-reported misconduct place its frequency in somewhere between 1% and 4% (Fanelli, 2009). In addition, up to one-third of respondents admitted to a variety of other questionable research practices, which do not constitute "scientific misconduct," including dropping data points based on a gut feeling and changing the design, methodology or results of a study in response to pressures from a funding source.[7] When asked about their colleagues, scientific misconduct had been personally observed by 14% of respondents.

"Scientific misconduct" is a proxy for something that is much harder to define and measure, namely "bad science." This has predictably led to the proxy taking over from the real problem and consequently we have all these well-meaning programs to fight misconduct by instructing students and their professors on the evils of plagiarism and making up data, while the people who fudge their research piecemeal can live in the happy knowledge that their activities conform to the standards of their chosen profession. Growing cognizance of the impotence of this particular proxy has led to calls for redefining scientific misconduct to encompass more actual wrongdoing (Fanelli, 2013). No doubt this would quickly induce the intended targets of the new rules to saddle up their new proxy and continue their happy rides in the bliss of conformity and success. However, as long as we do not really understand the scientific method(s), we cannot legislate it; and this means that every attempt to solve the problem of bad science by enforcing new rules of conduct will not only fail—it is likely to make things worse by substituting for the real thing, an easily winnable proxy battle.

7.6   SCIENCE AS A LOTTERY

If we cannot successfully measure important but fuzzy concepts like scientific worth or quality, maybe it is possible to find sets of parameters which can

---

7. Of course, changing the design of one's study in response to criticism need not always be a bad thing—here a lot hinges on the motivation of the funding source.

successfully predict them? In other words, from the parameters that we can measure today, which ones predict future success in science? To answer this Acuna et al. tried to predict scientists' future (5 years or 10 years) h indexes on a large sample of neuroscientists (Acuna et al., 2012). They found that a neuroscientist's current h index was not very good at predicting his or her future h index ($r^2 < 0.4$). When they added the number of published articles, the years since first publication, number of publications in five prestigious neuroscience journals, and the number of distinct journals to the mix and built a best-fit model, they were able to reach a more respectable fit ($r^2 = 0.66$). Unfortunately, further testing of the model in independent samples of life scientists showed that "generalizations to other fields within and outside of life science may be limited." Of course, even if your future h index could be accurately predicted, it would require a leap of fate to equate this proxy measure with the quality of science you will be doing.

Thus a more radical question arises: if presented with a bunch of professional scientists, can we predict *in principle*, which ones are going to do more interesting science in 5 or 10 years? This is the sort of question university hiring committees are faced with.

On the face of it, we may assume that this is a far from impossible task. We know that some people are more intelligent than others and know how to measure it, we know that some are more proficient at knowing the literature, at their ability to work with others, solving research puzzles, putting in more working hours, etc. Each specific ability that you have is quantifiable. Moreover, we know that in each ability which you might care to measure, most people are about average and a few have very high or low talent. In fact, distribution of human ability is approximately normally distributed, which means that it falls off relatively sharply at either side from the mean. For example, the distribution of general intelligence in adults centers at IQ = 100, and the standard deviation (SD) is 15. This means that your chances of ever meeting somebody with IQ = 160 (4 sigma) are slim and seeing an IQ = 200 near-impossible.

All this means that if we measured enough separate qualities of the candidate and put this information together then we could in principle have a very good measure of her scientific ability *vis-à-vis* other candidates. Because the components of this composite measure would be largely independent of each other (the first candidate likes to work 30 h weeks but is good in collaborating with others while the second candidate has a high IQ but low social intelligence, etc.) the distribution of candidates in our final ranking is still near-normal. Therefore, most candidates have about equal abilities and are presumably fit for the job, plus there is a very small fraction of outstanding candidates.

This is the way of scientific ability, but it is not the answer to our original question, which was not about ability but about success. And success behaves very differently. In the scale of success—and it doesn't matter if we measure it in science or in business or somewhere else entirely—most people are not moderately successful. Success is not described by normal distribution. It is described by a power law (an exponential distribution), which means that most people are in fact unsuccessful, relatively few people are moderately successful and a small minority are enormously successful. Exponential distributions tend to have "fat tails," meaning that it is much more likely to find extreme values under a power law than under a normal distribution. If ability would follow the same law as success, we would not be very surprised to meet a person 8 m tall, or somebody with an IQ of 400.

The presence of power laws in scientific success was first proposed by Lotka in 1926. According to Lotka's law the number of scientists producing n papers is approximately proportional to $1/n^2$. Similar power laws have been found in income distributions, in telephone and internet networks, and number of human sexual contacts (Newman, 2003).

The highly skewed distribution of citations to scientific papers was first described by de Solla Price (1965). A more recent very large study revealed a highly skewed distribution of citations where 9–10% of the most cited articles account for about 44% of all citations (Pedro Albarrán, 2011). Also, less than 1% of scientists who have published at least one paper every year from 1996 to 2011 (150,608 individuals) have between them managed to insert their names into author lists of 42% of the nearly 26 million papers published during that period; and into 87% of the extremely highly cited papers with over 1,000 citations (Ioannidis et al., 2014).[8]

Although the impact factors of different journals are widely different, within each journal the distribution of citations between papers is so unequal that it appears to make little sense to use journal impact factors to rank individual authors (Seglen, 1992). Importantly, highly skewed citation distributions hold even for the body of work of individual scientists. Does this mean that as we feel the need to prize individuals for doing well-cited work, so should we punish the same individuals for wasting public resources on their much more frequent poorly cited papers? If your gut feeling, as does mine, answers this question in the negative, ask yourself "why not"?

How does one explain these ubiquitous power laws of achievement, when the underlying ability is normally distributed? The original hypothesis of de Solla Price was that the observed steady state distribution arises simply if every year 10% of the less-cited papers randomly "die" (go out of collective consciousness permanently) (de Solla Price, 1976). However, as the skewedness of the distribution does not depend on the size of the time-window where accumulating citations are counted, it seems that differential rate of obsolescence of different papers cannot be behind the power law distribution of citations to them (Seglen, 1992).

---

8. Dependence of achieving top papers on literary prolificacy can be explained either by a mass effect whereby highly regular publishing is in itself necessary for raising the profile of authors' individual papers—the implication being that gathering 1,000+ citations is a poor predictor of the true value of the work—or by the theory that a scientific breakthrough is the product of perseverance and hard work, rather than genius and luck.

In his later classic paper de Solla Price (1976) proposed a simple urn model to explain the data. Here we have an urn with two types of balls, one labeled "win" and another labeled "lose." Now a ball is randomly drawn from the urn. If it happens to be a losing ball, the next ball is drawn. If it happens to be a winning ball, then N winning balls are added to the urn before the next draw. The result of such random experiment with added positive feedback is the kind of distribution seen in power laws. In effect, this is a limited form of Matthew effect where those who have, shall be given even more, while to those who have not, no punishment is meted. If there is a continuous inflow of fresh uncited papers, a steady state can be achieved, which is described by a power law.

Under this model a paper's first citation is essentially random and depends on the number of available citation slots in the citing papers and on the number of papers published. The assumption is that each year the same fraction of references randomly gives non-cited papers their first citation (Wallace et al., 2009). This simple random model turned out to be remarkably accurate in describing the actual steep reduction in the fraction of uncited papers that took place over a 100-year period. The effect was the strongest in medicine where about 80% of published work went uncited in 1900, while by 2000 this number had fallen to about 20%. The recent decrease in uncitedness can be explained as due to lengthening reference lists in modern scientific papers.

I can think of three models to explain the high impact of some papers but not others.

1. *Deterministic and predictable model.* Science is easy: competent scientists know their stuff and by working hard they obtain a predictable string of quality papers and a solid *h* index. A good scientist should therefore publish few low-impact papers and a mediocre scientist should publish few high-impact papers. Competent scientists' *h* indexes should therefore be close to their theoretical maximum (square root of total cites, assuming that this number is smaller than the total number of papers).

2. *Deterministic and unpredictable model.* Science is hard: most scientists' *h* indexes should stabilize near some value which is indicative of their scientific worth and is irrespective of the number of their total papers or citations. Therefore, *h* indexes of most established scientists would not be predictable from the number of total papers or citations. Under model 2 the age-adjusted *h* index would be worse than useless. Specifically, there would be no stable ratio of *h* index/theoretically maximal *h* index across many researchers.

3. *Random and predictable model.* Citations accumulate randomly with positive feedback (Matthew effect). Unlike in models 1 and 2, citations to one's papers do not predict real scientific excellence, but the fraction of high-impact papers is statistically predictable. The *h* index of a scientist, worthless as it is, can be predicted by the number of his papers and/or by his total citation count.

Indeed, consistent with the 3d model, an excellent correlation between the *h* index and the total citations has been described in the field of mathematics

(Yong, 2014). Yong proposes a rule of thumb that one's h index can be predicted from the total number of citations as follows:

$$\text{Predicted } h \text{ index} \approx 0.54 \times \sqrt{\text{total no. of citations}}$$

If this simple relationship holds in natural sciences, at a minimum this means that h index is no better proxy than total citation count. I therefore did a very small analysis of the true h indexes and predicted h indexes calculated through the equation above, using as my sample 22 Estonian natural scientists, who belong in the top 1% of scientists in their fields of study, according to their citation records (http://www.ut.ee/en/research/field-rankings; accessed 08.01.2015). If citations tell us anything, my sample consists of excellent scientists. I retrieved their total number of publications, citation counts, and h indexes from WoS and calculated their predicted h indexes. Figure 7.1A shows an excellent correlation between predicted and actual h index ($R^2 = 0.91$), and Figure 7.1B shows somewhat less correlation between the number of papers and h index ($R^2 = 0.72$). Also, the actual h indexes constitute a remarkably stable fraction of 0.50 (SD = 0.06) from the maximal possible h indexes of the researchers. These results support a random positive-feedback citation game and indicate that h index is not a good proxy for the overall quality of one's work (and neither are total citations or total publications).

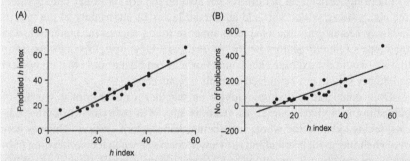

FIGURE 7.1   Concordance of the h index with (A) a predicted index calculated directly from the total number of citations and with (B) the total number of publications of a set of natural scientists, who are each within the top 1% of their respective fields (as measured by citations to their work).

The previous arguments are all very abstract. When we understand that many scientists try honestly and very hard to do good work and some are clearly more talented than others, why should we even contemplate models implying that most differences in scientific achievement are essentially random, "undeserved," and unpredictable? Based on his personal experience in managing commercial R&D, J. D. Bernal put it thus in 1956:

If work were done on a product about which we knew in advance, then it could not be called research; true research had to work into the unknown all the time.

Since, therefore, we did not know what we wanted to find no one could blame research workers for not finding the right answers, nor praise them for finding them. In running a research establishment one was really running a gambling concern and taking un-calculable risks for unassessible rewards ... The expenses and the rewards were totally unrelated"

(Shapin, 2008)

What could be the causes of such a discrepancy between effort and outcome? Firstly, although some scientists may be geniuses, there is no way of knowing if the nature of their genius matches the requirements of nature, for example, what is left to discover by available methods. The most important requirement for discovery is that there better be something to discover! The key to this requirement must always lie below the research horizon, and is therefore unknowable to science. Thus the question about the ultimate potential of science is metaphysical and must forever remain so. We simply cannot know who will be best suited for doing science 5 or 10 years from now, because we don't know what will be needed to do good science then. Even if a hiring committee could identify the best candidates, they couldn't possibly say which ones will be successful and which ones will fail. This is because success of a scientist will not only depend on her efforts but also on the efforts of her collaborators and competitors, which will add additional layers of uncertainty to the mix. It suddenly seems quite reasonable to suppose that a successful prediction of a candidate's future trajectory is a hopeless dream. Of course, this does not mean that we should not exclude clearly incompetent candidates; only that competent candidates should not be judged on their past success.

Even comparatively simple advice on whether a manuscript is worthy of publishing in a given journal, as given by specialist reviewers to journal editors, seems to be on the whole little better than random guesses. There is now overwhelming evidence that inter-reviewer concordance in recommending publishing or rejection is barely, if at all, above chance levels (Lee et al., 2012; Onitilo et al., 2013; Rothwell and Martyn, 2000). And manuscripts on which referee opinions diverge are 50- to 70-fold less likely to be published (Rothwell and Martyn, 2000). Available evidence indicates that scientific publishing is essentially a lottery, which is weakly biased against the worst (quality-wise) and the best (novelty-wise) manuscripts, and biased for prestigious institutions and groups (Lee et al., 2012).

Mathematical modeling of the peer review process suggests that any success of the peer review system presupposes a combination of high competency level (recognition of excellence) and a low level of rational behavior (suppression of one's competitors) on the part of reviewers (Thurner and Hanel, 2011). However, as the current system of peer review needs to annually process several millions of manuscripts and grant proposals, and the pool of high-quality specialists is in the hundreds of thousands at best, by necessity a lot of reviewing is done by under-qualified graduate students et al.

It is doubtful that much improvement in the quality of peer review is even possible, unless the volumes of reviewing are drastically reduced (and because of the rise of Asian scientific publishing and its emphasis on quantitative biblio-metrics proxies of excellence they are actually rising fast). A sensible alternative to peer review seems to be increased reliance on paid editorial committees over and above peer review (Neff and Olden, 2006); a fix, which would no doubt increase the monetary cost of publishing. Another possibility is to substitute pre-publishing peer review with post-publishing crowd-sourced comments and authors' responses on the quality of their work. This depends on the willingness of members of the scientific community to spend their time on commenting on others' work freely and without malice. Also, this model would probably result in a more fluid literature, where already "printed" papers are continuously changing and improving.

The randomness-oriented view of scientific publishing has some interesting implications. Firstly, since the number of available slots for citations is limited by the number of newly published papers, the fact that your paper is still uncited doesn't mean that it is unimportant for other scientists. It only shows that the available space for citations is limited. Every uncited paper is associated with a quantifiable probability of making it into the top.

Secondly, it makes little sense to punish the less-cited authors in the hope that they will go away and that the remaining authors will continue to do important science and be well cited. All such a policy can do is to reduce the total number of published papers, which means less slots for citations, which will lead to smaller science and the same fraction of well-cited papers. Paradoxically, if we remove less successful scientists from the game, the absolute number of successful scientists will go down sharply. Eventually we are left with a single scientist, all citations to whom will be self-citations. The fact, that a minority of the published articles collects the majority of citations cannot be altered by grant committees (Seglen, 1992). Science thrives on plurality, not on pre-planned success.

Thirdly, it no longer seems true that the best (most able) scientists publish the most, or the best, papers. This is, among other reasons, because we have no way of knowing which discoveries are still waiting for their champion; and of the ever-present threat from scores of scientists, most of whom you have never heard of, of scooping your work. Therefore the likelihood P(publishing | good scientist) is likely to be considerably lower than the 0.9 value assumed above, which in turn lowers the posterior probability P(good scientist | publishing).

And finally, a requirement behind a respectable h index is that its owner does steady well-cited work. Pauses in publishing due to developing new experimental systems or to some exciting idea not panning out cannot be tolerated. Here the advantage is (i) with good scientists who publish a stream of papers on a single important subject, never taking the risk of straying from the golden trail; (ii) with PIs, who have large groups, and who can therefore insert their name into each of the many manuscripts written by other people; and (iii) with people who unscrupulously manipulate the methods of science to always end up with important-sounding statistically significant results. Of course, the bigger the fraud, the bigger the advantages it brings. On the other hand, excellent but adventurous scientists are at a disadvantage and will be presumably sooner or later pushed out of the game. There is a good chance that the *h* index works exactly opposite to what it was created for—in the long run it increases conformity and herding; and it works against risk-taking and true innovation.

7.7 SCIENCE AS A CAREER

Science is a wonderful thing if one does not have to earn one's living at it.

Albert Einstein

In previous chapters we have seen that scientific funding and publishing are highly competitive and, at the same time, gameable. This seems like a recipe for trouble, but is not necessarily so. Whether there really are strong incentives to game the system depends to a great extent on the career security of the academic scientist. After all, don't we have a tenure system where a university scientist has complete freedom to pick and choose her own research avenues with the full knowledge of the security of her job? The answer seems to be that "yes, we do have tenure in most countries where high-visibility science is done—but it may no longer fulfill its original purpose."

University-produced PhDs have three main professional reasons for existence: to do basic science in the academic sphere, to teach scientific subjects at the university level, and to do applied science and development in industry settings. In addition, there are less numerous science journalism and bureaucracy roles for PhD-level people, plus opportunities to further "top-up" their education by doing a next round of studies on patent law or medicine. Outside these academic and advanced engineering spheres there seems to be very little demand for outspokenly critical, independent, and compulsively truth-seeking behavior for either private- or government-sector employees.

The golden era for scientific career building started with the huge economic boost after the end of World War II (although the exponential growth in scientific jobs started around 1910 in the United States (Shapin, 2008)), which coincided with the advent of mass university education in Europe in the 1950s–1960s (in America this process started in the 1930s), the advent of revolutionary advances in medicine and pharmaceutical and chemical industries, and the Cold War (Alberts et al., 2014). It was widely perceived that the war had been won thanks to technological innovations, like radar, antisubmarine technologies, and the atom bomb—and had introduced things like jet engines, rockets, and vastly improved submarines. (The most important advance in technology bought about by the war effort—electronic computers—was still under tight wraps.) This was done by lavishly funded, highly coordinated and cooperative factory-like efforts, which were later emulated in industry and government settings. At the same time, mobilization of scientists away from their areas of expertise and keeping an entire generation for entering universities led to a steep overall reduction in the volume of science that was not compensated, even after the war, by the novel methods and technological products of the war effort (de Solla Price, 1951).

There ensued a "perfect storm" of quickly developing industry, increasing need for university-level teachers, the birth of whole new sciences (molecular biology, computing) and militarization of the old ones (big chunks of physics). Also, the prestige of science and of scientists was very high on both sides of the Iron Curtain. All this meant the need for quickly increasing the number of scientists and the system of science education turned out to be up to the task.

Scientists are, of course, not made in lecture halls—they are made in the labs under close one-to-one supervision of established working scientists. This process of gradual molding by example may seem inefficient, but it isn't. Over his lifetime a successful biological scientist can easily train 10–15 graduate students to become highly competent researchers—and this means seriously exponential growth for science. Accordingly, graduate schools were well-funded and individual scientists were awarded not only for their publications, but also for the number of fresh PhDs they produced. This led to the system where many fresh graduates entered a PhD program for 3–4 years, followed by a 2–3-year postdoctoral stint, after which they could count on a safe academic or lucrative industry job when still in their late-20s. For talented and studious youths, entering the system was a safe bet for secure and interesting jobs for the rest of their lives.

Unfortunately, this wouldn't last. The economic boom found a sharp end with the 1974 Yom Kippur War and the subsequent OPEC oil boycott. By the same time the post-war university system was complete with more than half of American kids attending college and the very large post-war generation (the baby-boomers) had already rioted their way through it. This coincided with increasing economic and political stagnation in the Soviet block. Also, to the surprise of everyone, in a dozen years the Cold War was abruptly over, military expenditures down, and the globalization began to move the world's industrial base from the United States and Europe to the then science-poor developing nations of East Asia. Consequently the job structure of the developed countries found itself heavily tilted in the direction of serving and sales jobs which require empathy and social skills, but not engineering skills or analytical thinking. The scientists still have high prestige in the West (although less so in the former Eastern Block) and they have successfully lobbied the state for long-term science funding as a motor for future economic growth, but the jobs for young scientists are simply not there (unless one is prepared to move to Asia).

This reduction in the number of PhD-level jobs in the developed countries has not been accompanied by commensurate changes in the PhD production system. For instance, between 1999 and 2010 the number of awarded science and engineering doctorates grew by 35.5% in the United States (National Science Board, 2014). Professors still train as many PhD students as they can, and are incentivized to do so by outside financing of the lab work done by their graduate students. Also, the recent graduates still find their way to temporary postdoctoral positions, but increasingly there is no way out into the stability and relative prosperity of the tenure track. This has led to gradual lengthening of the postdoc period, which is now about 7 years on average and typically consists of two to three consecutive postdoc jobs. There are currently more than 40,000 postdocs in the US biomedical research system, and the number has been increasing rapidly in recent years (Alberts et al., 2014). This has led to a situation where many talented people spend their best years in doing science with no real independence in setting their research agendas.

Postdocs typically work at least 50 h per week and often over 60 h per week. By comparison in the United States the average working week of a private, non-farm employee is just over 34 h long.[9] They earn around half to two-thirds of a staff scientist's salary and compared to the scenario of not entering the graduate school in the first place will lose about $20,000 for each year spent as a postdoc (Stephan, 2012). The real wages of chemistry postdocs have actually fallen over the last decade or so and their average hourly wage was estimated at $16 in 2012, which is slightly over twice the US federal minimum wage (and the current trend is for states to raise their minimum wages to around $10) (Halford, 2013). Sadly, most postdocs go through all this for the uncertain perspective of landing a tenured academic job in the market that produces exponentially more candidates than there are jobs! This is how a *Nature* editorial (February 5, 2009) puts it:

> *The career crisis is especially stark in the biomedical fields, where the number of tenure-track and tenured positions has not increased in the past two decades even as universities have nearly doubled their production of biomedical doctorates. Those who do land jobs in academic research are struggling to keep them, because competition for grant money in biomedicine has grown at a steep rate. … beyond a certain point, the hyper-competition for grants, publication and tenure hurts everyone. The process ceases to select for only the very best young scientists, and instead starts to drive many of the smartest students out of research entirely. They realize that the risks outweigh the benefits in science.*

Accordingly, the proportion of biomedical PhDs going to be tenure-track faculty in the US had fallen below 10% and the total number of employed researches in the United States has remained stagnant since about 2003 (National Science Board, 2014). Meanwhile, from 2000 to 2010 a total of 599,100 newly minted PhDs entered the job market in the United States and the higher education teaching market, which makes up over 80% of the job market for PhDs, created 150,000 new jobs (Siegmund, 2014). According to the NSF, the US job market for PhDs in the biomedical sciences is as follows: 54,000 academic jobs (tenure-track and other), 24,000 non-research but science-related jobs, 22,500 industry

---

9. http://blogs.nature.com/naturejobs/2011/04/01/are-long-working-hours-inevitable-for-postdocs.

research jobs, and 7,000 government research jobs.[10] This total of 107,500 jobs gets an annual inflow of 9,000 freshly minted US PhDs plus 1,900–3,900 foreign-trained PhDs starting their postdocs in the United States, which means that for the Americans only, to get any kind of a science-related job would require an annual turnover of 8.4% of these jobs. The actual labor turnover rate in the US economy as a whole is around 3%. Another thing to notice is that although the private sector gets the vast majority of the R&D funding in the United States, only 13% of the relevant jobs for biomedical PhDs are in industry research.

Ensuing levels of competition puts scientists on a par with other professions, where the most important prerequisite for a career is not competence, but hypercompetence and/or luck. In this tournament career model many people compete for very few highly attractive jobs and only a few will be the winners, while the vast majority gets nothing (Lamar and Rosen, 1981). Other examples of the tournament model in action include professional sports, movie actors, writers, and drug dealers (Levitt and Dubner, 2005). What makes such tournaments worthwhile for the players is, firstly, the high stakes where the winner gets it all and, secondly, their relatively short duration, which allows the losers to pick up the pieces and move on without losing too much. Such a tournament is really a lottery where the win is so hugely greater than the entry cost that it makes sense to try the game. It might be a much safer bet to train as a surgeon (most conscientious students will end up with good jobs) than a football player, but the entry barrier to medical school is much higher than to high school sports. In other words, when there is a lot to win and not much to lose, entering a tournament can be a rational choice.

While an average Chicago drug dealer spends only a couple of (admittedly dangerous) years on the corner, before getting a better-paying minimum-wage job at a local fast food restaurant (Levitt and Dubner, 2005), a typical science student will have spent over a decade in intensive training before realizing that the tenure track will forever be closed to him. While our drug dealer will weigh a fairly high risk of getting killed against the near certainty of never amounting to anything if going for a legal job in his neighborhood on one hand, and the rather low probability of making it into a lavishly remunerated drug kingpin on the other hand, most budding scientists actually had other and better choices. By limiting his level of education to an ordinary college degree, he could have entered a well-paying job at 23 and by his 30th birthday would perhaps be well on the way to the top of his career ladder. Instead, as a scientist at 30, he probably finds himself in his first postdoc with no certainty as to in which walk of life his future career actually lies. The implication here is that, as long as there are ghettos, the future of drug dealing is safe under tournament economics, while the economics of science is inherently irrational for the participants, and thus unstable.

---

10. http://www.ascb.org/ascbpost/index.php/compass-points/item/285-where-will-a-biology-phd-take-you.

Another consequence of increasing competition is, paradoxically, the growth of individual labs. The bigger you are, the better: you will publish more, which increases your visibility, citation counts, and chances of occasionally getting into top journals. This will translate into more funding to the lab, into more graduate students and postdocs, and into further increases in your hitting power. On the downside, the heads of such super-labs, who spend most of their time in securing ever more financing, increasingly resemble politicos and businessmen, rather than creative scientists they once were. There seems to be an upper boundary on the size of labs, over which the scientific output will suffer (Youtie et al., 2013). As long as the overall science funding remains stagnant, such trends will translate into preferential closing of smaller labs and into reduced diversity and innovation. Increased competition leads to increased concentration of resources and to decreased diversity of science.

The third consequence of the heightened competition is that it provides an incentive for scientists to pervert the content of their science; or in simpler terms, to cheat. In a game where only a small fraction of players win, even with only a few cheaters they will get enough advantage to fill a disproportionately large part of the winning slots. Cheating then becomes a part of the job requirement. This happened in professional cycling and is quite possibly happening in science. While in cycling the problem was (at least temporarily) solved by new drug tests, in science there appears to be no technological fix to doping, as there are many more ways of different degrees of legitimacy of doing science than riding a bicycle. There is no methodological solution to the problem of how to do science "right."

In this view, the problems of reproducibility and of the general success of science are not created by "a few bad apples," nor by methodological shortcomings of individuals, nor by a technical failure of institutions to control the quality of the scientific product, but by a deep structural problem where the failure of science is hardcoded into the institutional system of science (Sovacool, 2008). Such structural flaws surface as insistence of journals in publishing only neat and quickly digestible storylines (the authors of which then are forced to bury the inevitable anomalies of fact and to disregard the potentially messy nature of the underlying reality itself), in the institutional overemphasis of publication rates, and in the alienation of the individual members of large research groups from their work product.

In the final analysis, nobody can truly police a scientist like he himself and if the cost of doing so entails loss of career, then the prospects for science as a whole don't look good.

REFERENCES

Abrams, I., Leung, G., Stevens, A.J., 2009. How are US technology transfer offices tasked and motivated-is it all about the money. Res. Manage. Rev. 17 (1), 1–34.

Acuna, D.E., Allesina, S., Kording, K.P., 2012. Future impact: predicting scientific success. Nature 489 (7415), 201–202.

Adams, J., 2012. Collaborations: the rise of research networks. Nature 490 (7420), 335–336.

Alberts, B., et al., 2014. Rescuing US biomedical research from its systemic flaws. Proc. Natl. Acad. Sci. USA. 111 (16), 5773–5777.

Allison, P.D., Long, J.S., 1990. Departmental effects on scientific productivity. Am. Sociol. Rev. 55 (4), 469.

Anon, 2013. Sparking Economic Growth 2.0. Available at: <http://www.sciencecoalition.org/reports/Sparking%20Economic%20Growth%20FINAL%2010-21-13.pdf>.

Berg, J.M., 2012. Science policy: well-funded investigators should receive extra scrutiny. Nature 489 (7415), 203.

Boettiger, S., Bennett, A.B., 2006. Bayh-Dole: if we knew then what we know now. Nat. Biotechnol. 24 (3), 320–323.

Bornmann, L., Daniel, H.-D., 2008. What do citation counts measure? a review of studies on citing behavior. J. Doc. 64 (1), 45–80.

Cable, D., Murray, B., 1999. Tournaments versus sponsored mobility as determinants of job search success. Acad. Manage. J. 42 (4), 439–449.

Chang, H., 2014. Is Water H_2O? Springer, Heidelberg.

Collins, F., Tabak, L., 2014. Policy: NIH plans to enhance reproducibility. Nature 505, 612–613.

de Solla Price, D.J., 1951. Quantitative measures of the development of science. Arch. Int. Hist. Sci. 4 (14), 85–93.

de Solla Price, D.J., 1965. Networks of scientific papers. Science 149 (3683), 510–515.

de Solla Price, D.J., 1975. Science Since Babylon. Yale University Press, New Haven, CT.

de Solla Price, D.J., 1976. A general theory of bibliometric and other cumulative advantage processes. J. Am. Soc. Inf. 27 (5–6), 292–306.

Dienes, Z., 2008. Understanding Psychology as a Science. Palgrave MacMillan, London.

Fanelli, D., 2009. How many scientists fabricate and falsify research? A systematic review and meta-analysis of survey data. PLoS One 4 (5), e5738.

Fanelli, D., 2010. "Positive" results increase down the hierarchy of the sciences. PLoS One 5 (4), e10068.

Fanelli, D., 2013. Redefine misconduct as distorted reporting. Nature 494 (7436), 149.

Fanelli, D., Glänzel, W., 2013. Bibliometric evidence for a hierarchy of the sciences. PLoS One 8 (6), e66938.

Feldman, R., Price, W.N., 2014. Patent trolling—why bio and pharmaceuticals are at risk. Stan. Tech. L. Rev. 773.

Feyerabend, P., 1975. Against Method: Outline of an Anarchistic Theory of Knowledge. Verso, London, New York.

Feyerabend, P., 1970. Consolations for the Specialist Criticism and the Growth of Knowledge. Cambridge University Press, pp. 197–230.

Feyerabend, P., 1995. Killing Time: The Autobiography of Paul Feyerabend. University of Chicago Press, Chicago, IL.

Goodman, D.S.N., Altman, D.G., George, S.L., 1998. Statistical reviewing policies of medical journals. J. Gen. Intern. Med. 13 (11), 753–756.

Goodstein, D.L., 1994. The Big Crunch. In: NCAR Symposium. Portland, OR, p. 23. Available at: <http://www.its.caltech.edu/~dg/crunch_art.html>.

Hacking, I., 1983. Representing and Intervening: Introductory Topics in the Philosophy of Natural Science. Cambridge University Press, Cambridge, New York, NY.

Halford, B., 2013. Postdoc pains and gains. Chem. Eng. News 91 (20), 41–44.

Hickey, T., 2014. Twentieth-Century Philosophy of Science: A History. Available at: <www.philsci.com>.

Horrobin, D.F., 1990. The philosophical basis of peer review and the suppression of innovation. JAMA 263 (10), 1438–1441.

Hoyningen-Huene, P., 2002. Paul Feyerabend und Thomas Kuhn. J. Gen. Philos. Sci. 33 (1), 61–83.

Ioannidis, J., 2006. Concentration of the most-cited papers in the scientific literature: analysis of journal ecosystems. PLoS One 1, e5.

Ioannidis, J.P., 2012. Why science is not necessarily self-correcting. Perspect. Psychol. Sci. 7 (6), 645–654.

Ioannidis, J.P.A., Boyack, K.W., Klavans, R., 2014. Estimates of the continuously publishing core in the scientific workforce. PLoS One 9 (7), e101698.

Ioannidis, J.P.A., Tatsioni, A., Karassa, F.B., 2010. Who is afraid of reviewers' comments? Or, why anything can be published and anything can be cited. Eur. J. Clin. Invest. 40 (4), 285–287.

Jones, B., 2009. The burden of knowledge and the "death of the renaissance man": is innovation getting harder? Rev. Econ. Stud. NBER Working Paper No. 11360, <www.nber.org/papers/w11360>.

Jones, B., 2010. Age and great invention. Rev. Econ. Stat. NBER Working Paper No. 11359, <www.nber.org/papers/w11359>.

Kaiser, D., 2012. In retrospect: the structure of scientific revolutions. Nature 484, 164–166.

Kavvoura, F.K., Liberopoulos, G., Ioannidis, J., 2007. Selection in reported epidemiological risks: an empirical assessment. PLoS Med. 4, e79.

Kuhn, T.S., 1962. The Structure of Scientific Revolutions. Chicago University Press, Chicage, IL.

Kuhn, T.S., 1970. Reflections on my critics Criticism and the Growth of Knowledge. Cambridge University Press, pp. 231–278.

Kyzas, P.A., Denaxa-Kyza, D., Ioannidis, J.P.A., 2007. Almost all articles on cancer prognostic markers report statistically significant results. Eur. J. Cancer 43 (17), 2559–2579.

Lamar, E., Rosen, S., 1981. Rank-order tournaments as optimal labor contracts. J. Polit. Econ. 89, 841–864.

Ledford, H., 2013. Universities struggle to make patents pay. Nature 501 (7468), 471–472.

Lee, C.J., et al., 2012. Bias in peer review. J. Am. Soc. Inf. Sci. Technol. 64 (1), 2–17.

Levitt, S.D., Dubner, S.J., 2005. Freakonomics: A Rogue Economist Explores the Hidden Side of Everything. William Morrow, New York, NY.

Li, Q.L., et al., 2002. Causal relationship between the loss of RUNX3 expression and gastric Cancer. Cell 109, 113–124.

Loise, V., Stevens, A.J., 2010. The Bayh-Dole Act turns 30. Sci. Transl. Med. 2 (52), 52cm27.

Long, J.S., Allison, P.D., McGinnis, R., 1979. Entrance into the academic career. Am. Sociol. Rev. 44 (5), 816.

Markel, H., 2013. Patents, profits, and the American people—the Bayh-Dole Act of 1980. N. Engl. J. Med. 369, 794–796.

Masterman, H., 1970. The Nature of a Paradigm Criticism and the Growth of Knowledge. Cambridge University Press, Cambridge, UK.

Merton, R.K., 1968. The Matthew effect in science: the reward and communication systems of science are considered. Science 159 (3810), 56–63.

Merton, R.K., 1974. The Sociology of Science: Theoretical and Empirical Investigations. University of Chicago Press, Chicago, IL.

Merton, R.K., 1988. The Matthew effect in science, II: cumulative advantage and the symbolism of intellectual property. Isis, 606–623.

Mervis, J., 2014. Peering into peer review. Science 343 (6171), 596–598.

Mynatt, C.R., Doherty, M.E., Tweney, R.D., 1978. Consequences of confirmation and disconfirmation in a simulated research environment. Q. J. Exp. Psychol. 30 (3), 395–406.

National Science Board, 2014. Science and Engineering Indicators 2014. National Science Foundation (NSB 14-01), Arlington, VA.

Neff, B.D., Olden, J.D., 2006. Is peer review a game of chance? BioScience 56 (4), 333–340.

Nelson, B., 2014. Synthetic biology: cultural divide. Nature 509 (7499), 152–154.

Newman, M., 2003. The structure and function of complex networks. SIAM Rev. 45 (2), 167–256.

Nicholson, J.M., Ioannidis, J.P.A., 2012. Research grants: conform and be funded. Nature 492 (7427), 34–36.

O'Brien, T.L., 2012. Change in academic coauthorship, 1953–2003. Sci. Technol. Human Values 37 (3), 210–234.

Onitilo, A.A., et al., 2013. Reliability of reviewer ratings in the manuscript peer review process: an opportunity for improvement. Account. Res. 20 (4), 270–284.

Ortega, Y., Gasset, J., 1932. The Revolt of the Masses. W. W. Norton and Company, New York.

Pedro Albarrán, A., 2011. The Skewness of Science in 219 Sub-fields and a Number of Aggregates. Working Paper, Departamento de Economía, Economic Series 11-09, Universidad Carlos III de Madrid, pp. 1–123.

Pfeiffer, T., Hoffmann, R., 2009. Large-scale assessment of the effect of popularity on the reliability of research. PLoS One 4 (6), e5996.

Polanyi, M., 1962. The Republic of Science: Its Political and Economic Theory. Minerva 1:54–74. Available at: <sciencepolicy.colorado.edu/students/envs_5100/polanyi_1967.pdf>.

Rasmussen, N., 2014. Gene Jockeys: Life Science and the Rise of Biotech Enterprise. Johns Hopkins University Press, Baltimore, MD.

Roberts, G.G., 2006. International Partnerships of Research Excellence: UK-USA Academic Collaboration, University of Oxford. Available at: <http://www.immagic.com/eLibrary/ARCHIVES/GENERAL/OXFORD/O060427R.pdf>.

Roessner, D., et al., 2013. The economic impact of licensed commercialized inventions originating in university research. Res. Policy 42 (1), 23–34.

Rorty, R., 1987. Science as Solidarity, in the Rhetoric of the Human Science. University of Wisconsin Press, Madison, WI.

Rothwell, P.M., Martyn, C.N., 2000. Reproducibility of peer review in clinical neuroscience. Brain 123 (Pt 9), 1964–1969.

Rzhetsky, A., et al., 2006. Microparadigms: chains of collective reasoning in publications about molecular interactions. Proc. Natl. Acad. Sci. USA. 103 (13), 4940–4945.

Schacht, W.H., 2012. The Bayh-Dole Act: Selected Issues in Patent Policy and the Commercialization of Technology. Congressional Research Service, Washington, DC.

Seglen, P.O., 1992. The skewness of science. J. Am. Soc. Inf. Sci. 43 (9), 628–638.

Shapin, S., 2008. The Scientific Life: A Moral History of a Late Modern Vocation. University of Chicago Press, Chicago, IL.

Siegmund, F., 2014. Is Getting a PhD Worth It? On Jobs, Salaries and the Market. Automatic Finances. Available at: <http://www.automaticfinances.com/is-getting-a-phd-worth-it/> (accessed 8.08.14).

Simmons, J.P., Nelson, L.D., Simonsohn, U., 2011. False-positive psychology: undisclosed flexibility in data collection and analysis allows presenting anything as significant. Psychol. Sci. 22 (11), 1359–1366.

Sovacool, B.K., 2008. Exploring scientific misconduct: isolated individuals, impure institutions, or an inevitable idiom of modern science? J. Bioeth. Inq. 5 (4), 271–282.

Stanford, P.K., 2006. Exceeding Our Grasp: Science, History, and the Problem of Unconceived Alternatives. Oxford University Press, Oxford, New York, NY.

Stephan, P., 2012. Research efficiency: perverse incentives. Nature 484 (7392), 29–31.

Stevens, A.J., 2004. The enactment of Bayh-Dole. J. Technol. Transf. 29 (93–99).

Stevens, A.J., 2010. Open Source Research in the Pharmaceutical Industry. Association of European Science and Technology Professionals (ASTP), Paris.

Su, X., 2009. Postdoctoral training, departmental prestige and scientists' research productivity. J. Technol. Transf. 36 (3), 275–291.

Thurner, S., Hanel, R., 2011. Peer-review in a world with rational scientists: toward selection of the average. Eur. Phys. J. B. 84 (4), 707–711.

Thursby, J., Thursby, M., 2004. Patterns of research and licensing activity of science and engineering faculty. Working Paper Series, Georgia Institute of Technology. Available at: <http://hdl.handle.net/1853/10723> (accessed 6.08.14).

van Dijk, D., Manor, O., Carey, L.B., 2014. Publication metrics and success on the academic job market. Curr. Biol. 24 (11), R516–R517.

Vinck, D., 2010. The Sociology of Scientific Work. Edward Elgar Publishing, Northampton, MA.

Wallace, M.L., Larivière, V., Gingras, Y., 2009. Modeling a century of citation distributions. J. Informetrics 3 (4), 296–303.

Yong, A., 2014. A critique of Hirsch's citation index: a combinatorial Fermi problem. Not. Am. Math. Soc. 61 (09), 1040.

Young, N.S., Ioannidis, J.P.A., Al-Ubaydli, O., 2008. Why current publication practices may distort science. PLoS Med. 5 (10), e201.

Youtie, J., et al., 2013. Career-based influences on scientific recognition in the United States and Europe: longitudinal evidence from curriculum vitae data. Res. Policy 42 (8), 1341–1355.

Chapter 8

What Can Be Done: A Utopia

If science is broken, then how to fix it? Can Francis Collins do it by reorganizing the principles of distribution of NIH grants (Collins and Tabak, 2014)? Should we dramatically increase the funding of science, or maybe do the opposite? Maybe we need to send the scientists to ethics classes? Or should we expand the definition of and increase the severity of punishments for scientific misconduct? Should we put our faith in increasing the reproducibility of raw results of biological experiments? Maybe we should do more power calculations and thus increase sample sizes in our experiments? Or should we spend more effort in teaching statistics to scientists and perhaps adopt a more stringent threshold for making type I errors? Or maybe we should ditch type I errors altogether and go Bayesian? Or are more controls needed in our experiments? Maybe we need better model organisms? Or more journals that publish scientific results without the customary gloss? Should we listen more to patients in assigning research priorities in biomedicine or should we leave the setting of priorities to individual scientists, who know their stuff best?

Don't worry if you don't know the answers, because nobody else does either. My guess is that there isn't a single fix. Instead we have a complex problem requiring a complex solution. A possible analogy is the problem of how to reduce traffic congestion and air pollution in large cities. Here the typical simple fix would be to make the roads wider and put more buses on them, but the fixes that have actually led to good results are distinctly more complex. They consist of a combination of the carrot and the whip, involving building safe cycling lanes, instigating rent-a-bike and share-a-car schemes and reserved bus lanes, in combination with higher taxes and slower roads for the people who prefer cars. But crucially, they also involve wider decisions concerning planning for new suburbs, city parks, schools, etc.

In this, our last chapter, we will enter into the realm of utopia and try to offer some fixes of our own. In doing so we will not care about the politics of it all and our understanding is that in the final analysis we really do not know what we are doing and merely offer possibilities that may or may not point to the right direction. Whether or not the reader should take these seriously, depends largely on whether she believes that science (or biomedical science, to be exact) is broken. If we really believe that over 75% of published preclinical cancer

Interpreting Biomedical Science. DOI: http://dx.doi.org/10.1016/B978-0-12-418689-7.00008-9

research is little more than contamination of literature, whose main effect will be to hide the minority of good research in plain sight, then we must admit that this belief has consequences. We should lose faith in the journals that publish such drivel and seem to actively engage in shaping the form and content of scientific communication to simplify its explanatory models, so that any anomalies are hidden away. We should lose faith in the universities that slavishly use the same publications as their major criteria in promoting people who are the best in producing such papers. And finally, we should lose faith in the inevitability of the progress of science. If bad science looks and feels and smells like good science, is published in the same journals and takes up the majority of the journal space, then the future scientists will stand on the shoulders of midgets—with the result that science ceases to be cumulative and relevant to society. The endless frontier of the post-WWII years will gradually turn into an endless bog where resources sink without a trace.

If this is what we believe, then our highest priority as scientists is to fix science, to make it work again. To do so, we must understand not only the symptoms, but also the causes of the problem. We must understand the incentives that really drive scientists to behave in certain ways and to move in certain directions; but we must also understand the assumptions and limits behind the methods that scientists often blindly use. Right now the major questions include the use of frequentist or Bayesian statistical methods in combination with very small sample sizes and clarification of the concepts of and relationships between statistical power and statistical evidence. As low power almost inevitably leads to high rates of false-positive results, we probably cannot escape the conclusion that sample sizes of many biological studies must be significantly increased (but at the moment we do not even have reliable published information on the level of variation in most experimental systems, which makes power calculations difficult). Also, one has the feeling that more could be learned of the relations between statistical and scientific hypothesis testing and statistical and scientific evidence.

Another bundle of problems that seems to cry out for new thinking and wider approaches is the silent multiplicities and the relations between exploratory and confirmatory studies. The received wisdom seems to be that to avoid the plight of multiplicities we must preregister our research questions and methods, and be careful of not letting ourselves be swayed by the actual data to change them in mid-study. This would decrease the frequency of false-positives but would make us willingly blind ourselves to any surprises that our data might try to send our way. A similar question arises about the exhortations that one-to-one replication of results by an independent (and maybe commercial) lab should serve as a gatekeeper to scientific publishing. Could we in this way create a situation that discourages the development of new groundbreaking experimental systems? There are conflicts between conservatism and pioneer spirit in science, for which we cannot even begin to offer a solution.

Another serious problem is that however well we build our data analysis methods, they will remain largely irrelevant to the most important problem in scientific methodology—bias. Bias is fought at the level of experimental planning, especially in planning controls. There is a strong and growing scientific basis for this fight in clinical experiments, which is currently pretty much ahead of molecular biology, methodologically speaking. Thus, the study of experiment—the science of experimentology, if you wish—is waiting for researchers to fill it with content.

Next we will discuss some possible remedies that could be used to improve the efficiency of science in doing what it is supposed to do—creating knowledge where previously there was none. My goal is not so much to convince the reader of the necessity of adopting some approach or other, but to provoke her either to agree or to violently disagree, and thus to provide some food for her own thought processes.

Our first recommendation is discussed in the following section.

8.1 TAKE METHODOLOGY SERIOUSLY

The most obvious fix to the problems facing science, especially if seen through the prism of reproducibility, would be to teach scientists to use scientific method(s) properly. Teaching methodology involves both experimental planning and data analysis and Part II of the book is largely devoted to this.

However, there are sources of uncertainty here. To start with, we have a multitude of separate and at least partially incompatible methods, statistical and otherwise, which are more-or-less successfully used by different scientists and in different branches of science. There are in fact at least two (and perhaps three) fundamentally different statistical paradigms—frequentism, Bayesianism, and (possibly) likelihoodism. This means—and I'm afraid that this book further underlies—that teaching the method in science is far from an easy task.

In teaching and using scientific methodology we must be very much aware that methodology itself is a science (and we shall shortly see why this is not necessarily a good thing). Even worse, scientific methodology is a relatively new science, which is now in great flux thanks to advances in computing, in Bayesian methods, and the exponential rise in big data, simple model (nonhypothesis-driven) kind of science. This means that in doing the science of biology we strongly depend on the science of methodology, which is, simply put, in its beta-version and may not be fit for sustained use. Therefore, a close understanding of the progress of methodology has gradually become necessary for a modern biologist. This requirement brings with it a need for greatly improved mathematical and philosophical education for biologists.

We can hazard a guess that the near future of development of scientific methodology will put pressure on the scientist to more formally express and use her background knowledge. This will transform not only how experiments are planned and analyzed, but also how they are presented in papers and talks. Such developments will be implemented for small sample size studies, prevalent in traditional biology and they have the potential to greatly increase the usefulness of individual papers in forming scientific consensuses on specific problems.

Another point is that we cannot be sure if teaching of the received wisdom of frequentist error-statistics (P values and the like) as the "scientific method" is even a good thing for the progress of science. This is because the function of error-statistics is quite limited: it is to increase the quality of the scientific corpus by limiting errors. In doing this it will necessarily weed out not only most of the errors but also some of our real findings. If it happens that the structure of the parts of the reality, which are still left to discover, is more complicated than currently depicted in our models, then it stands to reason that we will discard our most interesting discoveries by the correct and judicious implementation of statistical methods and will thus sacrifice Novelty on the altar of Quality. Obviously we need to find a balance here, but where the right balance lies is currently unknown and it is likely to be in different places in different fields of science.

Another problem with the traditional statistical approach to scientific methodology is that its reliance on black-and-white decisions (rejection of the null hypothesis) inevitably leads to inflated effect sizes, hopelessly tangled problems with silent (and not so silent) multiplicities, and troubles with the pseudocausality of the regression to the mean. The only real solution here would be to change the methodological paradigm from black-and-white to shades of gray: from decisions and long-run quality control to probabilities as degrees of belief about specific theories.

It can be argued that as the low-hanging fruit is being increasingly picked in biological discovery and the models needed to describe the as-yet-undiscovered parts of biology become increasingly complex, the need for complex statistical methods increases correspondingly and any intuitive approach to experimental science will increasingly be a fool's errand (Dougherty and Shmulevich, 2012). The history of the shifting fortunes of 40 years' worth of search for the right level of complexity for explanation in cancer studies is discussed in Weinberg (2014).

Unfortunately, sometimes the level of complexity of our models of reality is not determined by our prior understanding of the scientific problem, nor by our mathematical or computational capabilities, but rather by the experimental methods that we choose to use. For example, currently very popular massively parallel omics strategies need huge sample sizes to enable correction for multiple testing, which has largely limited their use to very simple statistical models consisting of individual odds-ratios with assorted P or q values (Dougherty and Shmulevich, 2012). Because of that, in the omics-based cancer studies, "[t]he coupling between observational data and biological insight is frayed, if not broken. We lack the conceptual paradigms and computational strategies for dealing with this complexity" (Weinberg, 2014).

If we are to put our trust in such arguments, it follows that we really cannot assume that the past successes of science are any guide to future discoveries, without assuming that we have access to a methodology, which is suitable for the current level of complexity of potential discoveries. Of course, we can never actually know what the *actual* level of complexity is in the parts of nature which we are currently studying.

The above should not be construed as an argument for the futility of methodology in science. Of course we need to teach methodology to our students. But we must also take great care to better understand the methodologies we have and to develop new ones. A crucial insight is that a good methodology in itself is not enough for scientific progress, and probably never will be. This leads us to the next part of our utopia.

8.2 BRING PHILOSOPHY BACK TO SCIENCE

A PhD stands for "Doctor of Philosophy." And yet many scientists are currently taught philosophy as a quaint historical subject; as something that may be good for your general level of education and intellectual stimulation, but is barely relevant to an actual education in science. Another oft-stated (by philosophers) reason for bringing together philosophers and scientists is to argue that it is the job of a philosopher to clarify the meaning of scientific concepts, which the scientist then tests empirically (see "Moritz Schlick, The Future of Philosophy", in Balashov and Rosenberg, 2002), or to teach the scientist how to think efficiently and logically (Casadevall and Fang, 2012). And yet, I suspect that, except for a few specialties, like the study of consciousness, few scientists would admit that their concepts or thinking need clarifying by nonscientists.

I will next try to argue that what the scientist needs most for arriving at a useful philosophy of science is not philosophy of language or the history of philosophy, or even philosophical logic—instead it is the branch of philosophy called metaphysics.

The methodology of science is all about working with theories and models of reality: there is nothing in it that tells us how to connect our data and theories to the true structure of the world. This is not to say that our data don't reflect reality—at least sometimes they do. However, we do not match data to reality but to our models and theories. Because there are infinitely more self-consistent theories than there are worlds (unless you are a believer in a multiverse), every dataset will support many possible theories, which are far from the truth and there is no guarantee that any theory under consideration right now, is reasonably close to the truth.

The true structure of reality is not the subject of science—it is the subject of metaphysics. Does the structure of the world correspond to the structure of our theories? Is there a way to find out? How can we separate data from theory? Should we be working with theories or models? Should we be chasing truth or usefulness by our scientific theories? What are the roles of truth and causality in science? How can we falsify a theory? How can we verify one? What does probability mean in science and in the physical world? Are these meanings connected?

To even begin to think on these thorny questions requires a philosophical education, not to mention an understanding of the history of science. The research methods and philosophies of those giants, whose shoulders we claim to have soiled with our footprints and whose discoveries have withstood the test of time, are much more than historical curiosities. Apparently these people were able to do something then, that we are unsure that we can do or even recognize now—namely good science. We should be very curious about these matters. Without a sound philosophical education a scientist will never be fully aware of the assumptions behind her work and methods; an awareness that she needs every day in the lab.

Thus the part of philosophy most relevant to science is not philosophy of language, which can help us to clarify the meanings of words. Instead it is metaphysics, a branch of philosophy that could help us to connect (or reconnect?) science with the world. Good science has an integral metaphysical component and I can see no indication that we could one day cure science from it. This may be unfortunate, but all that is left to us here and now is to face reality and work with what we have.

To conclude, here are some rather philosophical conclusions drawn from our understanding of methodology of science, especially Bayes theorem. It must be stressed once again that Bayes theorem is a logical necessity, derived directly and uncontroversially from the axioms of probability theory. We simply have no alternative, which would enable us to rationally assess probabilities of individual theories.

1. There is no room for certainty in science. Once we are sure of something, there is no rational way in which additional evidence can change our beliefs.
2. Evidence is always relative to more than one competing hypothesis. The strength and even direction of evidence change with our hypotheses.
3. Propositional logic does not work well in empirical science. Probabilistic logic, which does work, requires input of prior knowledge, which is always subjective. Therefore, the objectivity of science is a logical fallacy.
4. Science is not omnipotent. It can only test hypotheses, whose prior probabilities are higher than are the prior probabilities of systematic error (bias). This leaves great chunks of "hypothesis-free" massively parallel biology, as currently practiced, out in the cold.
5. Extraordinary claims require extraordinary proof. Bayes theorem can be used to quantify how much less trust we should put on surprising results that are published in the high-impact journals, as compared with ordinary results published in ordinary journals. Enrichment of the journals that matter with false results, in combination with increased competition for academic jobs, depending on publishing in those same journals, can kill science. Once we lose the working tradition of doing successful science, if history tells us anything, it is unlikely that science will ever be born again.

6. There is no criterion of truth. We can give no methodological advice on how to formulate a true theory, or how to recognize that a theory is true (or close to truth). Should basic science turn into a useless cadaver, the scientists will be the last to notice.

If our investigations into methodology and philosophy of biomedical science have taught us anything, it is that there is no semi-automatic method of doing science; there is no single scientific method, which can be taught at school. Scientific progress is a miracle that should never be taken for granted. Thus, combined with our recent difficulties, it seems like good advice that we should hedge our bets and...

8.3 STRIVE FOR MORE PLURALITY IN SCIENCE

Pluralism is strongly defended by the historian and philosopher of chemistry, Hasok Chang, who discovered the need for it from his study of the rapid conceptional shift from the phlogiston theory of chemistry to the oxygen theory (Chang, 2014). I will try to present the crux of his argument as best I can.

Phlogiston was a fundamental substance that, when combining with ordinary matter around 1775, gave it combustibility. A combustible substance was rich in phlogiston and its burning released phlogiston in the form of flame. Metals too were rich in phlogiston, which moved freely through metal and gave it its characteristics, like malleability and electrical conductivity. By losing its phlogiston, a chunk of metal lost some of its characteristic properties and was reduced to "calx" (rust). In a smelter calx could easily be rephlogisticated by mixing it with hot phlogiston-rich charcoal to turn it into proper metal again. Similarly, Joseph Priestly was able to turn mercury calx back into metal by making it absorb phlogiston from air, thereby producing dephlogisticated air. Because metabolism was known to release bodily phlogiston and animals breathed in order to extricate it, it was predicted that breathing dephlogisticated air would be an especially pleasant experience. And so it was: in modern terms, Priestly had discovered oxygen.

Also, inflammable air (hydrogen) was produced, by dropping metals into acids, and it was shown that mixing phlogiston-rich inflammable water[1] with dephlogisticated water produced normal water (phlogiston + dephlogisticated water → phlogisticated water).

Phlogiston theory was soon completely superseded by a novel system of chemistry, authored by Lavoisier, in which water was created by combining hydrogen and oxygen. In Lavoisier's winning but also false theory, all acids contained oxygen and heat was generated in combustion by the release from oxygen gas of a fundamental substance, called caloric (although combustion without oxygen was already a well-known phenomenon).

1. Hydrogen was equated with phlogiston for a while.

What Chang discovered was that firstly, the eighteenth-century phlogiston and oxygen theories are about equally wrong from the modern perspective, secondly, that in the eighteenth century there wasn't a good scientific reason to accept one theory and to discard the other, and thirdly, that with the benefit of hindsight we can say that it would probably had been the best if both theories had been allowed to mature in parallel and over many decades. Had phlogiston lived, it would have been easier to reach concepts, like the electron, and to increase our understanding of metals, nutrition, and geology. In these fields the oxygenists had nothing to say; while in the theories of heat, changes of physical state, and chemistry of salts they were the frontrunners. For studies of combustion, chemical reduction and acids, both paradigms had good potential.

While discussing modern science we do not have the benefit of hindsight. As science develops ever further from the common-sense world, which we can directly observe, we increasingly view our objects of study in terms of what we can do to them experimentally (corresponding to an operationalist theory of knowledge, see Section 2.1). The more experimental manipulations we can do, the more different properties our objects of study have for us. This leads to a more general realization that by having more experimental frameworks or paradigms to choose from, we will be better equipped to generate interesting experiments and thus a better chance of making interesting discoveries. Therefore, it should be in everybody's interest that not only for the best or most popular paradigm to be allowed to flourish, but a couple of alternatives as well. A hundred flowers can then bloom together and happily cross-fertilize each other.

Accordingly, universities and grant agencies should not shy away from supporting research programs that are incompatible with each other. This means that the focus must shift from truth to usefulness and progress. Admitting our current ignorance of what makes a good science (or a good scientist), we should strive to keep several kinds of scientific traditions in play: high-risk projects, slow scientific drudgery, foxes and hedgehogs, big and small data, hypothesis corroboration and hypothesis generation, small groups and large groups, interdisciplinary and highly disciplinary approaches, academic science, citizen science, government science, company science, and whatnot. We could even apportion a small proportion of science funding to what we think of as cranks, as long as they have interesting agendas.

Our immediate goal is to seek experimental contact with reality by any theory that can help to do just that, and we should therefore retain our older theories for what they are still good for, even when having given up on their veracity. Instead of reducing to as few general theories as possible, we should proliferate and find a new balance.

The pluralistic view is well described in the following passage, where incompatible scientific theories, possibly true and surely false, are integrated into a single working technological fix.

A nice ... example of such integration is the global positioning system (GPS): by means of satellites kept in place by Newtonian physics, and atomic clocks ruled by quantum mechanics and corrected by special and general relativity, this system maps the spherical surface of the round earth on a geocentric grid (or rather geostatic grid), and gives advice to people on the ground from a flat-earth point of view.

Chang (2014)

There is no logical consistency in this, but nevertheless it works. As long as our theories fail to cover everything, pluralism can be recommended.

For our work to be successful, we have to actually want to be successful. That is, individual scientists need to set their aims to science, as opposed to using science as a means to achieve other ends. To be able to do this, they need an ethos of science and they need the incentives to support and enforce this ethos. This leads us to our next radical recommendation.

8.4   REINTRODUCE MERTONIAN VALUES

A scientist must not only know her methodology and metaphysics, she also needs a value system to guide her work. If we take the current system to entail the values of the marketplace, which offers the scientist money for her discoveries and thus motivate intense competition and selecting the objects of study by their potential to lead to quick profit, then a clear-cut alternative would be Mertonian CUDOS.

The market-based system leads to scientists trying to satisfy the requirements of the market: if cancer medicine is needed, then many basic scientists will compete in using existing technologies to as quickly as possible find promising molecular targets, which they then try to sell on to drug developers. A quick sale leaves them or their universities with no additional monetary motivation to ensure that their discoveries actually succeed in the clinic. Such a short-term perspective encourages sloppy science and can divert valuable resources. In science at least, greed is not good.

Money Isn't Everything

In 1971 the US Congress declared war on cancer. This war was precipitated by the steep increases in the incidence of many forms of cancers observed from the early 1900s and the inspiration for it came from the great technological race for putting an American on the moon before the Soviets get there.

The modus operandi of the war on cancer was in 1971, as it is today, greatly increased competitive grants to basic research. Society provided the money and left it to many individual scientists to sort out how to best use it. Thus the war on cancer is a war with many generals, which over 40-plus years has led to a $300 billion investment (in adjusted USDs, counting both public and private sector contributions) and cancer research in the US currently uses about $16 billion annually (Leaf, 2013). Not surprisingly, this has resulted in a huge output of studies on basic cancer mechanisms.

In the meanwhile, the population-size-adjusted (but not age-adjusted) US death rates for all causes, except cancer, have decreased by 24%, while cancer death rates have increased by 14% (Leaf, 2013, pp. 37–38). During the same time period (1970–2010) death rates from heart disease fell by 47%.

Cancer is currently the major focus of drug discovery by the pharmaceutical industry. This is in large part due to profusion of potential target molecules coming from academic studies. As of 2012, there are nearly 1,000 anticancer drug candidates in the development pipelines of the pharmaceutical industry (LaMattina, 2013). Yet, while the average success rate from clinical trials is slightly over 10% and for cardiovascular disease about 20%, for cancer drugs it has fallen below 5%. Even from the cancer drugs that have successfully passed stage I and II clinical trials, about two-thirds fail to yield an approved treatment. There is no consensus about the reasons for this but insufficient grasp of the biology of cancer, hurdles of the clinical trial procedures (methodological and otherwise), and the sheer impossibility of finding a single-molecule cure for complex heterogeneous mixes of cancerous cells are popular explanations.

There is also this nagging feeling that in fighting cancer money isn't everything; that simply concentrating resources on the peer reviewed academic system of basic science will not emulate the success of NASA engineers in flying a man to the moon. Indeed, studies of the technology industry indicate that concentrated spending may not be the best way to real innovation. For example, in 2011, the ailing giant Microsoft spent $9.0 billion on R&D (12.9% of its sales), being just behind Pfizer. In contrast, the successful innovator Apple spent "only" $3.4 billion on R&D both in 2011 and in 2012, which is a meager 2% of its sales (Feldman and Price, 2014). The worlds top five R&D investing companies in 2013 were Volkswagen, Samsung, Microsoft, Intel, and Toyota (Pfizer had fallen to 10th, and Google was 13th) (http://www.businessinsider.com/50-biggest-companies-by-rd-investment-2013-12).

The CUDOS-based value system is based on a different set of prizes. Here the ultimate in recognition is not the money and power, but prestige, conveyed by the society to the best scientists. Society cannot see directly, who are the scientists worthiest of their adulation: this information they get from other scientists, who form the committees that give out the important prizes like memberships of National Academies or the Nobel Prize. The winners acquire a privilege of directly speaking to the public and are thus trusted to directly influence public perception on matters scientific. They will not be amongst the richest members of society, but are in the very top by the level of trust that the public bestows on them.

The first letter of CUDOS denotes communism, e.g., free dissemination of scientific knowledge. If knowledge is power, and the market economy of supply and demand rules, how does the communism of scientists, who are freely giving away valuable knowledge, benefit the otherwise capitalist society? Why not rather privatize the scientific knowledge by the currently popular patenting and exclusive licensing schemes, thus creating stronger incentives for the private sector to develop it into proprietary technologies?

As it happens, economic theory suggests that the communism of knowledge may be essential to the long-term well-being of capitalism (Piketty, 2014). Historically, it was the serial implementation of new technologies, i.e., unhindered dissemination of knowledge, that more than anything started a rise in the efficiency of labor in the second half of the nineteenth century, which in turn ended a half-century of wage stagnation of industrial workers and, starting from about 1910, led to steady improvements in workers' lot, thus averting the danger of communist revolutions in Western Europe and America. The communal nature of knowledge also offers the best hope of overcoming the current trend of radical income divergence:

> *The main forces for convergence [of incomes] are the diffusion of knowledge and investment in training and skills.... Knowledge and skill diffusion is the key to overall productivity growth as well as the reduction of inequality both within and between countries.... The technological convergence process may be abetted by open borders for trade, but it is fundamentally a process of the diffusion and sharing of knowledge—the public good par excellence—rather than a market mechanism.*
>
> Piketty (2014)

If this theory is correct, then by expunging the communism from science by privatization of its product, we may destabilize the capitalist society by allowing too great inequalities, and thus inviting the return of the specter of the true political communism!

According to Merton, scientists are tacitly socialized to CUDOS principles by their peers, thus beginning to see these norms as not only useful but also as morally good. A scientist, like a soldier, is "proud to serve" and like a soldier his valor is awarded by golden-hued medals, not by real gold.

The original CUDOS-ethos did not come about because somebody decreed it. It has been carried to us from the historical depths of science, where the original members of the Royal Society—a group of independently wealthy gentlemen who got their patronage, if not their funding, directly from their king—who tailored their work to the needs of the British Empire for astronomical (navigational) and botanical (growing food in tropical colonies) knowledge (Jardine, 2000). CUDOS may just be a historical accident whose reanimation would require the restoration of the original motivational structure of individual scientists.

Thus, should we wish to reintroduce Mertonian ideals to science, we could start by learning from history.

8.5 PUT SCIENTISTS BACK TO THE IVORY TOWER

But wait, you might say, there is a lot of talk about the societal impact of science, as anybody who has recently written a grant application should know. For an example of the (almost) accepted view of what a university should look like, here is a glowing description in the Orwellian tradition of the campus of University of Arizona, which has the largest faculty and student body in the United States, from a *Nature* feature story:

> Big placards in hallways announce "A New American University" with eight ambitious calls to action. "Fuse Intellectual Disciplines" is one, along with "Transform Society", "Value Entrepreneurship", "Enable Student Success", and "Conduct Use-Inspired Research". The campus itself has a modern, utilitarian look: large buildings with clean lines, many topped with solar panels. Construction cranes poke into the sky as they continue a building boom [...]. Throngs of students thread their way around them.
>
> Fishman (2014)

So, shouldn't we do the opposite and lock the scientist out from the ivory tower? This is exactly what the current system of science funding tries to do, and to see why this is not a good idea we shall look at a recent comprehensive review of literature on measuring the societal impact of science (Bornmann, 2012). Their major findings include: (i) there are well over 60 different indicators used to assess the societal impact of science, none of them well developed, and there is currently no cohesive scientific field or community to develop them; (ii) the scientific and societal impact barely correlate with one another, which means that few synergy effects will be produced, and that the two should not be evaluated together; (iii) the societal impact of research is much harder to assess than its scientific impact; (iv) the societal impact of research often takes many years to manifest itself, and often it is difficult to identify causal links between concrete research and its impact; (v) scientists often are unaware that their research even has a societal impact, which means that they may not be competent to assess the societal impact of other researchers' work; (vi) the lack of clarity in assessing social impact leads to the tragedy of the proxy with a vengeance, where people increasingly spend their effort in complying with bureaucratic criteria and indicators, rather than trying to do genuinely important work.

As a conclusion, it seems that even if we will one day learn to recognize the true societal importance of a scientific proposal, there will always be a gap between scientific importance (basic science) and societal importance (applied science, engineering) that cannot be easily bridged—and that, therefore, we must keep our basic science well isolated from societal pressures.

Our proposition is that a good way to increase the practical relevance of basic science is, paradoxically, to put the basic scientist back into the ivory tower and to keep him there under lock and key. This would entail reducing direct contacts between the working scientist and society. The gatekeepers and messengers between the two could include peer review and peer pressure, professional organizations like universities and academies of science, various governmental organizations and committees, and professional science journalists, who would be able to approach their subject critically.

An individual scientist should not be caught lobbying lawmakers for a direct share of the budget for his institute or subfield. Neither should he try to convince the public through media of the potential fruits of his work or of the groundbreaking results that he just got or will shortly obtain. We have to reestablish the social norm of disinterestedness. There is no harm in a scientist engaging the public on a question he cannot directly benefit from, but the opposite should be frowned on.

The reasons behind our proposed lock-up are twofold. Firstly, scientists are a danger to society. This is because they have a unique combination of high competiveness, the ability to ask society directly for money without having to give anything back for it in the short to medium term, and a specific competency lacking in general society. If you doubt this, please take a minute to imagine a world where the journalists deal in "popular politics," meaning that they uncritically repeat politicians' claims and, when in doubt, assume that it is the politicians, who know better. In this world a "critical" commentary is always taken from a member of the same party and the journalists are happy to confess that they themselves were trained as movie critics and know nothing of politics. Welcome to North Korea, the country where you really must trust the politicians.

Now look at the science pages of your local paper (or the op-ed pages of the *NY Times*) and my guess is that you will find scientists promoting their work through uncritical popularizing. The same goes for the interactions of scientists with politicians, where the object is to use one's personal charisma to convince them that giving money to science in general and the experts' institution in particular is a surefire method to increase one's country's competitiveness (and to brand a politician as a progressive). The outcomes are infantile claims of yet another breakthrough in the fight against cancer, earmarked funding to specific institutions, and political obsession with keeping science funding at some predetermined proportion of the gross domestic product (GDP).

All this has the potential of turning good scientists into highly motivated confidence artists, who are not above promising cures for terrible diseases (or even aging) and clean fixes for serious societal problems (think of cold fusion), if only their groups or institutes or universities get lavish funding. Such cons can be fairly subtle and opaque to everyone except a few specialists, i.e., the peers of the perpetrators. As always, there is a blowback. Namely, sooner or later the con artists will start to believe their own cons and, conversely, even the politicians might start to wonder about the nonarrival of the promised riches and breakthroughs. This leads to a state of confusion about what it really is that scientists and universities do, and what it is that they ought to do to justify their existence. For example, from my own small university there has recently been a newspaper article by a prominent professor, proposing that the university should have "zero tolerance" for principal investigators who are unable to bring in outside financing—i.e., it is not the quality of the scientific work but its cost to the society that should decide who gets to keep their jobs.

And secondly, we need to protect scientists from society. The norms of society are centered on hopeful and mystical thinking, self-interest, and free market—there is a premium in buying cheaply and selling dearly, but no premium for telling the truth. In fact, it is not hard to see how truth can be inimical in private life ("Honey, you look like a used piece of jet trash today"), in business ("Sir, we only paid x for this used car that I want to sell you for xxx, and its carburetor isn't really what it used to be"), or in politics (example not required). Indeed, dishonesty seems to be a social norm in the business world (Cohn et al., 2014; Ariely, 2013). Also it is noteworthy in this context that 84% of US parents and 98% of Chinese parents lie to their children to manipulate their behavior—typically by threatening to abandon their children in a public place (Heyman et al., 2013).

Meanwhile, our best bet for science is that scientists freely share their product (behave as communists), act disinterestedly, do not aim for the riches, are critical toward self and others, and passionately seek the truth. If there be any chance of these two contradictory set of norms co-existing (and remember that the scientist still goes home most evenings to say: "Honey, you look beautiful tonight!"), there has to be a clearly set boundary between them. And, of course, since the main condition for the long-term existence of the ivory tower is the relevance of basic scientific knowledge to society, there must also be mechanisms for communication through the barrier.

Perhaps we should evaluate the currently popular efforts to improve the market value of future PhD by teaching them additional marketable skills, like law, business, and leadership, in this context. For starters, imagine a fresh PhD starting his private sector employment by insisting on honesty toward clients, disinterestedness in the product margin, and open criticism of his boss. Now ask yourself, is it really in anybody's interest to invest four years in a PhD AND a business education? If this was not enough, next imagine that you have just employed a scientifically honest lawyer to represent your interests in a civil suit, while the opposing counsel is an ordinary professional working in the confines of official regulations.

One implication of our ivory-hued proposal is that scientific method will be practiced both inside and outside of the tower. Inside there are people doing the so-called Mode 1 science, that is basic science which is governed by CUDOS and peer review. This mode of doing science is independent and individualistic, it is aimed at knowledge creation and an important measure of its success lies in scientific publication. Meanwhile, in the outside, other people are using scientific method to work directly for the benefit of society—they do Mode 2 science, which is not directly concerned with creation of generalized knowledge but specialized knowledge in support of specific technologies (Ernø-Kjølhede and Hansson, 2011; Nowotny et al., 2001). Their projects are to a large extent initiated, controlled, and evaluated not by their peers but by their bosses, that is by the beneficiaries of their work. Mode 2 science is not independent, not individualistic, and its criteria of success come outside of science. Thus it cannot be assessed by the same criteria as basic science and any attempts to produce mixed criteria (e.g., by giving differing weights to publication record, scientific relevance, and criteria of societal relevance), for evaluating both modes of science as one, will end in systemic failure (Ernø-Kjølhede and Hansson, 2011).

Of course, mode 1 scientists need assurances from society that by forgoing the possibility of ever being truly rich, they receive as compensation a steady and well-respected, if not a well-paid, job. The society in turn needs to trust the scientists to do their best to provide useful insight into problems relevant to it. This leads to a picture of a communistic bubble inside a capitalist society, whose borders need to be maintained by a peacekeeping force.

In our utopia this force consists of scientific peers, who are bound by the CUDOS-ethos, in the inside and in the outside of bureaucrats and legislators, who trust the scientific institution as a whole to be working in society's interests. Maintaining the borders and channeling resources into the bubble is largely a matter for the democratically elected legislator. In contrast, directing the flow of resources inside the bubble is a more opaque politbureau type of thing where committees of scientists decide on the financing of themselves and of their colleagues. Outside we have the democratic rule of the masses, inside there is the Platonic rule of the wise.

Scientific committees perpetuate themselves—there is no democracy in communism. This can be good for efficiency in a Platonic sort of way, but creates the danger that the harshness of the competition for money leads to excessive concentration of the limiting resources to the institutions of the central committee members, thus starving the periphery and, ultimately, bursting the bubble.

8.6 CHANGE THE RULES OF THE TOURNAMENT

If we want scientists to be happy in their bubble, we have to ensure that the scientific profession is not overcrowded. We know that the resources that the society can spare for science are limited and ultimately it is in no-one's interest to pretend that investments into science will be quickly returned with interest.

The current system of education of scientists is premised on quick growth of science funding. It takes in many more PhD students than there are permanent jobs for professional scientists and pushes the bottleneck, where most candidates fail, a decade into the future to the postdoc level. For the young scientist this means an overly long period of working in insecurity, which leads to the tournament model of hiring and reluctance on the part of scientists to spend their best and mentally most active years working on risky projects. You having a decade-long postdoc experience is in the interest of the university that hires you—because of the hope of future tenure in some other university you are willing to forgo about half to a third of your salary—but it is probably not in the interest of science and it is definitely not in the interest of you.

To escape this trap, we must shift the career bottleneck from the end of postdoc (about 35 years of age) to the start of graduate studies (about 23 years). This we can easily do by dramatically lowering the entrance number of PhD candidates so that it would not be very much larger than the expected number of science jobs in academia and industry. The competition would still be high, but it would take place much earlier in life, would be for study positions rather than jobs, and both the winners and the losers would have real perspectives ahead of them. This solution goes against the institutional interests of the universities and of the "common sense" of state bureaucracy (currently the US government spends around $600 per schoolchild to raise their interest in science and technology careers (Macilwain, 2013)).

To give up the cheap workforce of graduate students and postdocs may seem like a very stupid idea to many lab heads of today, but some very recent developments in the automation of biology labs promise to change this economic balance (Check Hayden, 2014). As I write this at the end of 2014, it seems likely that in a few years it will be both economically and methodologically preferable to outsource standard wet-lab experiments (cloning, protein purification, Westerns, Northerns, qPCR, etc.) by sending directly machine-readable protocols to commercial vendors. The experiment will be done offsite, but will still be controllable in the real time by the scientist. This means that the function of a graduate student or postdoc will become fundamentally different: no longer will she serve as a cheaper—and slightly less qualified—version of a professional lab technician. Instead her main contribution will be in designing experiments, controlling their execution, and in analyzing the data. As the machine-readable protocols are by definition comprehensive, repeatable, and easily standardized, these developments can be expected to work wonders on the reproducibility of biology and to usher in a new era of metastudies in the areas where now there are none.

While in the current form of the tournament the winners get an academic job, whose main perk is rapid entry into the next tournament of getting highly competitive grants and renewing them every 5 years or so, in our utopia the winners would get their keys to the ivory tower and a reasonably high probability of stable financing for the rest of their working lives. The losers, who will still comprise the vast majority of contestants and each have 3–5 years of systematic academic education, would still be able to enter the marketplace with their heads held high, or to find satisfying work as teachers, science communicators, or government workers. They would not lose their ability to compete in a timely fashion outside the ivory tower for nontournament jobs, where "mere" competence can give you a good life—including the jobs in engineering, involving the implementation and development of the knowledge emanating from the commons of science. And, as an added bonus, the universities would now get a more homogeneous top-notch cadre of graduate students, who would ask for serious tutoring. This system would also reduce the postdoc period to a length needed for actually giving the postdocs their individualized education, which would make them suitable for independent research careers and give them the high level of competency needed for working in modern scientific teams. This means an overall reduction in the number of postdocs and a relative increase in the number of scientists who are not in training but hold steady jobs.

The overall result of such policies would be a greater freedom, financial and job security, and motivation to do good science. This would lead to a considerable added bonus: the dismantling of the incentive system, which currently leads successful scientists to assemble very large groups in order to beat the odds in financing. With a large group where many postdocs try to get the sort of extraordinary results that will allow publishing in top journals there is a greater probability that somebody will get lucky and the group has something to present for its efforts each and every year. This is not only significant in terms of introducing hidden multiplicities (see Chapter 3), but will translate into academic jobs for the lucky postdocs and into continuous financing of the host laboratory. Unfortunately, the laboratory head, who was once a lucky and industrious postdoc himself, must as a result control the monies (and by extension the research) of many people, which has an effect of reducing the diversity of thought in science. In addition, he must spend a lot of his time writing grant applications and sitting in committees, which pretty much guarantees that we honor our most successful scientists by removing them from doing actual science.

The problem here is not large groups of scientists working together as such; it is the hierarchical command structure of such endeavors. By limiting group sizes—which is easy to do by limiting the resources available to each individual scientist—we will end up with more lab heads for the same money, which will increase the diversity of approaches to existing problems and hopefully will lead to opening of new collaborations and new vistas for research. In addition, being a lab head will then be more of a scientific position than a bureaucratic one. The big work-intensive projects that need to be done for scientific reasons would still get done—but by collaborations and pooling of diverse resources. They would not be brought into existence simply because they are expensive and thus constitute a necessary evil that allows the lab head to finance his large group in a single go.

The financial size of science must be decided by the democratic society—not by the promises, threats, and arguments of committees of scientists, but based on the actual return that science provides to society. Society doesn't need scientists to tell it whether science works. The actual size of science depends on more than financing, and it is in here that the wise men (and women) of science can have their say. How to organize science, what to set as its priorities, to whom to redistribute the money bequeathed by society, and, of course, who gets the next Nobel Prize—these are the questions for our scientific peers, of which the politicians should stay clear. Where scientists and politicians should actively engage in is setting and policing the boundaries between science and society. Scientists should give their best to widen the sphere of critical inquiry into the general domain, including race, religion, ethics, and social policy. Society, in turn, should learn to accept the presence of inconvenient viewpoints, as long as it can trust that they are given honestly, without expectation of personal gain or aggrandizement. This does not mean that science should be seen as the only, decisive, or even the most important contribution on these matters—only that its advice should be listened to in the same good faith that it is given.

Our understanding here is that most people who complete their doctorates should be able to relatively easily land in either a steady academic job or a well-paying industrial research post.[2] The assumption behind our utopia is that PhD doesn't mean "can do everything," but is a more specialized affair, which qualifies the holder to work in the ivory tower of academia. PhD is not a ticket to success in the conventional society; it is a ride away from it, into the Mertonian world of communism, universality, disinterestedness, and organized skepticism—none of which will help you much in an ordinary workplace. Thus the society must rethink its belief that by producing a scientist it produces a job creator who can be of great value in many walks of life by bringing the scientific method to bear on any problem of interest. Supporting science may well be an economically sound choice for the society in the long term, but this doesn't mean that we should expect technological innovation from each individual scientist.

The scientists' primary job is to produce knowledge, not technological innovation. We should leave curing disease to physicians and building new gadgets to engineers, who need not be scientists. This leads us to our next proposition: if science is the scientists' business, then we should not allow outside influences to define the content of science. We must protect science from the market forces, which distort the content of science by awarding the use by scientists of certain types of models and interpretations of data, which can be presented as simple yet exciting storylines that lack loose ends and unexplained phenomena.

8.7   PROTECT SCIENTISTS FROM SCIENTIFIC JOURNALS

It is not only the scientists who are engaged in fierce competition for limited resources; the same goes for scientific journals. This may sound strange,

2. According to US census, in 2012 only 15% of biological sciences graduates worked in their chosen field, which in combination with stagnating real wages for PhDs, should deincentivize future scientists (Hira et al., 2014).

considering the ongoing strong growth in the number of scientific papers that all need to be published and the corresponding growth in the number of journals that publish them. There is clearly enough resource (fresh manuscripts) for everybody?! This view is not wrong as long as we consider the modern open-access publishing model, where the authors pay to the journals they publish in. However, in today's climate most authors, who need publications in order to secure a real academic job, need not only to publish in good journals—they need to publish at least a couple of papers in the top journals, which reject over nine-tenths of the manuscripts they receive. Even worse, they need to secure these papers over a relatively short time during their postdoc years.

This leads to an inordinate pressure at the doors of a dozen journals and therefore to outsize influence of collective preferences of a handful of journal editors over who will populate the principal investigator slots in the better universities. They, of course, will pretty much set the direction for science as a whole.

The editors of top journals are not volunteers; they are paid by their journals to work at the interest of their owners. From the top journals in biology most are for-profit enterprises with larger-than-average profit margins, which belong to trans-national companies. To stay at the top, they need to achieve maximal impact for their papers. To this end most top journals stringently ration their space for original research, which enables them to select the manuscripts most likely to gather readers outside their immediate fields, to catch the attention of the general media, and to collect citations. These journals compete amongst themselves and advertise themselves by their impact factors, which is the average number of citations per paper during a 2-year time frame. Currently (2014) the citation factors for the top ten journals in biomedicine (excluding the review journals) are as follows: *New England Journal of Medicine*—54, *Nature*—42, *Lancet*—39, *Nature Biotechnology*—39, *Cell*—33, *Science*—31, *Nature Genetics*—30, *Journal of the American Medical Association*—30, *Nature Medicine*—28, *Nature Methods*—25. Taken together, these journals published 3,610 papers in 2012. As a comparison, *PLoS One*, the largest open-access journal, whose main requirement is technical quality rather than the scientific impact of its papers, published an astonishing 31,496 articles—and still managed to secure an impact factor of 3.5.

To stay in competition and to make profit for their owners, these editors put a premium on (i) unexpected or important or otherwise newsworthy discoveries usually involving the elucidation of simple-to-describe causal mechanisms, (ii) novel and expensive big-data methods like genome-wide association studies (GWAS), (iii) consistency, meaning that there are no admissions of experiments pointing in awkward directions, and (iv) the level of technical quality customary in the field.

There are problems with each of the above and with combinations of them. Firstly, important discoveries imply intrinsic implausibility (or more formally, low prior probability), which means that the technical standard of the field as usual shouldn't apply. From the rejection rates alone we can guess that the customary prior probability for a surprising discovery published in *Nature* must be below 0.08, and probably significantly lower than that (Reich, 2013). From our experience of using Bayes theorem we know that we need dramatically more evidence and very low levels of bias to justify intrinsically implausible claims—but instead, as many a hopeful author has discovered to his or her dismay, top journals tend to ask for more experiments to fortify the discovery with a plausible causal theory of the biological or molecular mechanism behind the biological phenomenon. They seem to be more interested in widening the scope of the paper than addressing the, to us by now obvious, problem of low priors.

Secondly, while the novel expensive methods may have the charm of youth, they also tend to have the brashness of youth. They are interesting to the journals because they represent the moving front of science—and for this very reason they are more prone to error. It takes time to work in a method, to find its limits and biases—and in modern biology each new method seems to be progressively more complicated and thus harder to domesticate.

Thirdly, a simple story is good because it is easier to understand and understanding a theory is a prerequisite for criticizing it. Unfortunately there is no good reason to suppose that a simpler story should *ipso facto* be closer to truth than a more complicated one. This would be a metaphysical assumption best supported by introduction of a demiurge, the creator of worlds, who likes simplicity because complexity makes its head hurt. And yet people like stories they can quickly understand.

Recently *Cell* has begun to publish "graphical abstracts" for every paper it publishes, whose stated reason of existence is that it "should allow readers to quickly gain an understanding of the main take-home message of the paper and is intended to encourage browsing, promote interdisciplinary scholarship, and help readers identify more quickly which papers are most relevant to their research interests" (*Cell*, instructions to authors, accessed 15.04.2014). My analysis of 20 consecutive abstracts (vol. 157, issue 2) containing 33 graphical models of reality revealed that 80% of the abstracts contained causal schemes (the remainder being descriptions of cellular structures), 69% of which were linear having a median of three nodes (range 2–9). In addition, there were just two circular feedback models (one depicting positive and one negative feedback), four models where a single cause has two or three separate effects and four slightly more complicated causal models.

Fourthly, the requirement for consistency of interpretation of experimental results within the expounded model (although seemingly not always of the results themselves) makes the assumption that if a work is published in a top journal it should be regarded as a final word on its subject—at the $\alpha =$ 0.05 level, of course. For an example, here is an excerpt from a *Science* editorial about the level of risk the journal is happy to take of a paper being wrong:

> ... *publishing papers with some ... risk is a good thing. Of course, a journal would love for every paper it publishes to turn out to be perfectly correct—but not at the expense of publishing papers that are all perfectly "safe." Science moves forward by communicating findings that challenge old ideas and force us to test new theories against the evidence. The key is to contain that risk.*
>
> *The sources of risk and how they are best managed vary within scientific disciplines, notably in my experience between the observational fields versus the experimental fields. Experiments take advantage of controls, use multiple replicates, vary initial conditions and independent variables, and hold constant the factors that might otherwise confound results. Barring outright fabrication, results from the experiments should be reproducible to within known uncertainties. Because the experiments have been designed to test the authors' hypotheses, there is generally a relevant result.... The paper may still be risky, but the risk is generally quantifiable.*

McNutt (2013)

In this view there is little room for discrepancies between experimental results and the theories they test, be it due to random error or bias—there should be no loose ends. The published effects are expected to be accompanied by quality control measures, like P values or confidence intervals (CIs), but otherwise the experimental results must match the theory they test. And yet, considering the lack of power studies in molecular biology, the commonly used sample sizes (often $N = 3+$ or it is unmentioned) and the increasing number of different experiments required by the referees and editors for each paper, as well as the large and heterogeneous collaborations that are needed to provide them— the assumption of the correctness of scientific conclusions seems unfounded, to put it mildly.

I should like to propose that as a rule of thumb we should view biological studies that do not publicly state AND defend their sample size, as preliminary. It is all right to do preliminary studies and to publish them, but one has to expect a certain number of inconsistencies in their results simply emanating from the study design (or lack thereof). In addition, perfect consistency of models with experimental evidence strongly suggests that scientists have the uncanny ability to build models that perfectly mirror reality—and that they can do it every time! If we choose to ignore the above possibility, then the large experimental studies with dozens of separate experiments that we can publish in a top journal are either a result of luck (sometimes chance results and biases work in our favor), or of misrepresentation or selective burying of results, or of "scientific misconduct."

So, how can we save science from the clutches of scientific journals. The only answer that I can see is to deemphasize publishing as a career determinant.

8.8 JUDGE SCIENTISTS BY THEIR PROMISES, NOT THEIR DEEDS

A Moral Question

To be a successful scientist in the biomedical field you need at least (i) to get funded, which is a prerequisite for doing research, and (ii) to publish papers that your colleagues will take notice of. It is currently in the vogue to dispense funding by the quality of candidates' previous work, rather than by the quality of their current research ideas (Daniels, 2015). This system seems to lead to higher future impact of grantees' papers and is thus likely to spread further in the near-future.

In this light I would like to report a recent letter from a low-impact but peer-reviewed and open-access journal, which stated that thanks to my high scientific standing (citing a well-published paper of mine) they invite me to publish a free paper in their international US-based journal. Importantly, for my paper "the standard by which ... (the journal) will accept it for publication will be based upon the fact that you have completed multiple fund-supported projects and published several articles in the field of biochemistry and molecular biology." Good-bye peer-reviewers—a good publication record can now guarantee both semi-automatic funding and semi-automatic publishing!

Here comes the moral question: if you accept the morality of funding by the criterion of past success, but not the morality of publishing by the same criterion, please explain, why.

There is no question that as long as there will be scientists, they will keep on checking each others' publication lists with various degrees of envy. There is also no doubt that scientific discoveries need to get published to be of any use to science (and, arguably, to society). A scientist who has made a discovery and refuses to publish it should be reprimanded for wasting public resources.

Alas, such behavior seems to be rare—the much more common case being that scientists who have not made a real discovery nevertheless seek to publish one. Such activity is far from benign. To see why, perhaps we should return to the very beginning of this book, where science is likened to a slowly simmering soup pot, with scores of cooks adding ingredients and at the same time eating from the pot. If a few toss in high-quality discoveries, which will then have to simmer in the company of dead cats and bodily fluids contributed by others, what would you expect the soup to taste like?

Should we choose to improve our cooking by deincentivizing the cooks who lack access to the good ingredients, then the volume of the soup will decrease dramatically and its thickness will improve. But then, how can we decide, who are the best scientists? We cannot anymore use publication records to make this decision because by doing so we would invite the contestants to "cook their records" and to "publish like a pro." A theoretically workable solution would be to limit the number of papers researchers are allowed to publish each year and hope that they will choose to publish their best work in an honest manner. This scheme would probably backfire by decreeing that now everybody must fill his or her allotted quota, even if they don't have much to say in a given year. Another proposed solution is to remove the peer review from the equation, thus making it trivial to publish in biomedicine in a manner that physicists do in arXiv (Nosek et al., 2012). This way everyone could publish as much as he or she wanted to and it would no longer make sense to compare the lengths of publication lists (but we would need to make a conscious decision to ignore the fact that some papers are still published in peer reviewed high-impact journals).

My proposal is to leave peer review as it is and simply ignore publications in hiring decisions. If it is possible to lead a successful scientific career in the pharmaceutical business without being associated with development of a single blockbuster drug, because everybody understands that drug development is a risky business (LaMattina, 2008), why shouldn't this be true in academic science? For the pharmaceutical firms the calculation is simple: if developing a single blockbuster is financially worth the risk of a dozen failures, then the risk is taken knowingly, and individual researchers or departments are not punished for the failures that are, in a way, expected of them. What is the right balance between success and failure in science? Do we want to get true novelty from scientists and are we willing to risk a lot of failure? Or do we prefer to tread on the safe ground and abhor failure as a vice? If so, doesn't this mean that research and development have changed places in the scale of innovativeness and biological scientists should now be compared to engineers or physicians, who obviously are not allowed to lose too many patients or have too many bridges collapsing on them?

If we decide that failure is the expected outcome for a science project, how then should we decide who gets funded? We cannot use past success as a yardstick because in this system it is probably a better predictor of cheating than of future real discoveries. All that is left then, is to assess the intrinsic value of each proposal and the technical competence and level of motivation of the people behind it. A failed project that results in zero publications (or should we require a notification of negative results in some online database?) is then completely irrelevant for future funding, as is a successful ending with good publications. A necessary condition for this particular utopia is that the number of applications for each funding slot be relatively small—this is because the selection process cannot be automated and will thus take many more highly qualified man-hours per candidate than do most current systems, which typically do the first sweep semi-automatically by publication history and then interview the top candidates at length. Therefore a prerequisite seems to be that we must first reduce the competiveness in science at the level where professional scientists are tenured and their work is funded, by placing the bottleneck onto a lower career level (see above).

Now that we have removed the incentives of wrongdoing we can finally come to the central point of this chapter and, perhaps, this book.

8.9 TEACH HONESTY AS THE GUIDING PRINCIPLE OF SCIENCE

Teaching a young scientist to be honest is not difficult, is it? All you need is to instruct her not to tell lies and everything will be fine. Or maybe not. Our everyday lives are filled with situations where we are lied to by telling the truth. Think of the advertising industry, for instance, which is strictly regulated not to send out false messages and yet manages to never tell the truth. My current favorite is the sign in front of my local shoe shop:

All
winter boots on sale up to
−40%

You can see no lie, but there is no truth either. Does this remind you of some scientific communications you have read, or perhaps even written?

Scientist's truth must somehow be different from advertiser's truth or lawyer's truth, or, most definitely, a mother's truth. It must be a special kind of truth that is built like a brick: not prone to stretching and bending.

The story of what happens when scientists fail to see their mission differently from laypeople was ingeniously told by Richard Feynman in his Caltech commencement lecture of 1974. He began by describing the cargo cults of the South Seas where the locals developed a liking for US air force bases of the Second World War, which for them were associated not with terror-bombing, but with an influx of Western goods. When the war ended and the American airmen left, the locals begun to build their own airfields, complete with military-style parades and bamboo air control towers with traffic controllers sitting in them wearing bamboo headsets; all designed to make the planes land once more.

According to Feynman, the expectation behind a cargo cult is that when the form is perfect and everything looks exactly as before, the desired outcome will follow. Of course, we in our great Western wisdom know that it doesn't matter how well one recreates a WWII airfield on some little island—to make the American bombers come again and bring you chocolate you need big events thousands of miles away, rather than an airfield in some tiny Oceanian island.

Feynman next points out that neither is such a situation foreign to modern science: "... I call these things cargo cult science, because they follow all the apparent precepts and forms of scientific investigation, but they're missing something essential, because the planes don't land." When in a real cargo cult we have the requisite knowledge to easily point out the missing ingredient, which is war—the real reason for the planes to land—in science the answer is not so obvious:

> *But there is one feature I notice that is generally missing in cargo cult science. That is the idea that we all hope you have learned in studying science in school— we never explicitly say what this is, but just hope that you catch on by all the examples of scientific investigation. [...] It's a kind of scientific integrity, a principle of scientific thought that corresponds to a kind of utter honesty—a kind of leaning over backwards. For example, if you're doing an experiment, you should report everything that you think might make it invalid—not only what you think is right about it: other causes that could possibly explain your results; and things you thought of that you've eliminated by some other experiment, and how they worked—to make sure the other fellow can tell they have been eliminated.*
>
> Feynman (2010)

Does this sound sensible to you? If yes, then you probably agree that science and scientists must be kept at a safe distance from the lawyers, advertisers and, by implication, businessmen and politicians:

> *The easiest way to explain this idea is to contrast it, for example, with advertising. Last night I heard that Wesson oil doesn't soak through food. Well, that's true. It's not dishonest; but the thing I'm talking about is not just a matter of not being dishonest, it's a matter of scientific integrity, which is another level. The fact that should be added to that advertising statement is that no oils soak through food, if operated at a certain temperature. If operated at another temperature, they all will—including Wesson oil. So it's the implication which has been conveyed, not the fact, which is true, and the difference is what we have to deal with.*
>
> Feynman (2010)

The difference between science and the rest lies in the difference between telling the truth and not telling lies; ultimately it lies in *quo bono*—who benefits. It lies in the ability of the individual scientist to admit that he is wrong: that his results fail to support a spectacular theory—and to get away with it. As long as the premium in science is placed on truth and the premium in a scientific career is on sales, the incentives of scientists will be skewed, and no amount of methodological or ethical training can change that.

8.10 CONCLUSION

What is the take-home message from this discussion, and, perhaps, from the book? In my mind it is that it is not technological competence which defines a scientist. We can never educate a successful scientist the way we can a lawyer or a physician to follow a set of well-defined technical rules. Science as a process is simply too complex, too poorly understood, and too metaphysical to be amenable to a strict deterministic analysis, which is a prerequisite for the development of its rules of conduct.

The key to success of the science as a whole must therefore lie within the mindset of individual scientists. It seems that for science to be a success, most scientists must fail consistently and always. This, of course, is an evolutionary view of science. Darwinian theory of evolution has two premises, (i) that there must be a constant source of heritable variation in the population, and (ii) that natural selection uses this (essentially random) variation to select for useful phenotypes (Mayr, 1982). Thus individual variation, most of which is of slightly negative adaptive value, is absolutely necessary for the occurrence of natural selection—without variation there could be no evolution.

What is true for Darwinian evolution holds for the evolution of science. In science the part of variation is played by the diversity and pluralism of thought of many individual scientists and the part of natural selection is played by Mother Nature (not the journal or the hiring committee), red in tooth and claw, which allows some theories to live and propagate with modification, while killing the vast majority of them by experiment. This means that, as with the biological evolution, the failure of most scientific theories should not be construed as a failure for the scientists who created them. Instead, it is absolutely necessary for the wellbeing and evolution of science.

Mother Nature selects for theories, not theorists. On the other hand, when a university or a grant agency or *Nature* (the journal, this time) selects the scientists, before nature has had a chance to refute their theories, then we have effectively starved the evolution from its fuel, which is diversity. A result is that evolution of science gradually grinds to a halt, which seems to be exactly what we are observing in biomedical science in this day and age. When real progress is inhibited, false prophets will rise to ply their false effects, confirming a huge web of false theories. In the end everything will be connected with everything else, every interaction heralds a biological relevance, every molecule is a potential drug target, every proposed medical treatment works (at least in the published literature), and every false-positive association confirms somebody's theory of life.

So, what can be done about it in concrete terms? As for institutional reforms, I don't know. I'm sure there are people who get paid to think about these things and who may have answers. As for the individual scientist, I do have a proposal. If a scientist decides that what she wants to do is simply the best science that she can, as honestly and with as little bias as she knows how to, then the scientific enterprise should be in no danger. Let us end with this thought.

REFERENCES

Ariely, D., 2013. The (Honest) Truth About Dishonesty: How We Lie to Everyone—Especially Ourselves. Harper Perennial, New York.

Balashov, Y., Rosenberg, A., (Eds.), 2002. Philosophy of Science. Contemporary Readings. Routledge, London.

Bornmann, L., 2012. What is societal impact of research and how can it be assessed? A literature survey. J. Am. Soc. Inf. Sci. Technol. 64 (2), 217–233.

Casadevall, A., Fang, F.C., 2012. Reforming science: methodological and cultural reforms. Infect. Immun. 80 (3), 891–896.

Chang, H., 2014. Is Water H_2O? Springer, Heidelberg.

Check Hayden, E., 2014. The automated lab. Nature 516 (7529), 131–132.

Cohn, A., Fehr, E., Maréchal, M.A., 2014. Business culture and dishonesty in the banking industry. Nature 516 (7529), 86–89.

Collins, F., Tabak, L., 2014. Policy: NIH plans to enhance reproducibility. Nature 505 (7485), 612–613.

Daniels, R.J., 2015. A generation at risk: young investigators and the future of the biomedical workforce. Proc. Natl. Acad. Sci. USA 112 (2), 313–318.

Dougherty, E.R., Shmulevich, I., 2012. On the limitations of biological knowledge. Curr. Genomics 13 (7), 574–587.

Ernø-Kjølhede, E., Hansson, F., 2011. Measuring research performance during a changing relationship between science and society. Res. Eval. 20 (2), 130–142.

Feldman, R., Price, W.N., 2014. Patent Trolling—Why Bio and Pharmaceuticals Are at Risk. Stan. Tech. L. Rev. 773.

Feynman, R.P., 2010. "Surely You're Joking, Mr. Feynman!": Adventures of a Curious Character. W. W. Norton & Company, New York.

Fishman, J., 2014. The research rethink. Nature 514, 292–294.

Heyman, G.D., et al., 2013. Instrumental lying by parents in the US and China. Int. J. Psychol. 48 (6), 1176–1184.

Hira, R., et al., 2014. Bill Gates' Tech Worker Fantasy: Column. USA Today. Available at: <http://www.usatoday.com/story/opinion/2014/07/27/bill-gates-tech-worker-wages-reforms-employment-column/13243305/>.

Jardine, L., 2000. Ingenious Pursuits. Random House Digital, Inc, New York.

LaMattina, J.L., 2008. Drug Truths: Dispelling the Myths About Pharma R&D, first ed. Wiley, New York.

LaMattina, J.L., 2013. Devalued and Distrusted: Can the Pharmaceutical Industry Restore Its Broken Image? Wiley, New York.

Leaf, C., 2013. The Truth in Small Doses. Simon & Schuster, New York.

Macilwain, C., 2013. Driving students into science is a fool's errand. Nature 497 (7449), 289.

Mayr, E., 1982. The Growth of Biological Thought. Belknap Press, Harvard.

McNutt, M., 2013. Risk. Science 341 (6142), 109.

Nosek, B.A., Spies, J.R., Motyl, M., 2012. Scientific utopia: II. Restructuring incentives and practices to promote truth over publishability. Perspect. Psychol. Sci. 7 (6), 615–631.

Nowotny, H., Gibbons, M., Scott, P., 2001. Re-thinking Science: Knowledge and the Public in an Age of Uncertainty. Wiley, Hoboken, NJ.

Piketty, T., 2014. Capital in the Twenty-First Century. Harvard University Press, Cambridge, MA.

Reich, E.S., 2013. Science publishing: the golden club. Nature 502 (7471), 291–293.

Weinberg, R.A., 2014. Coming full circle-from endless complexity to simplicity and back again. Cell 157 (1), 267–271.

Statistical Glossary

Measures of the Average

Mean, arithmetic (μ) — $\dfrac{\Sigma(X)}{n}$ is the sum of a collection of numbers divided by the number of numbers in the collection.

Mean, geometric the geometric mean is the nth root of the product of n numbers. For instance, the geometric mean of the three numbers 4, 1, and 3 is the cube root of their product: $\sqrt[3]{4 \times 1 \times 3}$.

Median the middle value of a ranked list if n is odd (or the mean of two middle values if n is even).

Probability Distributions

Normal distribution a specific class of symmetrical distributions whose shape is uniquely determined by mean and standard deviation.

Standard distribution a normal distribution with mean = 0, SD = 1.

Lognormal distribution an exponential distribution that can be turned into a normal distribution by log-transforming the data.

Binomial distribution a distribution that describes experiments with binary outcomes (like tossing a coin). Converges on normal distribution if n is large.

Beta distribution on the interval [0, 1] is uniquely defined by two positive shape parameters, α and β. It models the behavior of random variables limited to intervals of finite length. The beta distribution is used in Bayesian analysis to describe prior knowledge concerning binary outcomes (like probability of success). It models the random behavior of percentages and proportions.

Sampling distribution describes the frequency distribution of the sample statistics (means, t statistics, etc.) calculated from many hypothetical samples drawn from a population.

Measures of Variation and Uncertainty

Standard deviation (SD) $\sqrt{\dfrac{\Sigma(X - X_i)^2}{n - 1}}$ where X is the sample mean, X_i is an experimental value, and n is the number of measurements (sample size). One SD defines 68% of a normal distribution, 2 SDs define 96%, and 3 SDs 99%. Describes the variation of measurements around the mean.

Average deviation $\dfrac{\Sigma \mid X - X_i \mid}{n}$ is the arithmetic mean of the absolute deviations.

Standard error of the mean (SEM) $\dfrac{SD}{\sqrt{n}}$ describes the uncertainty about the placement of the population mean.

95% confidence interval a frequentist procedure that produces intervals, which contain the population value 95% of the time.

95% Credibility interval a Bayesian interval that contains the population value with 95% probability.

1/8 likelihood interval contains the population values that are consistent with the observations (the data do not provide strong evidence otherwise).

Probability, Population, and Sample

Population, statistical any collection of individuals to which we draw inferences from the data.

Population, biological a collection of actually or potentially interbreeding biological individuals.

Sample, statistical a subgroup drawn randomly and independently from the statistical population.

Sample, biological a quantity of biological material.

Randomness postulate the probability of each individual in a population to be drawn to the sample is equal.

Independency postulate the probability of drawing an individual from the population does not depend on the previous draws (there is no correlation between results of any two draws).

Frequentism the philosophy behind classical statistical inference, which is based on viewing probabilities as objective long-run frequencies of observing some outcome or event.

Bayesianism the philosophy behind statistical inference using Bayes theorem.

Probability, frequentist a long-run relative frequency of an event.

Probability, Bayesian a degree of belief in a hypothesis.

Propensity a physical tendency of an experimental system to behave in a certain manner.

Null Hypothesis Testing

H_0 null hypothesis (mathematically fully defined).

H_{alt} (for the purposes of this book) the alternative hypothesis, which is mutually exclusive and exhaustive with H_0.

H_1 (for the purposes of this book) any other statistical hypothesis, except H_0, that is mathematically simple.

Simple hypothesis mathematically fully defined function, for instance a normal distribution with known mean and SD is fully defined.

Composite hypothesis mathematically not fully defined function, for instance a normal distribution with defined mean and undefined SD (a collection of all possible normal distributions with the given mean).

NPHT the Neyman-Pearson procedure of statistical hypothesis testing, which involves H_0 and H_1.

P **value** the probability of seeing these data or more extreme, given that the H_0 is true. $P(D \mid H_0)$.

q **value** the probability of H_0 being true, given that we see these data or more extreme. $P(H_0 \mid D)$.

FDR false discovery rate = false discoveries at a significance level/all discoveries at that significance level.

Type I error rejecting a true H_0.

Type II error accepting a false H_0.

α the long run relative frequency of type I errors; the significance level.
β the long run relative frequency of type II errors.
Specificity (medical usage) $1 - \alpha$.
Size (statistical usage) $1 - \alpha$.
Sensitivity (medical usage) $1 - \beta$.
Power (statistical usage) $1 - \beta$.
Significance, statistical the relative frequency of type I errors.
Significance, scientific minimal scientifically interesting effect size.

Likelihoodism and Bayesianism

Conditional probability $P(D \mid H)$ probability of the data, given the truth of hypothesis.
Conditional probability $P(H \mid D)$ probability of the hypothesis, given observation of the data.
Likelihood $P(D \mid H)$ probability of exactly the observed data, given the truth of hypothesis.
Likelihood function a continuous function, which defines likelihoods for all parameter values.
Likelihood principle all the evidence present in data is described in the likelihood function.
Likelihood ratio (LR) $P(D \mid H_1)/P(D \mid H_2)$.
Strength of evidence $k = $ LR.
Prior probability $P(H)$ a degree of belief in a hypothesis that is informed by all relevant evidence, except for that contained in the likelihood.
Posterior probability $P(H \mid D)$ a degree of belief that integrates the prior probability with likelihood.

Index

Note: Page numbers followed by "*b*," "*f*," and "*t*" refer to boxes, figures, and tables, respectively.